NOOMAIR AHMER ZUBERI

D1309018

The 68HC11
Microcontroller

*The Saunders College
Publishing Series
in Electronics Technology*

The 68HC11 Microcontroller

Joseph D. Greenfield
Rochester Institute of Technology

SAUNDERS COLLEGE PUBLISHING
Harcourt Brace College Publishers

Fort Worth   Philadelphia   San Diego   New York   Orlando   Austin
San Antonio   Toronto   Montreal   London   Sydney   Tokyo

Text Typeface: Times Roman
Compositor: Maryland Composition
Acquisitions Editor: Barbara Gingery
Assistant Editor: Laura Shur
Managing Editor: Carol Field
Project Editor: Laura Maier
Copy Editor: Elaine Honig
Manager of Art and Design: Carol Bleistine
Associate Art Director: Doris Bruey
Text Designer: Rebecca Lemna
Cover Designer: Lawrence R. Didona
Text Artwork: GRAFACON
Director of EDP: Tim Frelick
Production Manager: Charlene Squibb

Cover Credit: Cover photo © 1991 by Michael Simpson; cover artwork by Lawrence R. Didona.

Printed in the United States of America

THE 68HC11 MICROCONTROLLER

0-03-051588-2

Library of Congress Catalog Card Number: 91-050630

5678901   016   987654

This book is dedicated to my wife,
Gladys,
and
to my grandchildren,
Chana, Miriam, and Yosef Chaim.

Preface

Recent advances in electronic semiconductor technology have resulted in more powerful integrated circuits. Modern microprocessors, such as the **68HC11** discussed in this book, are far more capable and flexible than earlier microprocessors, like the Intel **8085** or the Zilog **Z80**, which dominated the 1980s but are now essentially obsolete. The literature, however, has not kept pace with the technology. At the time of this writing there are very few textbooks whose subjects are the more advanced microprocessors, especially the often-used and easily understood 8-bit microprocessors. I believe that the **68HC11** is the best and most powerful 8-bit microprocessor in general use, and its study provides an ideal introduction to the topic of microprocessors.

This book has a dual purpose: it can be used for a one-semester introductory microprocessor course in an engineering or technology program, and it can also be used by more experienced engineers who must study specifics of the **68HC11**. The book is divided into three basic sections. The first three chapters form an introduction to microprocessors, which is necessary for students who have had no previous exposure to the topic (the book has no prerequisites). The middle chapters discuss the details of the register and instruction sets of the **68HC11**, so that the reader can competently program it. The later chapters discuss the many Input-Output (I/O) capabilities of the **68HC11**, which make it a superior microprocessor.

Pedagogical Features

Each chapter contains Instructional Objectives, Self-Evaluation Questions, examples, programs, and many problems for the students. Solutions to some of the problems are provided in Appendix D. An instructor's manual, which has solutions to all the problems and several suggested laboratory exercises, will be available to the faculty. A diskette containing the AS11 assembler and some programs will also be available. After reading this book the student should be able to write programs for the microprocessor and incorporate it in a system where its I/O capabilities can be used.

Text Philosophy

I believe in teaching microprocessors that are both state-of-the-art and commonly used. The **68HC11** is one of Motorola's newest and most powerful 8-bit microprocessors. Not only does the **68HC11** have a more advanced Central Processing Unit (CPU), with features such as two index registers that greatly simplify programming, but it is also a *microcontroller*. This means that it has internal memory and I/O capability, including both parallel and serial data transmission and A/D conversion. These features make it ideal for use as a controller in industrial processes. Millions have been used in the electronic control modules of automobiles, for example. In most mass-produced items that contain a microprocessor, the superior computational capability of a 16-bit microprocessor is unnecessary, and 8-bit microprocessors, such as the **68HC11**, continue to outsell them. I also maintain that 8-bit microprocessors are simpler and easier to understand than the 16-bit microprocessors; this makes them a better choice for an introductory course on microprocessors.

This book is intended for the *user* of the **68HC11**. Topics such as the internal circuitry of the **68HC11**, or its testing, are not generally of interest to computer users, although they are necessary for the designers of microprocessors. These topics are only alluded to briefly in this book.

Content Overview

Chapters 1 and 2 introduce the hexadecimal and 2s complement number systems and the basic operation of a computer. These chapters are intended as an introduction for students who have never taken a microprocessor course. More advanced students, especially those conversant with another microprocessor, can skim these chapters and start with Chapter 3.

Chapter 3 introduces the **68HC11** and describes its register set and some of its basic instructions. Chapter 4 is on the Evaluation Board (EVB), a very inexpensive kit manufactured by Motorola for use in the laboratory. I believe that some "hands-on" experience in the laboratory is essential to understand a topic, and the EVB is the best way for students to obtain that experience. I have also found that programs rarely work the first time they are run, and there must be a way to test them.

Chapters 5 and 6 cover the instruction set of the **68HC11**, including stacks and subroutines. Many programming examples are presented to help clarify the concepts.

Chapter 7 introduces assemblers. The reader is shown how to write a source code, assemble it into list and object files, and download it to a **68HC11**. Simulators are also discussed.

Chapter 8 describes the internal hardware of the **68HC11** and its important and versatile interrupt structure. It discusses topics such as the crystal clock that drives the microprocessor, RESET, and the memory interface.

Chapter 9 starts the discussion of the I/O capabilities. Parallel I/O using ports B and C, and the two control lines, STRA and STRB are covered. The use of port A for I/O is discussed in Chapter 10, where the use of input captures, output captures, and the pulse accumulator are explained.

Chapter 11 starts with a discussion of the A/D converter on the **68HC11**. This is one of its most valuable features. The chapter then covers serial I/O including both the SCI and the SPI.

Chapter 12 covers features such as the operational control of the **68HC11**. It also covers the memories within the **68HC11**. Finally, the hardware of the EVB, a system using the **68HC11**, is discussed.

Acknowledgments

I would like to thank Gordy Carlson, a former student who now works for Motorola in their Rochester office, Gordy Davis, and other Motorola personnel for their cooperation. I appreciate the help of my students, including the attendants at the seminars on the **68HC11** I held at RIT. My reviewers, Billy Wood of Arizona State University, Jack Foster of Broome Community College, and my former student, Jeff Muehl, of the Lord Corporation, contributed many valuable suggestions. I would also like to thank Barbara Gingery, Laura Shur, and Laura Maier of Saunders College Publishing for their help. Last, but certainly not least, I must thank my wife for both her help and her understanding.

Joseph D. Greenfield
December 1991

Contents

Chapter 9
Basic Input/Output—Ports B and C   238

Chapter 10
The Timing System and Port A   267

Chapter 11
Ports D and E   302

Chapter 12
Microprocessor Control and Memories   341

Appendix A
Instruction Set Details   A-1

The Mathematics of Computers

1

INTRODUCTION

The **MC68HC11**, the microprocessor that is the topic of this book, is one of the most powerful and widely used 8-bit microprocessors in current use. Since the Motorola Corporation introduced its first microprocessor, the **M6800**, in 1974, Motorola has been constantly upgrading and improving its line of microprocessors. As the years progressed, and the technological processes for producing microprocessors were improved and refined, new microprocessors were introduced with enhanced capabilities. Indeed, comparing the **6800** to the **68HC11** is like comparing the Ford Model T to a Thunderbird. The **6800**, and most of its successors, such as the **6801**, the **6802**, and so on, have been made obsolete by the appearance of the 68HC11.*

The purpose of this book is to enable the reader to use the **68HC11** to monitor and control processes *external* to itself. This is what the microprocessor (hereafter abbreviated as μP) was designed to do. One example of these external processes is the control of an automobile. The **68HC11** is the "brains" in the Electronic Control Modules (ECMs) in many automobiles.

Because the **68HC11** is primarily a *computer*, it must be presented as such. Therefore the first chapter of this book consists of an introduction to the **68HC11** and a discussion of the arithmetic system used by computers. Chapter 2 explains, conceptually, how computers work. Chapters 3, 5, and 6 cover the programming of the **68HC11**, while Chapter 4 introduces the EValuation Board (EVB), a kit that allows the student to write and test programs in the laboratory. Chapter 7 discusses the assemblers available for the **68HC11**, and Chapter 8 explains its hardware configuration. The rest of the chapters are devoted to the various *input/output (I/O)* capabilities of the μP. The I/O capabilities of the **68HC11** are some of its strongest features.

---

* In the 1970s Motorola sold an **M6800-D2** kit that was designed for laboratory experiments using the **6800**. Motorola no longer sells or supports this kit.

This chapter starts with a brief description of the physical aspects of a computer—what it consists of and how it is built. It then introduces the reader to the arithmetic of computers: the binary, hexadecimal, and 2s complement number systems. After reading the chapter, the student should be able to

- Convert decimal numbers to binary.
- Convert binary numbers to decimal.
- Add and subtract binary numbers.
- Convert numbers to their 2s complement representation.
- Add, subtract, and negate 2s complement numbers.
- Convert numbers to their hexadecimal representation.
- Add, subtract, and negate hexadecimal numbers.
- Combine hexadecimal and 2s complement arithmetic.

SELF-EVALUATION QUESTIONS

Watch for the answers to the following questions as you read the chapter. They should help you to understand the material presented:

1. Why do computers basically use binary arithmetic?
2. Why can't a computer use a minus sign? How can a number, such as −25, be represented?
3. Why are computer numbers a fixed number of bits?
4. What are the advantages of the 2s complement number system?
5. What is the difference between complement and negate?
6. What is the advantage of the hexadecimal system over the binary system?

1-1 The Physical Construction of the 68HC11

A *computer* is a complex electronic circuit that performs all the many operations required in a computer. A **microprocessor** (μP) is a computer shrunk down to the size of a single **Integrated Circuit (IC)** chip.

A computer is built from a large number of digital electronic circuits called **gates**. The reader may already be familiar with the basic electronic gates: AND gates, OR gates, and inverters. Each gate, in turn, is constructed using one or more transistors. There are several types of transistors, such as BJTs (Bipolar Junction Transistors), FETs (Field Effect Transistors), and CMOS (Complementary Metal-Oxide Silicon). Transistors, in turn, are made up of tiny areas of *p-doped* and *n-doped silicon*. The **68HC11** uses *high-speed* CMOS transistors. This is what the letters HC in the title stand for.

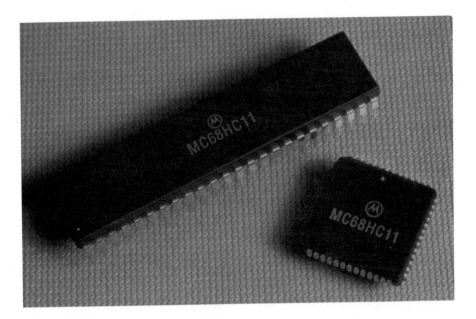

Figure 1-1   **MC68HC11 PLCC and DIP packages.** *(Reprinted Courtesy of Motorola, Inc.)*

To build a μP, a prototype is first constructed, using transistors and discrete gates (such as small-scale ICs). It is then tested to be sure it performs all the required functions. When it has been determined that the discrete computer is operating satisfactorily, the entire μP is placed on a small *silicon wafer* by a photolithography process. The silicon wafer is then bonded to the necessary I/O pins and encapsulated in a package.

There are several versions of the **68HC11**. The different versions contain varying amounts of memory and different I/O capabilities.* The most popular versions are the **M68HC11A8**, which uses internal ROM (Read Only Memory), and the **M68HC11A1**, which uses external ROM and is used on the EVB. These versions are otherwise identical and will be emphasized in this book. They come in one of two packages: a Plastic Leadless Chip Carrier (PLCC) or a Dual-In-line Package (DIP). The packages are shown in Figure 1-1 and the pinouts are given in Figure 1-2.

This introductory book is meant for those technicians, engineers, and programmers who need to *use* the **68HC11**; it is not intended for *designers* of μPs. Although this section has provided the reader with an introduction to, and a feel for, the internal workings of the μP, considerations of the type of transistor used or the manufacturing process utilized will not be discussed further. The reader who requires more information on such matters can refer to the References in Appendix B.

---

* The capabilities of some of the versions are discussed in Chap. 8.

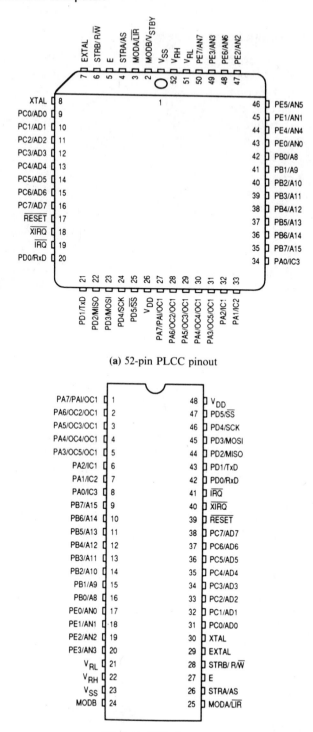

(a) 52-pin PLCC pinout

(b) 48-pin DIP pinout

Figure 1-2   MC68HC11A1 pin assignments. (*Reprinted Courtesy of Motorola, Inc.*)

1-2   Binary Arithmetic

The basic unit of information in a computer is called a **bit**. *A single bit can assume one of only two values: 0 or 1*. Bits are the *only* information a computer can use; the computer achieves its power by organizing and using many bits so that they can represent numbers, letters, and other information.

Groups of bits of a particular size have special names, as given in Table 1-1. The **68HC11** is an 8-bit μP because it uses data in 8-bit groups. It can

Table 1-1   Bit Sizes

4 bits = 1 hexadecimal digit (sometimes called 1 nibble)

8 bits = 2 nibbles = 1 byte*

16 bits = 2 bytes = 1 word†

32 bits = 2 words = 1 longword or 1 doubleword

* Although 1 byte usually refers to a group of 8 bits, the IBM-PC is sometimes considered to consist of 9-bit bytes because it uses a 9-bit wide memory. Each memory byte, however, consists of only eight information bits, plus one parity bit.
† The word *word* is often used generically to refer to any number of bits (see Chap. 2). When used in conjunction with 16-bit computers, however, it refers specifically to a group of 16 bits.

be considered as a byte-sized μP. *BYTE* is also the name of a popular computer magazine which, alas, costs considerably more than 8 bits per issue.

1-2.1   Binary Numbers

A single *decimal* digit is restricted in value; it can only represent the numbers 0 through 9. A single *binary bit* is even more restricted; it can only represent the numbers 0 and 1. But several decimal digits can be combined to represent larger numbers, and the same is true of binary numbers.

In general, a number in *any* number system can be represented by the mathematical equation

$$N = A_0 r^0 + A_1 r^1 + A_2 r^2 + \cdots \qquad (1\text{-}1)$$

where the *A* numbers are any of the numbers allowed in the number system and *r* is the *radix* or *base* of the system. The numbers allowed in a number system usually run from 0 to $r - 1$. Equation (1-1) is not nearly as complicated as it may look. In the **binary system**, where the radix or base is 2, only the numbers 0 and 1 are allowed. In the **decimal system**, which everyone uses daily, each *A* number is one of the digits 0 through 9 and the base is 10.

E X A M P L E   1 - 1

Consider the decimal number 7535. What are A_0, A_1, and so on?

Solution

A_0 is the *Least Significant Digit* (LSD). In this example it is 5. A_1 is the next digit, and so on. For the number 7535, $A_3 = 7$, $A_2 = 5$, $A_1 = 3$, and $A_0 = 5$.

Indeed, the number 7535 can be considered as a *sum* in accordance with Equation (1-1).

$$
\begin{aligned}
7535 = \quad 7 \times 10^3 &= 7000 \\
+5 \times 10^2 &= 500 \\
+3 \times 10^1 &= 30 \\
+5 \times 10^0 &= \underline{5} \\
&\ 7535
\end{aligned}
$$

Binary numbers can also be calculated using Equation (1-1). Now all the *A* numbers must be 1 or 0 and the radix or base is 2. Thus a binary number is a string of 1s and 0s.

Everyone knows the first few powers of 2—1, 2, 4, 8, 16, and so on, and the "binary boat" of Table 1-2 gives the additional powers. The binary boat also gives negative powers of 2, which can be used in floating point arithmetic.

1-2.2 Binary-to-Decimal Conversion

Any binary number has an *equivalent* decimal value. A binary number, a string of 1s and 0s, can be converted to its decimal equivalent by using the following algorithm:

The decimal value of a binary number is determined by finding the decimal value corresponding to each 1 in the number, ignoring the positions that contain 0, and then adding them.

From the powers of two, or by using the binary boat, the decimal value associated with each 1 in a binary number can be found. Then they can be added.

E X A M P L E   1 - 2

What is the decimal equivalent of the binary number 11001011?

Solution

The number 11001011 is an 8-bit number, but, as is usually the case with computers, counting starts at 0, not 1. So 8-bit positions are considered to

Table 1-2 Powers of 2*

2^n	n	2^{-n}
1	0	1.0
2	1	0.5
4	2	0.25
8	3	0.125
16	4	0.062 5
32	5	0.031 25
64	6	0.015 625
128	7	0.007 812 5
256	8	0.003 906 25
512	9	0.001 953 125
1 024	10	0.000 976 562 5
2 048	11	0.000 488 281 25
4 096	12	0.000 244 140 625
8 192	13	0.000 122 070 312 5
16 384	14	0.000 061 035 156 25
32 768	15	0.000 030 517 578 125
65 536	16	0.000 015 258 789 062 5
131 072	17	0.000 007 629 394 531 25
262 144	18	0.000 003 814 697 265 625
524 288	19	0.000 001 907 348 632 812 5
1 048 576	20	0.000 000 953 674 316 406 25
2 097 152	21	0.000 000 476 837 153 203 125
4 194 304	22	0.000 000 238 418 579 101 562 5
8 388 608	23	0.000 000 119 209 289 550 781 25
16 777 216	24	0.000 000 059 604 644 775 390 625
33 554 432	25	0.000 000 029 802 322 387 695 312 5
67 108 864	26	0.000 000 014 901 161 193 817 656 25
134 217 728	27	0.000 000 007 450 580 596 923 828 125
268 435 456	28	0.000 000 003 725 290 298 461 914 062 5
536 870 912	29	0.000 000 001 862 645 149 230 957 031 25
1 073 741 824	30	0.000 000 000 931 322 574 615 478 515 625
2 147 483 648	31	0.000 000 000 465 661 287 307 739 257 812 5
4 294 967 296	32	0.000 000 000 232 830 613 653 869 628 906 25
8 589 934 592	33	0.000 000 000 116 415 321 826 934 814 453 125
17 179 869 184	34	0.000 000 000 058 207 660 913 467 407 226 562 5
34 359 738 368	35	0.000 000 000 029 103 830 456 733 703 613 281 25
68 719 476 736	36	0.000 000 000 014 551 915 228 366 851 806 640 625
137 438 953 472	37	0.000 000 000 007 275 957 614 183 425 903 320 312 5
274 877 906 944	38	0.000 000 000 003 637 978 807 091 712 951 660 156 25
549 755 813 888	39	0.000 000 000 001 818 898 403 545 856 475 830 078 125
1 099 511 627 776	40	0.000 000 000 000 909 494 701 772 928 237 915 039 062 5
2 199 023 255 552	41	0.000 000 000 000 454 747 350 886 464 118 957 519 531 25
4 398 046 511 104	42	0.000 000 000 000 227 373 675 443 232 059 478 759 765 625
8 796 093 022 208	43	0.000 000 000 000 113 686 837 721 616 029 739 379 882 812 5
17 592 186 044 416	44	0.000 000 000 000 056 843 418 860 808 014 869 698 941 406 25
35 184 372 038 832	45	0.000 000 000 000 028 421 709 431 404 007 434 844 970 703 125
70 368 744 177 664	46	0.000 000 000 000 014 210 854 715 202 003 717 422 485 351 562 5
140 737 488 355 328	47	0.000 000 000 000 007 105 427 357 601 001 858 711 242 675 781 25
281 474 976 710 656	48	0.000 000 000 000 003 552 713 678 800 500 929 355 621 337 890 625
562 949 953 421 312	49	0.000 000 000 000 001 776 356 839 400 250 464 677 810 668 945 312 5
1 125 899 906 843 624	50	0.000 000 000 000 000 888 178 419 700 125 232 338 905 334 472 656 25
2 251 799 813 685 248	51	0.000 000 000 000 000 444 089 209 850 062 616 169 452 667 236 328 125
4 503 599 627 370 496	52	0.000 000 000 000 000 222 044 804 925 031 308 084 726 333 618 164 062 5
9 007 199 254 740 992	53	0.000 000 000 000 000 111 022 302 462 515 654 042 363 166 809 082 031 25
18 014 398 509 481 984	54	0.000 000 000 000 000 055 511 151 231 257 827 021 181 583 404 541 015 625
36 028 797 018 963 968	55	0.000 000 000 000 000 027 755 575 615 628 913 510 590 791 702 270 507 812 5
72 057 594 037 927 936	56	0.000 000 000 000 000 013 877 787 807 814 456 755 295 395 851 135 253 906 25
144 115 188 075 855 872	57	0.000 000 000 000 000 006 938 893 903 907 228 377 647 697 925 567 626 953 125
288 230 376 151 711 744	58	0.000 000 000 000 000 003 469 446 951 953 614 188 823 848 962 783 813 476 562 5
576 460 752 303 423 488	59	0.000 000 000 000 000 001 734 723 475 976 807 094 411 924 481 391 906 738 281 25
1 152 921 504 606 846 976	60	0.000 000 000 000 000 000 867 361 737 988 403 547 205 962 240 695 953 369 140 625
2 305 843 009 213 693 952	61	0.000 000 000 000 000 000 433 680 868 994 201 773 602 981 120 347 976 684 570 312 5
4 611 686 018 427 387 904	62	0.000 000 000 000 000 000 216 840 434 497 100 886 801 490 560 173 988 342 285 156 25
9 223 372 036 854 775 808	63	0.000 000 000 000 000 000 108 420 217 248 550 443 400 745 280 086 994 171 142 578 125
18 446 744 073 709 551 616	64	0.000 000 000 000 000 000 054 210 108 624 275 221 700 372 640 043 497 085 571 289 062 5
36 893 468 147 419 103 232	65	0.000 000 000 000 000 000 027 105 054 312 137 610 850 186 320 021 748 542 785 644 531 25
73 786 976 294 838 206 464	66	0.000 000 000 000 000 000 013 552 527 156 068 805 425 093 160 010 874 271 392 822 265 625
147 573 952 589 676 412 928	67	0.000 000 000 000 000 000 006 776 263 578 034 402 712 546 580 005 437 135 696 411 132 812 5
295 147 905 179 352 825 856	68	0.000 000 000 000 000 000 003 388 131 789 017 201 356 273 290 002 718 567 848 205 566 406 25
590 295 810 358 705 651 712	69	0.000 000 000 000 000 000 001 694 065 894 508 600 678 136 645 001 359 283 924 102 783 203 125
1 180 591 620 717 411 303 422	70	0.000 000 000 000 000 000 000 847 032 947 254 300 339 068 322 500 679 641 962 051 391 601 562 5
2 361 183 241 434 822 606 848	71	0.000 000 000 000 000 000 000 423 516 473 627 150 169 534 161 250 339 820 981 025 695 800 781 25
4 722 366 482 869 645 213 696	72	0.000 000 000 000 000 000 000 211 758 236 813 575 084 767 080 625 169 910 490 512 847 900 390 625

go from 0 to 7. The given number has 1s in bit positions 0, 1, 3, 6, and 7. Table 1-3 can then be constructed.

Table 1-3   Decimal Equivalent of 11001011

Bit Position	Decimal Value
0	1
1	2
3	8
6	64
7	128

By adding the numbers in the table, we find that $(11001011)_2 = (203)_{10}$. The subscripts indicate the base of each number.

1-2.3   Attributes of Binary Numbers

There are two *attributes* of binary numbers that the student should keep in mind when using the binary system. They will help prevent errors.

Attribute 1

An n-bit binary number can represent at most 2^n different numbers.

This should be apparent when one considers that a single bit can represent one of two numbers or choices (0 or 1). Two binary bits can represent four choices (00, 01, 10, 11). As the bits are strung together each additional bit *doubles* the number of possibilities. If n bits form a binary number, the 2^n combinations are generally the numbers from 0 to $2^n - 1$.

E X A M P L E   1 - 3

What numbers can be represented by an 8-bit binary number?

Solution

Because n is 8 and $2^8 = 256$, 256 different numbers can be represented. Using only positive binary numbers (negative numbers are discussed in Sec. 1-3), the numbers range from 0 (represented as 00000000) to $2^8 - 1$, or 255 (represented as 11111111).

Attribute 2

The second attribute concerns 0s in the Least Significant Bit (LSB) positions of the binary number. Even binary numbers end in 0; those that end in 00 are divisible by 4, and so on.

If a binary number has 0s in its n LSBs, it is evenly divisible by 2^n.

E X A M P L E   1 - 4

What is the decimal equivalent of 10000000? How does this relate to Attribute 2?

Solution

The number has only a single 1 in bit position 7, so this binary number equals $(128)_{10}$. Note that the seven LSBs of the given number are 0. Therefore the number must be divisible by 2^7 or 128.

1-2.4 Decimal-to-Binary Conversion

Decimal numbers can be converted to their equivalent binary numbers by using the following algorithm:

1. Examine the LSD of the given decimal number. If it is even, write 0 as the LSB of the binary number and go to step 3.

2. If the LSD of the decimal number is odd, subtract 1 from it and write 1 as the LSB of the binary number.

3. At this point the remaining decimal number must be even. Divide it by 2. Go back to step 1 and repeat until the decimal number is reduced to 0.

E X A M P L E   1 - 5

Find the binary equivalent of $(604)_{10}$.

Solution

Before plunging in and simply applying the algorithm, you should first think about the *characteristics* of the solution; this will prevent you from making a gross error. In this problem two characteristics of the solution are immediately present:

1. The number 604 is between 2^9 and 2^{10} ($2^9 < 604 < 2^{10}$). Therefore, in accordance with attribute 1, the answer should consist of ten binary bits.

2. The number 604 is divisible by 4, but not by 8. In accordance with attribute 2, this means that the two LSBs of the binary result must be 0.

Now that you have some feeling for the correct result, the algorithm can be applied to determine it.

Decimal No.	Procedure	Binary
604	Even. Write 0.	0
	Divide by 2.	
302	Even. Write 0.	00
	Divide by 2.	
151	Odd. Subtract 1 from decimal number.	
150	Write 1 in binary number—divide by 2.	100
75	Odd. Subtract 1 from decimal number.	
74	Write 1 in binary number—divide by 2.	1100
37	Odd. Subtract 1 from decimal number.	
36	Write 1 in binary number—divide by 2.	11100
18	Even. Write 0.	011100
	Divide by 2.	
9	Odd. Subtract 1 from decimal number.	
8	Write 1 in binary number—divide by 2.	1011100
4	Even. Write 0.	01011100
	Divide by 2.	
2	Even. Write 0.	001011100
	Divide by 2.	
1	Odd. Subtract 1 from decimal number.	
0	Write 1 in binary number.	1001011100

The decimal number has now been reduced to 0, which indicates that the problem is finished. Therefore $(604)_{10} = (1001011100)_2$. Note that the binary result complies with the attributes discussed at the start of this problem.

With a little practice, you will probably be able to combine several steps at once and do these conversions much faster.

1-2.5 Binary Addition

All the operations that can be performed with decimal numbers (addition, subtraction, multiplication, square root, and so on) can be performed with binary numbers. Addition and subtraction, the two most important operations, will be covered in this chapter. For the **68HC11**, addition and subtraction are necessary to calculate *offsets* (see Sec. 3-7.2) and for other operations. The **68HC11** is also capable of multiplication and division (see

Sec. 5-4.1), but more complicated mathematical operations are not considered here.

This section discusses the addition of binary numbers. By *addition* we mean adding an *n*-bit binary number $(A_{n-1}, A_{n-2}, \ldots, A_0)$, where the *A*s are the 1s and 0s of the number, to another *n*-bit binary number $(B_{n-1}, B_{n-2}, \ldots, B_0)$ to produce an *n + 1 bit sum*. The sum may require an extra bit to accommodate a carry-out.

Basic binary addition is shown in Figure 1-3. If we consider the mth state, its inputs are A_m, B_m, and a possible *carry-in* from the *previous stage*.

Figure 1-3 Binary addition.

Its outputs are the mth bit of the sum, S_m, and a possible *carry-out* to the *next stage*. The addition can be performed by using Table 1-4. The table is

Table 1-4   Adding Binary Numbers

	Inputs		Outputs	
	A_m	B_m	S_m	C_{out}
No Carry-in from Previous Stage	0	0	0	0
	1	0	1	0
	1	0	1	0
	1	1	0	1
Carry-in from Previous Stage	0	0	1	0
	0	1	0	1
	1	0	0	1
	1	1	1	1

divided into two sections, depending on whether or not there is a carry-in to the stage. Binary addition should start with the LSBs because there will never be a carry-in to stage 0.

E X A M P L E   1 - 6

Given the binary numbers $A = 110011001100$ and $B = 10011011010$, find their sum.

Solution

The numbers and their sum, S, are shown below, along with their stages:

```
Stage  12  11  10  9  8  7  6  5  4  3  2  1  0
  A         1   1  0  0  1  1  0  0  1  1  0  0        (3276)
  B             1  0  0  1  1  0  1  1  0  1  0      +(1242)
  S      1   0   0  0  1  1  0  1  0  0  1  1  0       (4518)
```

Each stage has been added in accordance with Table 1-4 and is discussed below. If you feel competent, you may skip to the next section:

Stage 0 $0 + 0 = 0$

Stage 1 $0 + 1 = 1$

Stage 2 $1 + 0 = 1$

Stage 3 $1 + 1 = 0$ This stage generates a carry-out.

Stage 4 $1 + 0 = 0$ Because of the carry-in from stage 3. This stage also generates a carry-out.

Stage 5 $0 + 0 = 1$ Because of the carry-in. This stage does not generate a carry-out.

Stage 6 $1 + 1 = 0$ Same as stage 3.

Stage 7 $1 + 1 = 1$ Because of the carry-in from stage 6. This stage also generates a carry-out.

Stage 8 $0 + 0 = 1$ The same as stage 5.

Stage 9 $0 + 0 = 0$ The same as stage 0.

Stage 10 $1 + 1 = 0$ The same as stage 3.

Stage 11   Because the B number only had 10 bits while the A number had 11 bits, a leading 0 is added to the B number. Now stage 11 becomes $0 + 1$ with a carry-in. The sum is 0, but there is a carry-out.

Stage 12   Neither number has a bit in stage 12. Now leading 0s are added to both stages so that A_{12} and B_{12} both equal 0. The result is the same as stage 5. Because there are no more bits and this stage does not generate a carry-out, the problem is complete.

The decimal equivalents of the given numbers are also shown in parentheses. The results of any binary addition can be checked by converting each binary number to its decimal equivalent as shown here.

1-2.6 Binary Subtraction

Binary subtraction starts with an n-bit number, A (the *minuend*) and a second number, B (the *subtrahend*) is subtracted from it. The result is usually called the *difference*. Binary numbers can be subtracted by using the following rules:

1. 0 minus 0 = 0

2. 1 minus 0 = 1

3. 1 minus 1 = 0

4. 0 minus 1 = 1 with a borrow

To *borrow*, the next 1 in the minuend must be changed to a 0. If there are 0s between the stage where the borrow occurred and the next 1, those 0s have to be changed to 1s. The procedure is best illustrated by an example.

E X A M P L E   1 - 7
_____

Subtract 1011100 from 1000010110.

Solution

This example is of the classic subtraction form (*A* minus *B* equals the difference, *D*). It is best analyzed stage-by-stage as follows:

```
Stage  9 8 7 6 5 4 3 2 1 0
  A    1 0 0 0 0 1 0 1 1 0      (534)
 -B          1 0 1 1 1 0 0    −  (92)
  D      1 1 0 1 1 1 0 1 0      (442)
```

Stage 0 0 minus 0 = 0
Stage 1 1 minus 0 = 1
Stage 2 1 minus 1 = 0
Stage 3 0 minus 1 = 1. This generates a borrow. Therefore the 1 in stage 4 of the minuend must be changed to a 0.

Stage 4 Because the 1 in the minuend has been changed to a 0, this now becomes 0 minus 1, or 1 and a borrow. The next 1 in the example occurs in stage 9. This is changed to a 0, but all *intervening 0s* (in stages 5, 6, 7, and 8) must be changed to 1s.

Stage 5 After the change this is now 1 minus 0 or 1.
Stage 6 After the change this is now 1 minus 1 or 0.
Stages 7 and 8 These are the same as stage 5.

Again, the problem can be checked by converting all the binary numbers to decimal numbers as shown in parentheses.
_____

1-3 Twos Complement Numbers

Until now we have had considerable freedom in presenting binary numbers. We used the numbers whose bit sizes were as long as necessary; negative numbers were not considered. Within a computer, however, this freedom is lost. There are two restrictions that must be observed:

1. *The number of bits that can be used to express a number is fixed.* For most 8-bit computers, including the **68HC11**, this number is 8.

2. *There must be a way to express negative numbers.* Remember that a computer contains only 1s and 0s, not plus and minus signs.

To meet these restrictions, *all modern computers use the 2s complement number system.* This system may be used with numbers of any bit length, but the bit length must be specified. For the rest of this section we will consider 8-bit (byte-sized) numbers, which are used in 8-bit μPs.

1-3.1 Expressing Numbers in the Twos Complement System

One of the major characteristics of the 2s complement number system is its ability to express both positive and negative numbers. In the 2s complement system *all numbers that have a Most Significant Bit (MSB) of 0 are positive, and all numbers with an MSB of 1 are negative.*

E X A M P L E   1 - 8

Which of the following numbers are positive?

a. 10000000

b. 00000000

c. 11111111

d. 10101010

e. 01010101

Solution

Only (b) and (e) have MSBs of 0, and are therefore positive. Numbers for (a), (c), and (d) have MSBs of 1 and are therefore negative.

Positive 2s complement numbers are identical to binary numbers, except that the MSB must be 0.

E X A M P L E   1 - 9

Express $+(53)_{10}$ as an 8-bit 2s complement number.

Solution

Using the methods of Section 1-2.4, we find that $+(53)_{10} = (110101)_2$. This number can be expanded to an 8-bit number by adding two leading 0s. Therefore

$$+(53)_{10} = (00110101)_2$$

in 8-bit 2s complement form. Note that the MSB of the number is 0, indicating it is positive.

E X A M P L E   1 - 10

What is the largest positive number that can be expressed as an 8-bit 2s complement number?

Solution

The largest positive 8-bit number is 01111111 or $(127)_{10}$. Note that if 1 were added to this number, the result would be 10000000, which has an MSB of 1, and is therefore negative.

In general, if numbers with n bits are available, they can express any one of 2^n different numbers. In the 2s complement system these numbers consist of 2^{n-1} positive numbers and 2^{n-1} negative numbers. The numbers that can be expressed by an n-bit number are

All positive numbers from 0 to $2^{n-1} - 1$

All negative numbers from -1 to -2^{n-1}

Specifically, for 8-bit numbers the 256 different numbers consist of the 128 positive numbers from 0 to 127 and the 128 negative numbers from -1 to -128. Note that *0 is considered as a positive number*.

1-3.2   Negating Twos Complement Numbers

Negation means finding the *negative equivalent of a positive number* or finding *the positive equivalent of a negative number*.

To negate a number in the 2s complement system

1. Complement the number (invert all its bits).

2. Add 1.

These two steps can be used to transform a positive number into its negative equivalent. The *identical* steps can also be used to transform a negative number into its positive equivalent.

E X A M P L E   1 - 11

Find the negative equivalent of

a. $+53$

b. $+1$

Solution

a. In Example 1-9 we found that $+53 = 00110101$. To negate it, we first complement it to give 11001010. Then 1 is added to give the final result:

$$-53 = 11001011$$

b. $+1 = 00000001$. This complements to 11111110. Adding 1 gives

$$-1 = 11111111$$

From Example 1-11 we saw that *−1 equals a solid string of 1s*. This is a characteristic of the 2s complement system. Observe that −1 can be negated by complementing it, giving all 0s, and then adding 1 to give +1 (00000001).

Another feature of the 2s complement system is that Attribute 2 of Section 1-2.3 applies to both positive and negative numbers; that is, all even numbers end in an LSB of 0, all numbers divisible by 4 have their two LSBs as 0, and so on.

As stated earlier, negative numbers can be converted to their positive equivalent by negating them.

E X A M P L E   1 - 12

What is the decimal value of the number 11001100?

Solution

Inspection reveals that this is a negative number because the MSB is 1. It can be converted to its positive equivalent by negating. The complement of the number is 00110011, and by adding 1 we obtain 00110100. Using the binary-to-decimal conversion techniques of Section 1-2.2, we find that 00110100 = 52. Therefore

$$11001100 = -52$$

Note that both +52 and −52 have two LSBs of 0.

1-3.3   Adding Twos Complement Numbers

Twos complement numbers can be added just like binary numbers, but 2s complement arithmetic has some advantages over decimal arithmetic. Consider the simple equation $C = A + B$. In decimal arithmetic we must be concerned about the signs of A and B. If A and B are both positive, they can simply be added. If they are both negative, they can also be added and preceded by a minus sign. If they are of different signs, however, they must be subtracted.

It is much simpler when using 2s complement arithmetic. For addition and subtraction the signs of the numbers do not have to be considered.

> *In 2s complement arithmetic the sign of the result will always be correct, regardless of the signs of the addend and augend.**

* This is true provided *overflow* is ignored. The examples presented in this section will not involve overflow, which is discussed in Sec. 3-5.4.

E X A M P L E   1 - 13

Using 2s complement arithmetic, add the number $+53$ and -1.

Solution

In Example 1-9 we found that $+53 = 00110101$, and Example 1-11 showed that $-1 = 11111111$. Adding the numbers, we obtain

$$
\begin{array}{r}
00110101 \\
+ 11111111 \\
\hline
00110100
\end{array}
$$

This is indeed $+52$ (see Example 1-12).

Observe that in this example there was a carry-out of the MSB. *In 2s complement arithmetic carries-out of the MSB are ignored.* Therefore, don't worry about it.

In Example 1-12 we found that -52 was 11001100 and that its negation, $+52$, was 00110100. To check this, we can add the two numbers. The sum, of course, must be 0. Again, the addition will produce a carry-out of the MSB, which is ignored.

1-3.4   Twos Complement Subtraction

Subtraction in the 2s complement system is probably best performed by *negating the subtrahend* and then adding it to the minuend. As with addition, the sign of the answer will be correct.

E X A M P L E   1 - 14

Subtract 52 from 25.

Solution

Twenty-five is a positive number equal to 00011001. We can subtract by negating 52 and adding. It has already been shown that $-52 = 11001100$. Therefore

$$
25 - 52 = \quad
\begin{array}{r}
00011001 \\
+ 11001100 \\
\hline
11100101
\end{array}
\qquad
\begin{array}{l}
(25) \\
+(-52) \\
\hline
(-27)
\end{array}
$$

The result is obviously negative. Using the methods of Example 1-12 it can be shown that $11100101 = -27$. This addition also produced a carry-out of the MSB, which was ignored.

1-4 Hexadecimal Arithmetic

By now you probably have observed a problem with binary arithmetic; there are too many 1s and 0s. In all modern computer literature the **hexadecimal system** (*base 16*) is used. This is actually a variant of the binary system. *Each hexadecimal digit represents four binary bits*, which condenses the numbers considerably. Printouts of computer code or portions of memory (sometimes called memory dumps) are almost always in hexadecimal.

The base 16 system requires 16 different *single-digit* numbers. The first ten of these are the ordinary decimal digits, 0 through 9, but because the number 10 requires two digits, it cannot be used. Therefore the letters A through F are used as the six additional digits. The bit configuration for each hexadecimal digit is given in Table 1-5. Observe that the 16 hexadecimal digits include all possible combinations of four binary bits. From the table it can be seen that the hexadecimal number A corresponds to a decimal 10, B corresponds to 11, and so on, until F corresponds to 15.

1-4.1 Binary-to-Hexadecimal Conversion

Binary numbers can be converted to hexadecimal by starting at the LSB, dividing the binary number into 4-bit groups, and then representing each group by its proper hexadecimal number as determined from Table 1-5.

Table 1-5 Hexadecimal-to-Binary Number Conversions

Hexadecimal* Digit	Decimal Value	Binary Value
0	0	0000
1	1	0001
2	2	0010
3	3	0011
4	4	0100
5	5	0101
6	6	0110
7	7	0111
8	8	1000
9	9	1001
A	10	1010
B	11	1011
C	12	1100
D	13	1101
E	14	1110
F	15	1111

* The word *hex* is often used as an abbreviation for hexadecimal.

E X A M P L E   1 - 15

In Example 1-5 (the binary subtraction problem) the difference was 110111010. Express this number in hexadecimal.

Solution

The number is divided into three 4-bit groups, as shown. Note that there is only a single 1 in the most significant group. This becomes a 4-bit group by adding three 0s in the most significant locations.

$$\underbrace{0001}\,\underbrace{1011}\,\underbrace{1010}$$

$$1 \quad B \quad A$$

Therefore $(110111010)_2 = (1BA)_{16}$.

1-4.2 Converting Hexadecimal Numbers to Binary

Hexadecimal numbers can be converted to binary simply by using Table 1-5, as Example 1-16 shows.

E X A M P L E   1 - 16

Convert 1F4C to binary.

Solution

Each hex digit is converted to its 4-bit equivalent by using the table:

$$1 \quad F \quad 4 \quad C$$

$$\overbrace{0001}\ \overbrace{1111}\ \overbrace{0100}\ \overbrace{1100}$$

or

$$(1F4C)_{16} = (1111101001100)_2$$

Notice how much more compact hex notation is; here a four-digit hex number replaces a 13-bit binary number.

1-4.3 Hexadecimal-to-Decimal Conversion

Hexadecimal (hex) numbers can be converted to decimal by using Equation (1-1) with $r = 16$. To simplify, the least significant hex digit (or nibble) has a value of 1, the next digit has a value of 16, the third has a value of 16^2 or 256, the fourth has a value of 16^3 (4096), and so on.

E X A M P L E   1 - 17

In Example 1-6 the answer was found to be 1000110100110, which converts to 11A6 in hex. This number can be converted to decimal by starting at the least significant hex digit and applying the algorithm discussed earlier.

Hex Digit		Value		Net
6	×	1	=	6
A (10)	×	16	=	160
1	×	256	=	256
1	×	4096	=	4096
				4518

The sum shows that $(11A6)_{16} = (4518)_{10}$. This agrees with the decimal answer given in Example 1-6.

1-4.4   Decimal-to-Hexadecimal Conversion

Decimal numbers can be converted by using the following algorithm:

1. Divide the decimal number by 16.

2. Write the remainder as the least significant hex digit.

3. Remove the remainder.

4. Go back to step 1 and repeat until the number is reduced to 0.

E X A M P L E   1 - 18

Convert the decimal number 7335 into hex.

Solution

Once again, it is wise to think about the answer before attempting the problem. In this example the number is between 2^{12} (4096) and 2^{13} (8192), so the answer contains 13 bits, or requires $13/4 = 3.25$ hex digits. Since fractional hex digits are not allowed, the answer contains four hex digits.

The problem is started by dividing the number, 7335, by 16. The answer given by a calculator is 458.4375. The remainder, .4375, is equal to 7/16. Therefore the least significant hex digit is 7.

We continue by dividing the quotient, 458, by 16. The result is 28.625. Of course, $.625 = 5/8 = 10/16$. So the remainder is 10, or A in the hex system.

Now 28 is divided by 16 to give 1 and a remainder of 12, or C in the hex system. Of course, dividing 1 by 16 will give 0 with a remainder of 1. This ends the problem. The results are

$$(7335)_{10} = (1CA7)_{16}$$

1-4.5 Adding Hexadecimal Numbers

While studying the **68HC11** the student will often need to add, subtract, or negate hex numbers. Hex numbers are added by mentally replacing any letters, such as A, B, and so on, with their numeric equivalent and adding. During addition *any sum that is equal to or greater than 16 is expressed by subtracting 16 from it. In this case a carry is sent to the next more significant place.*

E X A M P L E   1 - 19

Add the numbers 3BC5 and 66E7.

Solution

The addition is somewhat similar to the binary addition of Example 1-6.

Stage	3	2	1	0	
A	3	B	C	5	(15301)
B	6	6	E	7	(26343)
S	A	2	A	C	(41644)

Stage 0 5 + 7 = ? In decimal, of course, the answer is 12. Here the answer is the hex equivalent of 12, C. There is no carry-out because the sum is less than 16.

Stage 1 C + E = ? Here C (12) plus E (14) = 26. Because this number is greater than 16, there is a carry into the next stage. The sum is 16 less than 26 or 10 or A in hex.

Stage 2 6 + B = ? 6 + 11 = 17 The carry-in from the previous stage increments this to 18. The 18 becomes 2 with a carry-out.

Stage 3 3 + 6 = ? The sum is 3 + 6 + 1 (the carry-in from the previous stage). This equals A in hex. As before, the equivalent decimal numbers are given in the parentheses.

1-4.6 Subtracting Hexadecimal Numbers

Hex subtraction is similar to normal subtraction, with the stipulation that the numeric equivalents are substituted for the numbers represented by letters.

As in normal subtraction, when the subtrahend is larger than the minuend in any stage, a *borrow* occurs. In hex arithmetic *a borrow means that 16 is added to the minuend, and to compensate for this, the minuend in the next more significant stage is decremented (reduced by 1).* If this stage cannot be reduced because it contains 0, the minuend in the stage is changed to F, and the next stage is decremented.

E X A M P L E   1 - 20

Subtract 33F2 from 603F.

Solution

The solution proceeds as follows:

Stage		3	2	1	0	
A (minuend)		6	0	3	F	(24639)
$-B$ (subtrahend)		3	3	F	2	$-$(13298)
D (difference)		2	C	4	D	$\overline{(11341)}$

Stage 0   F $-$ 2 = (15) $-$ 2 = (13) = D

Stage 1   3 $-$ F = ? Here the subtrahend is greater than the minuend, so a borrow must occur. The borrow allows us to add 16 to the minuend, so stage 1 becomes (19) $-$ (15) = 4. We cannot borrow from stage 2 because it contains 0, so the minuend in stage 2 is changed to F. The borrow occurs by decrementing the minuend of stage 3 to 5.

Stage 2   The borrow in stage 1 has changed the minuend in stage 2 from 0 to F. The subtraction becomes F $-$ 3 = (15) $-$ 3 = (12) = C.

Stage 3   The borrow in stage 1 changed the minuend in stage 3 to 5. So the subtraction is simply 5 $-$ 3 = 2.

The subtraction can be checked by adding the difference and subtrahend to get the minuend, as in normal subtraction, or by converting the numbers to decimal, as shown in the parentheses.

1-4.7   Negating Hexadecimal Numbers

Negating hex numbers means finding the negative equivalent of a given hex number. Therefore, if the given hex number is positive (having an MSB of 0), its negative will have an MSB of 1, and vice versa. In general, *hex numbers whose most significant hex digit is between 0 and 7 are positive* because they have an MSB of 0, and *hex numbers whose most significant digit is between 8 and F are negative.* Hex numbers can be negated by converting them to binary and applying the procedure of Section 1-2, but we feel that the following method is more efficient.

To negate a hex number, examine the least significant hex number. Write the number that, when added to this hex number, will make the hex number total 10 hex or 16 decimal. For example, if the least significant digit is A, write 6. This will be the least significant digit of its negative.

Now examine the remaining hex digits. For each digit write the difference between that digit and F. This should complete the negation.

E X A M P L E 1 - 21
_____

Negate A0FF.

Solution

In accordance with the previous procedure, the least significant hex digit is examined first. Here it is an F. We write a 1 in its negation because F + 1 = 16. Now the next digit is examined; it is also F. Now, however, a 0 is written because F + 0 = F. The next digit is 0, so an F is written because 0 + F = F. Finally, the most significant digit is A, so a 5 is written. The results are

$$-A0FF = 5F01$$

_____

The reader should note some characteristics of the results of Example 1-21.

1. The given number was negative (it had a most significant digit of A). The negation, therefore, had to be positive and it was.

2. The given number was odd (odd numbers end in 1, 3, 5, 7, 9, B, D, F). Its negation was also odd. Attribute 2 of Section 1-2.3 applies.

3. The sum of the number plus its negation must equal 0. The reader can determine this by adding them. There will be a carry-out of the most significant digit but that can be ignored.

The preceding procedure works for all numbers except when the least significant digit is 0. This is because there is always a digit that will make the hex sum 15. But if the digit is 0, there is no digit available that will make it sum to 16.

When the least significant hex digit is 0, write 0 in the least significant digit of its negation. Find the first nonzero digit and make that add to 16; then make the rest of the digits add to F.

E X A M P L E 1 - 22
_____

Negate 3A00.

Solution

Here the given number ends in two 0s. Therefore its negative will also end in two 0s. The first nonzero digit is A. This must add to 16, so the negative digit is 6. Finally, the most significant is 3, so the most significant digit of the negation must be C (C + 3 = F).

$$-3A00 = C600$$

Again, this example can be checked by adding the two numbers to give a sum of 0.

Observe that both the original number and its negative have nine LSBs of 0. This means that they are both evenly divisible by 2^9, or 512.

1-4.8   Hexadecimal Numbers and Twos Complement Arithmetic

In Section 1-3 the use of 2s complement arithmetic with binary numbers was discussed. Twos complement arithmetic can be shortened considerably by using hex numbers. Although the hex representation and 2s complement arithmetic work for numbers of any bit length, we will confine our discussion to 8-bit 2s complement numbers because these are the numbers used predominately in the **68HC11** and other 8-bit μPs.

Because we are considering only 8-bit numbers, each hex number will consist of *exactly two hex digits or nibbles*. In hex representation they can be added, subtracted, or negated in accordance with the procedures developed previously in this section.

Suppose, for example, that we have to add the decimal numbers +59 and −87. First, each of these numbers must be expressed in its 2s complement form and then they can be added. Probably the easiest way is to mentally calculate how many 16s are in each number; then the hex representation is easily determined. Take +59 first. Because +59 equals 3 times 16 plus 11, the most significant nibble is 3 and the least significant nibble is B so that $(+59)_{10} = (3B)_{16}$.

To find the hex representation of a negative number, first find the hex equivalent of its positive value, and then negate. To find −87, we first find +87. Now +87 equals 5 times 16 plus 7, or 57. To find $(-87)_{10}$, we can negate $(+57)_{16}$, just as we did in Section 1-4.7. The result is A9. Now we can add the numbers:

$$\begin{array}{r} 3B \\ +\ A9 \\ \hline E4 \end{array}$$

Once again, the results are correct and with the proper sign. The number E4 is negative. To find its positive equivalent, it can be negated to $(E4)_{16}$ $= -(1C)_{16}$. But $(1C)_{16}$ equals 1 times 16 plus 12 or $(+28)_{10}$ decimal. Therefore $(E4)_{16} = (-28)_{10}$, which is correct.

Subtraction is performed similarly, as Example 1-23 shows.

E X A M P L E   1 - 23

Subtract $(-59)_{10}$ from $(03)_{10}$.

Solution

The number $(03)_{10}$ transforms into $(03)_{16}$. We have previously found that $(59)_{10} = (3B)_{16}$. Therefore -59 will be C5. The results are

$$
\begin{array}{rr}
03 & (03) \\
-\ C5 & -\ (-59) \\
\hline
3E & +\ \overline{(62)}
\end{array}
$$

In the least significant stage 5 was subtracted from 3. This requires a borrow. Do not be disturbed by the fact that there is no place to borrow from; this is 2s complement arithmetic, so just borrow. Borrowing allows us to add 16 to the 3 and the subtraction becomes $19 - 5 = E$. Because of the borrow, the 0 in the more significant stage of the minuend is changed to F and the subtraction becomes $F - C = 3$. The result is correct with the correct sign. The decimal numbers are shown in parentheses.

Summary

This chapter started with a brief introduction to the **68HC11**, its construction and manufacture. It then progressed to the study of the arithmetic used in computers. The binary, 2s complement, and hexadecimal number systems were discussed. These arithmetic systems are used by all modern computers, and the reader should be familiar with them before progressing further.

Glossary

Bit   A single unit of information. A single bit can be a 1 or a 0.

Binary system   A number system based on the number two. It consists only of bits (1s and 0s).

Decimal system   The most commonly used number system based on the number or base 10.

Gate   A logic electronic circuit; an electronic circuit that accepts binary inputs and produces binary outputs as a function of those inputs.

Hexadecimal system   A number system based on the number 16.

Integrated Circuit (IC)   A logic circuit or circuits encapsulated in a single package.

Microprocessor   A computer in a single IC.

Negation   Finding the negative equivalent of a given number.

2s complement system   A binary-based number system that allows for the expression of both positive and negative numbers.

Problems

1-1 Convert the following binary numbers to decimal:
  a. 10111
  b. 110011101
  c. 100110001111

1-2 Consider your Social Security number. For each of the nine digits that are odd, write 1; for each even digit write 0. Now convert the 9-bit number to decimal.

1-3 How many bits would be required to express the following decimal numbers in binary?
  a. 56
  b. 97
  c. 734
  d. 10,003

1-4 Find the binary equivalent of each of the numbers in Problem 1-3.

1-5 In the table you are given A and B. Fill in the third and fourth columns of the table.

	A	*B*	*A + B*	*A − B*
a.	11001	1000		
b.	1101101	10111		
c.	11100011101	1100111010		
d.	1010111011	*		

* Use your "binary Social Security number," as found in Problem 1-2.

1-6 By inspection, state which of the numbers in Problem 1-5 are divisible by 2 and which are divisible by 4.

1-7 Convert the following decimal numbers to 8-bit 2s complement numbers, and then negate them.
  a. 12
  b. 43
  c. 105
  d. −77

1-8 The following numbers are all decimal. Convert them to 8-bit 2s complement form and perform the indicated operation. In each case check the results by converting the answer back to decimal.

   a. 75 + 23
   b. 75 − 23
   c. 77 − (−23)
   d. −77 − (−87)
   e. 77 + (−12)

1-9 What range of numbers can be expressed in 16-bit 2s complement form? (*Note:* Because the BASIC language reserves 16 bits for an integer number, this is the range of numbers that can be expressed in BASIC.)

1-10 Convert the following numbers to hexadecimal.
   a. 10111000
   b. 110011111110
   c. 1001000011010110
   d. 110011111110011110

1-11 Convert the following hex numbers to binary.
   a. AB
   b. E01FA
   c. BCD123

1-12 Convert the hex numbers in Problem 1-11 to decimal.

1-13 Convert the following decimal numbers to hex.
   a. 98
   b. 1235
   c. 45,567
   d. 106,787

1-14 Perform the following hex additions.
   a. AD + 3D
   b. 12B5 + 45FF6
   c. 33EE48 + 1A7E8DC

1-15 Perform the following hex subtractions.
   a. A5 − 77
   b. 5A05 − 1EEE
   c. 345AA − F34F
   d. 7 − ABCD

1-16 Given the following 2s complement numbers in hex form, state by inspection which are positive and which are divisible by 4.

<u>8-bit numbers</u>
   a. A4
   b. 77
   c. F0
   d. 36

<u>16-bit numbers</u>

e. DDDC

f. 1223

g. 3A00

h. 887B

1-17 Negate the numbers in Problem 1-16.

1-18 Rework Problem 1-8 after converting the numbers to 2-digit hex numbers.

After finishing the problems, go back and check the self-evaluation questions. If any of them seem unclear, reread the appropriate sections of the chapter.

Elementary Computer Operations | 2

INTRODUCTION

This chapter explains, in general terms, how a computer functions, so that when the specifics of the **68HC11** microcontroller are discussed in Chapter 3 and subsequent chapters, the reader will have a better understanding of the operation of the microprocessor (μP) contained within the **68HC11** chip.

INSTRUCTIONAL OBJECTIVES

After reading the chapter, the student should be able to

- List the major components of a computer and explain their functions.
- Explain the function of each bus connecting the μP and the memory.
- Explain the steps required to read or write a memory.
- Write simple programs using the rudimentary instruction set presented in this chapter.
- Draw simple flowcharts.

SELF-EVALUATION QUESTIONS

Watch for the answers to the following questions as you read the chapter. They should help you understand the material presented:

1. What is a register?

2. What are access and cycle times? Why are they important?

3. What does nondestructive reading mean? Why is it important in a computer?

4. What are the advantages and disadvantages of ROMs compared to RAMs? Where are ROMs used?

5. Why is documentation important?

2-1 Overview of the Operations of a Computer

The *block diagram* of a computer is shown in Figure 2-1. To understand how a computer operates, one must first understand the function of each component of the computer.

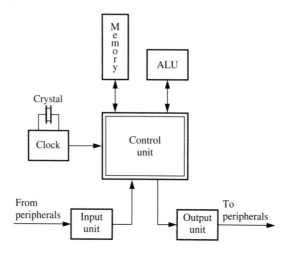

Figure 2-1 Block diagram of a computer.

The figure shows the following components:

1. Memory   The memory retains all the information necessary for the operation of the computer. This includes both data and instructions. Memories are discussed further in Section 2-2.

2. Arithmetic-Logic Unit (ALU)   The ALU acts as a calculator for the computer. It performs all the necessary arithmetic and logic operations. The ALU is discussed further in Section 2-3.

3. Input unit   This is the part of the computer that handles information coming in from external devices (also called peripherals). One example of an input device is a keyboard, which sends the operator's keystrokes to the computer.

4. Output unit   This is the part of the computer that sends information to output devices connected to the computer, such as a printer or the video display. Input and output are sometimes referred to collectively as I/O.

5. Clock   The clock is a *crystal-controlled oscillator* that controls the timing and the speed of the μP. The frequency of the clock must be very stable, which is why crystals are used. The most powerful modern μPs run at clock rates of 25 to 35 MHz. The clock rate determines the time it takes for the μP to execute each instruction; therefore a higher clock rate means the μP can execute more instructions per second and is more powerful.

6. Control unit   The control unit regulates and synchronizes the operations of the computer. It controls the flow of data between the other units of the computer.

One of the major components of the control unit is its set of *registers*. A **register** is a part of a μP that holds a group of related bits usually containing a character, number, or other data that the computer needs. The **8086** μP from Intel, and its smaller version, the **8088**, which is used in the IBM-PC, the PC/XT, and compatibles, both contain fourteen 16-bit registers. Some of these registers can be subdivided into two 8-bit registers. Motorola's **68000** μP contains 32-bit registers, which can be used for 8-, 16-, or 32-bit data. The **68HC11** uses 8-bit registers predominantly, although a 16-bit register can be used.

Perhaps the most important component of a computer is not shown in Figure 2-1. This is the **software** or the set of instructions that control the operation of the computer. Both engineers and programmers must be capable of writing the set of instructions, the program, that will make the computer do what they require.

2-1.1   Personal Computers

The readers should be able to relate something they are probably familiar with—the personal computer (IBMs, Apples, Commodores, etc.)—to the block diagram of Figure 2-1. If the system unit (the main box in the computer system) is opened, the user will see a **motherboard**, a large board full of IC chips. Among these chips are

1. The μP

2. Memory ICs

3. ICs that control the input and output devices

4. Expansion slots for I/O drivers

5. Many small-scale ICs that perform various functions

On most computers the disk drives are also contained in the system unit.

The IBM-PC is a typical personal computer (PC). The module chart of the motherboard is shown in Figure 2-2. The motherboard is inside the systems unit. It contains

1. The microprocessor   This is a single IC. The ALU and control registers are within this chip. Many PCs contain an extra socket so that the user can add a *math coprocessor* if needed. A math coprocessor is another IC that acts as a very powerful ALU and is used in problems that require much mathematical manipulation (number-crunching). The socket for a coprocessor (the Intel **8087** in this case) is also shown. Most PCs get along without coprocessors and rely on the ALU that is already within the μP.

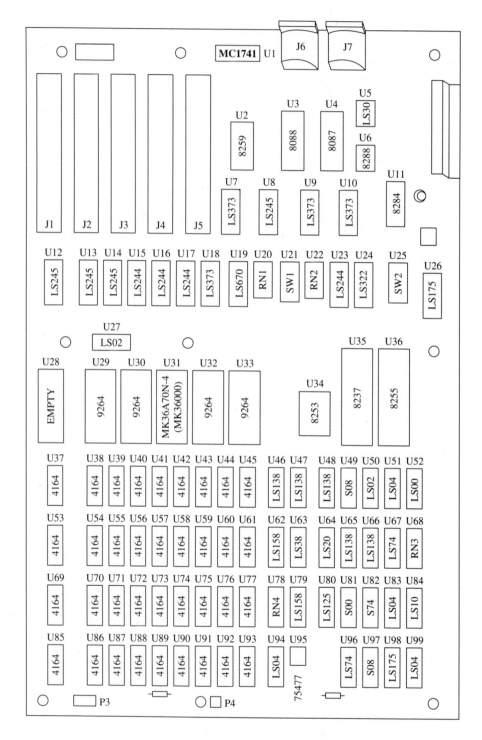

Figure 2-2 Module chart of the IBM PC.

2. The memory This is a set of ICs. For an IBM-PC there are four groups of nine memory ICs. Memories are often *expandable*; additional chips can be purchased to increase the size of the memory.

3. The crystal clock The crystal in an IBM-PC oscillates at 14.31818 MHz.

The rest of the motherboard consists of smaller-scale ICs for timing, buffering, and other functions, and expansion slots, labeled J1 through J5. Electronic drivers for the video display, printer, and modem (if one is used) mount in these expansion slots.

The disk(s) and power supply are also inside the system unit but not on the motherboard. The disk is both an input and an output unit; it retains files and sends them to the computer's memory as needed. In BASIC, for example, the command SAVE sends a program from the computer's memory to the disk for permanent storage. This is *output*, or *writing* to the disk. The command LOAD brings the program back into memory from the disk. This is *inputting* to the computer, or *reading* the disk.

PCs also contain I/O devices or peripherals that are external to the systems unit. These are as follows:

Keyboard This is an input device that transmits the user's keystrokes to the computer.

Video display or monitor The computer sends the user information by displaying it on the CRT (cathode ray tube) screen. It is an output device.

Printer A printer is *predominantly* an output device that prints the user's files, but the printer must be capable of sending some information back to the PC. *Status* information relating to the printer must be transmitted to the μP. The μP should not send data if the printer is offline or out of paper, so the μP must be able to read these status indicators (input) and respond accordingly.

Modem Modems enable the computer to communicate with remote devices, usually over telephone lines. During such communication the computer both sends and receives data; therefore modems are both input and output devices.

2-2 Memories

The *memory*, which has been alluded to in previous sections, is an indispensable component of any computer system. Memories retain large amounts of information for use by the computer as needed. Memories, in response to commands from the computer's control unit, perform two basic operations: *read* and *write*. During a *write* operation the computer stores information in the memory for future use; the memory must remember it. When the computer needs information, it issues a *read command*, which causes the memory to recall the proper information and sends it to the computer.

There are two basic types of memories: **Random Access Memories (RAMs)** and **Read Only Memories (ROMs)**. Without getting too technical at this time, RAMs are memories that can be *both* read and written to, whereas ROMs, as the name implies, can only be read and not written to. The information in a ROM is permanent and cannot be changed. Most of the following information applies to RAMs. ROMs are discussed in Section 2-2.6.

2-2.1 Memory Organization

As explained in the previous section, memories are *accessed* by the computer being read or written to. Memories are organized in groups of bits called **words.** Here word is used in its generic sense; it can refer to any number of bits.

A word is the number of bits involved in each memory access. In the IBM-PC, for example, there are 9 bits per word. They consist, however, of only eight information bits. The ninth bit is for parity (see Example 5-21). The reader has already seen that the IBM-PC's memory consists of rows of nine ICs. On any memory access each IC contributes 1 bit to the word.

Most microprocessors are classified according to the number of bits (the word length) on their data bus. The **68HC11**, the Intel **8085**, and the Zilog **Z80** are all 8-bit µPs with an 8-bit data bus and 8-bit data registers. The Intel **8086** and the Motorola **68000** are 16-bit µPs. Sometimes the distinction blurs, however. The Intel **8088** is generally regarded as a 16-bit µP because it has 16-bit internal registers and an instruction set identical to the 16-bit **8086**, but the **8088** only has an 8-bit data bus.

2-2.2 The Structure of a Memory

The structure of a memory is shown in Figure 2-3. It consists of n words by m bits per word. In Figure 2-3 the memory is shown inside the double solid line, and the connections between the memory and the microprocessor are also shown. The memory in Figure 2-3 consists of 65,536 words ($n = 65,536$) by 8 bits ($m = 8$). This is the largest size memory that a **68HC11** and most other 8-bit µPs can use.

Each word in memory has a specific *address*, just as each house on a street has. In the memory there are 65,536 addresses, but they are numbered 0 to 65,535. This is a 64K-byte memory. Byte is the appropriate size because there are 8 bits at each address, but one might think that 64K means 64,000 words. The abbreviation K (for kil) refers to 1000 when dealing with electrical quantities such as ohms or voltage, but it is only *approximately* 1000 when it refers to memory sizes. Memories are almost always constructed so that the number of words can be expressed as 2^n. When the term K refers to memory sizes, $K = 1024$ instead of 1000 because $1024 = 2^{10}$.

Memories are connected to a µP as shown in Figure 2-3. Microprocessors and memories have a *master-slave* relationship. Because the µP contains the "brains" of the system, it is the master. It tells the memory what to do, and the memory slavishly obeys.

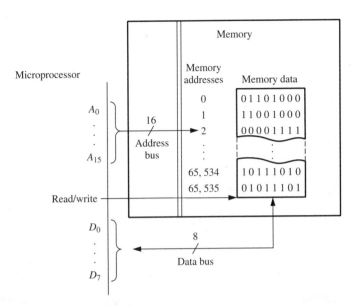

Figure 2-3 The structure of a memory.

A μP is connected to a memory by three *buses*. A **bus** is a group of wires that carry specific information. The three buses are

1. The address bus As explained in Chapter 1, *K bits are required to distinguish between n different things, such that* $2^K = n$. Here the *n* different things are the number of words in the memory, and the *K* bits to distinguish between them are the bits on the address bus. The memory shown consists of 65,536 (2^{16}) words; therefore the address bus must carry 16 bits and requires 16 lines.

Observe that the address bus is *unidirectional*; addresses go *from* the μP *to* the memory. This is because the μP is the master; it determines which word in memory it wants to access.

2. The data bus This bus transfers the data between the μP and the memory. It contains as many lines as there are bits in each memory word (8 in the figure). The data bus is *bidirectional*. During a read access it carries data from the memory to the μP, but data flows from the μP to the memory during a write operation.

3. Read/Write (R/W) bus This bus is only one line for Motorola μPs. It allows the μP (the master) to tell the memory (the slave) to perform a read or a write operation. A logic 1 on this line indicates a read and a 0 indicates a write.

E X A M P L E   2 - 1

How many address lines and how many data lines are required for a 128K word by 20-bit memory?

Solution

First, the number of words in the memory is calculated; 128K words means 128 (2^7) times K (2^{10}) or 2^{17} words. Therefore this memory size requires 17 bits for address and 17 lines on the address bus. The data bus must be as large as the word size of the memory or 20 lines in this example.

2-2.3   Reading Memory

In order to read the memory, the microprocessor must do the following:

1. Issue a read command.

2. Place the address of the word it must read on the address bus.

The memory will then place the data on the data bus, where the μP can read it. In Figure 2-3, for example, if the command is "read the word at location 1," the memory will place 11001000 (the contents of location 1) on the 8-bit data bus.

The memory cannot respond to command instantly; it will have to wait one *access time*. **Access time** is the time it takes the memory to respond or the time between the issuance of a command and the appearance of firm data on the data bus. Access time is illustrated in Figure 2-4a. Access time for typical memories is about 100 nanoseconds (ns).

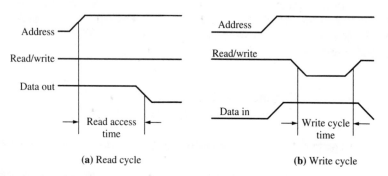

(a) Read cycle   (b) Write cycle

Figure 2-4   Memory timing. (*Greenfield/Wray, USING MICROPROCESSORS AND MICROCOM-PUTERS: The Motorola Family, 2/E,* © *1988, p. 53. Reprinted by permission of Prentice-Hall, Inc., Englewood Cliffs, N.J.*)

In any computer, reading must be **nondestructive**. In other words, the memory cannot destroy the data in the process of reading it; the memory must retain the data as well as present it on the data bus. Memories may be read as often as desired (memories are often read continuously) because *reading must not change the contents of the memory*.

E X A M P L E   2 - 2

In Figure 2-3, what data will appear on the bus if the address is FFFF?

Solution

This memory requires 16 address bits. They are usually represented by four hex digits. Because $(FFFF)_{16} = (65535)_{10}$, the data that will appear on the bus is 01011101 or $5D_{16}$.

2-2.4 Writing Memory

To *write* a memory, the computer must

1. Place the address of the word to be written on the address bus.

2. Place the data to be written on the data bus.

3. Issue a WRITE command.

The memory will then receive a WRITE pulse, which must be low for one *cycle time* as shown in Figure 2-4b. **Cycle time** is the time that the WRITE command must be active to allow the memory to write. For most modern memories this time is equal to the access time. The data and address must be stable throughout the write cycle.

Writing a memory is *destructive*. Writing alters the contents of memory by overwriting the addressed word with the new data. The old data is irretrievably lost; therefore the user should be very careful when writing memory. In most computers over 90 percent of the memory accesses are reads, so the data is retained.

2-2.5 Physical Memories*

Like μPs, most modern memories are etched onto silicon chips and then packaged in Dual-In-line Packages (DIP). The memory ICs in the IBM-PC is one example. The capacity of each memory IC has been increasing as time progresses; currently 1 megabit (MB) memories are readily available and 4 MB memories are coming online. A 4 MB memory contains 2^{22} bits and requires 22 address lines. Memory sizes are usually measured in *bytes*, regardless of whether an 8- or a 16-bit data bus is used. Thus eight 4 MB ICs are required for a 4 megabyte memory.

There are two types of RAM memories: *static* and *dynamic*. **Static RAMs** store their bits on flip-flops and are generally faster (shorter access and cycle times). **Dynamic RAMs** store their information on capacitors. This information tends to leak off, however, so that dynamic RAMs require the additional complication of a **refresh** cycle to rewrite the data before it vanishes.

* A more thorough discussion of memories is given in J. D. Greenfield, *Practical Digital Design Using ICs*, 2nd ed. Prentice-Hall, Englewood Cliffs, N.J., 1988.

Dynamic RAMs have one big advantage over statics, however; the manufacturer can place more bits on a single IC. The capacity of both static and dynamic RAMs is constantly increasing as the technology of fabricating memories improves, but at any time dynamic RAMs seem to contain more bits than static RAMs by a factor of 4. For this reason PC manufacturers tend to favor dynamic RAMs. Another advantage of dynamic RAMs is that they *time multiplex* their address lines so that a 4 MB memory only has 11 address pins instead of 22. These details need not concern a user of the **68HC11,** but more information is available in books on memories. (See References, Appendix B.)

Dynamic RAMs

TMS4416-12	120ns, 16K × 4	$ 2.25
TMS4416-15	150ns, 16K × 4	$ 2.00
4116-12	120ns, 16K × 1	$ 1.49
4116-15	150ns, 16K × 1	$ 1.09
4116-20	200ns, 16K × 1	$ .89
4164-100	100ns, 64K × 1	$ 2.75
4164-120	120ns, 64K × 1	$ 2.39
4164-150	150ns, 64K × 1	$ 2.15
4164-200	200ns, 64K × 1	$ 1.75
41256-60	60ns, 256K × 1	$ 5.25
41256-80	80ns, 256K × 1	$ 3.75
41256-100	100ns, 256K × 1	$ 3.15
41256-120	120ns, 256K × 1	$ 2.95
41256-150	150ns, 256K × 1	$ 2.59
41464-80	80ns, 64K × 4	$ 5.95
41464-10	100ns, 64K × 4	$ 4.95
41464-12	120ns, 64K × 4	$ 3.95
41464-15	150ns, 64K × 4	$ 3.59
511000P-70	70ns, 1M × 1	$13.95
511000P-80	80ns, 1M × 1	$12.95
511000P-10	100ns, 1M × 1	$12.35
514256P-80	80ns, 256K × 4	$13.45
514256P-10	100ns, 256K × 4	$12.95

Static RAMs

6116P-3	150ns, 16K × 1 (CMOS)	$ 2.79
6264LP-10	100ns, 64K × 1 (CMOS)	$ 6.95
6264LP-15	150ns, 64K × 1 (CMOS)	$ 4.95
43256-10L	100ns, 256K × 1	$10.95
43256-15L	150ns, 256K × 1	$ 9.95
62256LP-15	150ns, 256K × 1 (CMOS)	$10.95

Figure 2-5 Part of an advertisement showing memories for sale.

A typical advertisement for memories is shown in Figure 2-5. It identifies whether the RAMs are static or dynamic and gives the part number for each available memory, its speed (access time) in ns, its size (words times bits per word), and its price.

2-2.6   Read Only Memories

The problem with read/write memories is that they forget; they lose their information when power is turned off. This condition is known as being **volatile.** Consider someone using BASIC on a PC, for example. He or she can write and execute a BASIC program, but the program is now in RAM. If it must be preserved for future use, it must be SAVED. The SAVE command writes the program on a disk because the RAM will lose the program as soon as the computer is turned off. Disks are *nonvolatile* and can retain information indefinitely. At the next computing session the user can retrieve the program by LOADing it back into memory and then editing or running it.

Disks can be read or written to, but they cannot be used in place of RAMs because their access time is too long. Random access really means that any word in the memory can be accessed in about the same access time. Disks are not random access. If a word on a disk is required, the user must wait until the disk spins so that its read or write head is over the proper disk sector. This can take several milliseconds, as compared to the 100-ns access time for a RAM.

Read Only Memories (ROMs) are memories that can only be read. They are, however, random access memories whose access time is approximately equal to the access time of RAMs. To read a ROM, the user simply supplies the address for the designated word and the ROM places the data at that address on the data bus. Observe that no read/write (R/W) line is required because the memory will only be read.

ROMs have two advantages over RAMs to compensate for the inability to write them:

1. The programs or data will never change.

2. They are nonvolatile; they retain their data when power goes down.

ROMs often contain computer programs that must not change. One of their most common uses is for **monitor** programs. These are the programs that take control of a computer when it is first turned on so that the operator can use it. In the IBM-PC and compatibles, for example, the monitor program starts by checking memory and then searching for a disk that contains DOS (the Disk Operating System). ROMs are also ICs in DIPs. In Figure 2-2 sockets U29–U33 are for ROMs.

ROMs cannot be written during the normal course of computer operation. Nevertheless, there must be some way of entering information in them. The act of entering the information is called *programming the ROM*. There are three types of ROMs:

1. **Masked ROMs** These are ROMs programmed at a manufacturer's facility by using a *mask* that contains the proper bit pattern. Producing the mask is fairly expensive and has a long lead time. After the mask is made, however, copies are very inexpensive. Masked ROMs are used when many

identical ROMs are required. A PC manufacturer, for example, who uses the same monitor program in each PC, would use a masked ROM.

2. EPROMS EPROMS stands for *Erasable Programmable ROMs*. A special device, called a *PROM programmer*, is used to write the program or data into the ROM by overvoltaging certain inputs. These ROMs can also be erased by exposing them to ultraviolet light. Because EPROMS can be programmed and erased at the user's facility, they can save much time and money. EPROMS are used when small quantities of special purpose ROMs are needed.

3. EEPROMS These are *Electrically Erasable Programmable ROMs*. The information in EEPROMS can be erased and rewritten without removing them from their sockets.

The **68HC11A8** is a version of the **68HC11** that contains an 8K-byte masked ROM within the IC. The **68HC11A1** is almost identical to the **'A8** except that it has no internal ROM. On the board used in the laboratory the **'A1** uses an external ROM that holds the Buffalo monitor that allows the μP to communicate with a data terminal and lets the user examine memory and write instructions (see Chap. 4). The **'A8** and **'A1** also contain 512 bytes of EEPROM, so the user can store a small program permanently in this area.

2-3   The Arithmetic-Logic Unit

As Figure 2-1 shows, there is an Arithmetic-Logic Unit (ALU) within each μP. The ALU in the **68HC11** is capable of the following arithmetic and logical operations:

Arithmetic	Logical
Addition	AND
Subtraction	OR
Multiplication	EXCLUSIVE OR (EOR)
Division	Complementation
Shifts	Negation (2s complement)
Rotations	

These operations are discussed in detail in Chapters 3 and 5.

The ALU also provides inputs to the *Condition Code Register* (*CCR*). This is a register that retains information about the *characteristics of the results* of the last operation. Typical characteristics are as follows:

1. Were the results 0?

2. Were the results negative?

3. Did the last operation generate a carry?

The CCR is also discussed in detail in Chapter 3.

2-4   The Control Unit

Every computer contains a group of instructions that it is capable of executing; this is called its **instruction set**. The function of the computer is to execute these instructions properly. The function of the **programmer** is to provide the proper program (a **program** is a sequence of instructions) to make the computer do what is required. The advantage of a computer over a piece of dedicated electronic hardware, like a digital clock, is that the computer can execute a vast variety of different tasks. What the computer actually does depends simply on which program it is running.

Computers fall into two categories: *dedicated* or *special purpose*, and *general purpose*. A dedicated computer has a program that never changes (it is written in ROM) and it can perform a limited number of tasks. The μP in a VCR, for example, is dedicated. The user's commands tell the μP which function of the VCR it must perform (updating the clock, playing, recording, etc.), and the VCR conforms.

A *general purpose computer* can perform an unlimited number of tasks, depending on what program is loaded into it. A PC, for example, can run BASIC, do word processing, run a spreadsheet, or so on, depending on whatever the user requires.

All computers execute their instructions in approximately the following manner:

1. Fetch   The computer must fetch the instruction. *Fetching* means reading the instruction from memory.

2. Decode   The computer must decode the instruction so that it knows what it must do.

3. Execute   The computer must execute; it must do what the instruction tells it to do. After the execution phase the instruction is finished. The computer then fetches the next instruction. Computer instructions are executed *sequentially*.

The *control unit* in a computer or μP controls the execution of the μP's instructions. To do so, the control unit uses the following registers that are within the μP:

1. Memory Address Register (MAR)   The MAR contains the address of the word that the computer is reading or writing to memory.

2. Memory Data Register (MDR)   The MDR contains the data going to the memory on a write and the data coming from the memory on a read.

3. Program Counter (PC)   During a fetch cycle the computer reads the next instruction from memory. At which address? *The PC contains the address of the next instruction to be fetched.*

4. Instruction register   This register retains the instruction while it is in the process of being executed.

5. Instruction decoder   This is a logical decoder whose inputs come from the instruction register. It activates a particular line that corresponds to the instruction currently being executed. This line, being active, causes the control unit to perform all the tasks necessary to execute the particular instruction.

6. Accumulator The accumulator is a register in the control unit that typically holds an *operand*, a number that is part of an arithmetic or logical operation. Most arithmetic operations require two operands. Generally one of these operands is in an accumulator and the other is in memory. The results of the operation are usually placed in the accumulator. In subtraction, for example, the minuend is found in the accumulator and the subtrahend is found in memory. The result, the difference, is placed in the accumulator. When an operation requires only one operand, such as an INCREMENT or a SHIFT, that operand is usually found in the accumulator.

2-4.1 Control Unit Flip-Flops

There are several flip-flops (FFs) inside a control unit. An FF is a digital electronic circuit that is capable of retaining a single bit of information. One of the most important FFs is the FETCH/EXECUTE flip-flop. A computer operates by alternating between FETCH and EXECUTE modes. First, it *fetches* the instruction from memory, so it can determine which instruction it is. Then, knowing the instruction, it *executes* it. The computer then fetches the next instruction. The FETCH/EXECUTE FF is in FETCH mode when the instruction is being fetched and is in EXECUTE mode when the instruction is being executed.

We have now covered all the components of the computer as shown in Figure 2-1, except the Input-Output. The discussion of I/O is deferred until Chapter 9.

2-5 Execution of a Simple Routine

To give the reader a feel for the operation of a computer, the execution of a simple program segment is described in this section. This program will take two numbers from memory, add them, and then write them back into memory. It requires three instructions:

1. LOAD instruction The LOAD instruction reads a word (the first number in this example) from memory and copies or *loads* it into the accumulator.

2. ADD instruction This instruction takes the second number from memory and adds it to the number that is already in the accumulator. Observe that two operands are being added; one in the accumulator and one in memory. The sum is placed in the accumulator.

3. STORE instruction This instruction takes the contents of the accumulator and writes it (*stores* it) in memory.

To continue this discussion, we make two assumptions that, while not true in all cases, simplify this introductory explanation.

Assumption 1 Each instruction consists of two parts: an *Operations code* (Op code) and an *address*. The **Op code** part of the instruction tells the computer what to do. LOAD, ADD, and STORE are all examples of Op codes. We will further assume that each Op code requires 8-bits (1 byte).

Of course, a computer can only read a series of bits, not a word like L O A D. Each Op code has a corresponding bit pattern that the computer will recognize and respond to. For the **68HC11** the following Op codes apply:

Instruction	Op code
LOAD	B6
ADD	BB
STORE	B7

Note that the 8 bits of each Op code are expressed in hexadecimal.

The *address* portion of an instruction tells the computer *where the memory operands* are. We will assume that all memory addresses are 16 bits (2 bytes). This is true for the **68HC11** and is consistent with dimensions of the memory (64K bytes) that a **68HC11** can use.

Assumption 2 We assume that memory is divided into two areas: a *program* area and a *data* area. The program area contains the instructions and the data area contains the data on which these instructions must operate. Both data and instructions reside in the same physical memory. The user must designate a certain set of addresses (a memory area) for instructions and another set of addresses for data.

2-5.1 Instruction Execution in Detail

Let us assume that our first problem is to write a program segment to add the numbers 23 and 3A. To start, these numbers must be placed in memory. Let us arbitrarily select the addresses around C100 as our data area and assume that the number 23 is placed in C101, the number 3A is placed in C102, and C103 is reserved for the result.

The program consists of a LOAD, an ADD, and a STORE instruction. These instructions must also be placed in memory. Let us begin the program area at location 25, well away from the data area. Then the program would look like the following:

Location	Instruction	Code		
25	LOAD C101	B6	C1	01
28	ADD C102	BB	C1	02
2B	STORE C103	B7	C1	03

Address	Data	
0025	B6	← Op code of LOAD instruction.
0026	C1	} Address.
0027	01	
0028	BB	← Op code of ADD instruction.
0029	C1	} Address.
002A	02	
002B	B7	← Op code of STORE instruction.
002C	C1	} Address.
002D	03	
.	.	
.	.	
.	.	
C101	23	
C102	3A	
C103		

The leftmost labels: **Program Area** (for 0025–002D rows) and **Data Area** (for C101–C103 rows).

Figure 2-6   The memory and its contents.

Observe that each instruction requires three locations: one for the Op code and two for the address. The memory and its contents are shown in Figure 2-6.

Before execution can start, the starting address of the program (25) must be placed in the PC. The step-by-step execution of the LOAD instruction is shown in Figure 2-7. It proceeds as follows:

Step 1   The contents of the PC, 0025, are sent to the MAR. The µP provides this address to the memory and reads back its contents, B6. Because this is an instruction fetch, the memory contents are placed in the instruction register. The instruction decoder decodes B6 as the Op code for a LOAD instruction and activates the LOAD line as shown. The µP now knows that it is executing a LOAD instruction and that the address is contained in the next two memory locations.

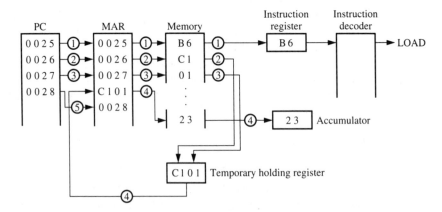

Figure 2-7   Execution of a LOAD instruction.

Step 2   The PC has been incremented during step 1. Its new contents, 0026, are sent to the MAR and the μP reads C1. The μP knows that this is the most significant part of the address and stores it in the top half of a 16-bit temporary holding register. The PC is incremented again.

Step 3   The contents of the PC, 0027, are again sent to the MAR. The μP now reads the contents of memory location 27, which is 01, and stores it in the lower half of the holding register. The PC is incremented to 0028 to prepare it for the next instruction.

Step 4   This is the step where the actual loading takes place. The proper address is now in the temporary holding register. It is moved into the MAR. The memory now reads the contents of C101, 23, and places it in the accumulator.

Step 5   This is the start of the next instruction. The contents of the PC are placed in the MAR and the μP can begin to fetch the next instruction.

This completes the LOAD instruction. It required four memory read cycles: one to fetch the Op code, two more to obtain the address, and a fourth cycle to load the contents of that address into the accumulator. Notice also that the PC contains the address of the Op code for the next instruction in the program.

The ADD instruction is executed similarly. With the PC at 0028, the Op code for an ADD is fetched and the next two memory reads place the address in the temporary holding register.

Step 4 of the ADD instruction is illustrated in Figure 2-8. The MAR is loaded from the temporary holding register, and the memory reads out the

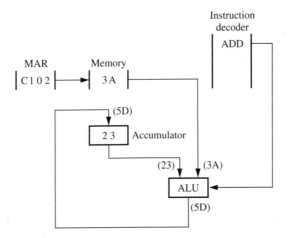

Figure 2-8   Step 4 of an ADD instruction.

contents of C102, which is 3A. Because this is an ADD instruction, these contents become one input to the ALU, and the contents of the accumulator, 23, become the other input. Also, the ADD line of the instruction decoder is active, which commands the ALU to add. The sum, 5D, is written into the accumulator at the end of the instruction.

The STORE instruction also requires four memory cycles. First, the μP reads the Op code to find that it is a write command. Then the next two locations are read so that the μP can determine the memory address into which it must write. The fourth memory cycle is now a write cycle; the μP writes the contents of the accumulator into the memory. In this example 5D is written into C103. Notice that it was not necessary to erase C103. Whatever data was there before has been overwritten and is lost.

It may seem that we have gone to a lot of trouble to show that 23 + 3A = 5D, but this is how the computer determines it, and by following the procedure, the student will have a better idea of how computers function.

2-6   Introduction to Programming

By now it should be obvious that in order to use a computer one must be able to program the computer. There are three levels that a computer can be programmed on: machine language, assembly language, and higher level language. Each level becomes progressively harder for the computer to understand and progressively easier for the human.

2-6.1   Machine Language

Machine language instructions are the only instructions a computer can actually understand. This is because they are written the only way computers can understand anything—as numbers. In the previous section we saw some examples of machine language instructions; B6 was the number (Op code) for LOAD, BB was the number for ADD, and B7 was the number for STORE. Thus writing in machine language means giving the computer the proper numerical Op code for each instruction it must execute.

2-6.2   Assembly Language

The next step up is **assembly language.** Using assembly language, the programmer can write *mnemonics* for each instruction. It is easier for humans to write LOAD, ADD, and STORE without translating them into their machine language equivalents.

The problem, of course, is that *the computer cannot read assembly language statements.* If a program is written in assembly language, it must be placed into a computer, along with another program called an *assembler*, that translates the words, such as LOAD, into their proper numbers. The output of the assembler is the *assembled code*, which contains the instructions in machine language so that the computer can understand them. The actual computer being used is called the *target computer*, and the computer that runs the assembler is called the *host computer*. They are often two different computers.

The EVB, which is the kit manufactured by Motorola for experimenting with the **68HC11**, uses a *one-line assembler*. It can translate instructions such as LDAA $C101 into machine language for the **68HC11**. Here LDAA is the mnemonic for "Load Accumulator A," but this is much more descriptive and easier to remember than B6.

Most assemblers have additional features and simplify programming in a variety of ways. Assemblers for the **68HC11** are discussed in Chapter 7.

2-6.3 Higher Level Languages

In assembly language each statement produces only one machine language instruction. Higher level languages allow the programmer to write one statement that produces *several* machine language instructions. Examples of higher level languages are BASIC, FORTRAN, PASCAL, C, and so on. In BASIC the statement C = A + B + C might produce the following assembly language statements:

LOAD A

ADD B

ADD C

STORE C

Even in this simple example a single BASIC statement produced four machine language instructions. Higher level language statements must be put through a *compiler* or an *interpreter* (programs similar to assemblers) to produce machine language for the computer.

Research on programmer productivity has shown that programmers can write about ten lines of code a day, regardless of whether each line is written in machine, assembly, or higher level languages. Because each higher level language statement can produce several lines of machine code, programming is more efficient when using these languages. When working with a microcontroller, however, where the programs are often short and sometimes involve much I/O, assembly language is often used. We will use assembly language predominantly in this book.

2-7 Looping Programs

Now that the reader is capable of adding 23 + 3A, we conclude this chapter by introducing slightly more complex programs. For a first problem, assume that the numbers 1, 5, 9, 13, 17, and 21 must be added. This is no more complex than adding 23 + 3A; it is only longer.* If the numbers are stored

---

* In this section we will assume these numbers are decimal.

in memory locations C100, C101, and up, the program becomes

LOAD	C100	(1)
ADD	C101	(5)
ADD	C102	(9)
ADD	C103	(13)
ADD	C104	(17)
ADD	C105	(21)

The numbers in parentheses are the contents of each memory location and the sum, 66, is in the accumulator at the end.

2-7.1   Program Documentation

Before discussing more difficult problems further, we need to digress to a very important topic—documentation. *Program documentation* means providing a *written description* of what a program is actually doing so that *someone other than the person who first wrote the program might understand it*. Poor documentation has often been the bane of many programs. A popular quip is that programmers can gain absolute job security by writing useful programs and then failing to document them. Then no one else will ever know how they work.

There are three ways of documenting programs:

1. A written description of how the program works

2. Comments

3. Flowcharts

A *written description* is simply an attempt to explain in words how the program works. For a long program it can get very long. After writing the programs for the laboratory exercises, the instructor may ask for a report containing a written description.

The instructions in any program should always be accompanied by *comments*. A cogent comment is perhaps the best way to explain the function of an instruction or group of instructions. Poor comments should be avoided, however. One of the most common mistakes is to state in a comment what the instruction is doing when it is obvious from the code. For example, comments on the program in the previous section might be "load the number in C100, add the next number," and so on. These comments are redundant and unenlightening. The code makes it obvious that the program is doing these things. A better comment, which would apply to all the instructions in the program, might be "add all the numbers in the buffer." A **buffer** is an area of memory reserved for specific data. Here the buffer would be memory locations C100–C105.

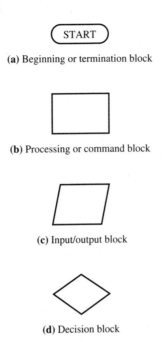

(a) Beginning or termination block

(b) Processing or command block

(c) Input/output block

(d) Decision block

Figure 2-9 The most common standard flowchart symbols. (*Greenfield/Wray, USING MI-CROPROCESSORS AND MICROCOMPUTERS: The Motorola Family, 2/E © 1988, p. 62. Reprinted by permission of Prentice-Hall, Inc., Englewood Cliffs, N.J.*)

Flowcharts are used to illustrate the flow of a program by using specific symbols. The symbols used in this book, which are the most commonly used symbols, are shown in Figure 2-9. They are

1. The *oval* symbol   This is either a beginning or a termination box. It is used simply to denote the start or the end of a program.

2. The *command* block   Inside the rectangular block there are statements that tell what must be done at this point in the program.

3. The *decision* box   This diamond-shaped box contains a question. Typically, there are two output paths. The program will follow one path if the answer to the question is yes and the other path if the answer is no. There might be three output paths when a comparison between two numbers is made. The paths correspond to the greater than, less than, and equal possibilities.

In a flowchart the lines and arrows between the boxes indicate the flow of the program. Flowcharts also use other symbols, but it is not necessary to discuss them here.

E X A M P L E   2 - 3

Draw a flowchart for the problem of the previous section.

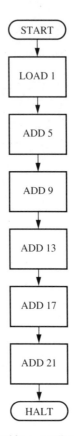

Figure 2-10   A flowchart for the problem: ADD 1, 5, 9, 13, 17, and 21.

Solution

The solution is shown in Figure 2-10. It consists simply of a start box, six command boxes, and a stop box. A decision box is not used because no decisions were made in this program.

2-7.2   JUMPs and BRANCHes

JUMPs and BRANCHes are instructions that *alter the flow* of a program. They do so by placing a new number in the PC; this causes the program to execute the next instruction at the new address in the μP rather than at the next sequential address.

A JUMP instruction causes the program to go to an *absolute address*. A BRANCH instruction causes the program to go an address *relative to its present location*. An analogy might help clarify the difference. Suppose you live at number 6 Main Street and a stranger knocks at your door and asks, "Where do the Smiths live?" There are two possible answers:

1. The Smiths live at number 12 Main Street.

2. The Smiths live three houses up the block.

The first answer is like a JUMP. We are directing the stranger to an absolute address. The second answer is like a BRANCH; we are telling the stranger to go to a house *relative* to where he or she is presently. While JUMPs may seem more precise, suppose the post office were suddenly to change the numbering system on Main Street by adding 50 to each house number. Now the first answer would have to change, but the second answer would still be correct. This may seem farfetched for Main Street, but the situation occurs often in programming. For this reason BRANCHes are usually preferred to JUMPs.

There are two types of JUMPs and BRANCHes: *conditional* and *unconditional*. The program will *always take an unconditional* JUMP or BRANCH. Conditional JUMPs or BRANCHes will be taken *only if certain conditions are met*. They correspond to the exits from a *decision box* in a flowchart. Conditional JUMPs or BRANCHes are really computer decisions.

In the **68HC11** there is only one JUMP instruction; it is unconditional. There are many BRANCH instructions. The most common BRANCH instructions are listed in the following table:

Mnemonic	Instruction	Action
BRA	BRanch Always	Unconditional branch
BPL	Branch PLus	Branch only if the results of the last instruction were positive.
BMI	Branch on MInus	Branch only if the results of the last instruction were negative.
BEQ	Branch on EQual	Branch only if the results of the last instruction were equal to 0.
BNE	Branch on Not Equal	Branch only if the results of the last instruction were *not* equal to 0.

E X A M P L E   2 - 4

Consider the following program segment:

```
LOAD   C101
ADD    C102
BEQ
```

Under what conditions will the program branch?

Solution

The program will branch only if the number in C102 is the negative equivalent of the number in C101. Only then will their sum be 0 and only then will the BEQ cause a branch.

2-7.3   Program Loops

A **loop** is a sequence of instructions that will be executed several times as the program progresses. Almost all programs use loops. A loop is characterized by having a START (where the loop begins), an END (where the loop terminates), and a *loop counter*. Typically, the loop counter is loaded with the number of times the program is to traverse the loop before the loop is entered. Then it is reduced by one (decremented) each time through the loop. When the loop counter reaches 0, the program exits the loop and proceeds to the next instruction.

The operation of a loop is shown in Figure 2-11. Observe that the last two instructions are a DECREMENT of the loop counter and a branch back

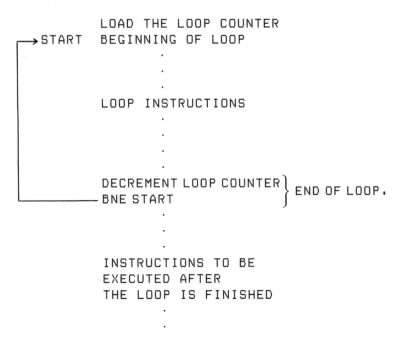

Figure 2-11   A typical program loop.

to the start of the loop if the counter is not 0. When the counter is 0, the program progresses to the next instruction. The use of arrows to show where the BRANCH goes helps clarify the flow of the program.

2-7.4   A More Advanced Problem

In Section 2-7 the numbers 1, 5, 9, 13, 17, and 21 were added using straight-line programming with no loops. Now consider the following problem:

Add the numbers 1, 5, 9, ..., 20,001.

Theoretically, it can be done by using the same methods as before, but the program would be long, tedious, and impractical. This problem must be attacked by using thought instead of brute force. The first thought might be, "How would this problem be handled in BASIC?"*

A simple BASIC program to solve this problem is shown in Figure 2-12. First, the program is *initialized*. **Initialization** means *setting the variables*

```
LIST
3   S = 0
5   N = 1
7   DEFDBL S
10  FOR I = 1 TO 5001
20  S = S + N
30  N = N + 4
40  NEXT I
50  PRINT N,S
60  END
Ok
RUN
    20005           50015001
Ok
```

Figure 2-12   A BASIC program for the problem in Section 2-9.4.

to the numbers they should have at the beginning of the program. In this problem the first number to be added is 1 and the sum, at the start, must be 0. Lines 3 and 5 set these values. Line 7 simply declares the sum, S, as a double precision variable, so it will print out in integer notation.

The addition is done in a loop. The loop starts with the FOR statement (line 10). First, the number to be added, N, is added to the sum. Then N is increased by 4. The loop ends at the NEXT statement. The program goes through this loop 5001 times before falling through and printing the results.

The number 5001 is the number in the loop counter. It is determined as follows:

1. At the start of the first loop $N = 1$.

2. At the start of the second loop $N = 5$.

3. At the start of the third loop $N = 9$, and so on.

4. With a little algebra we find that

$$N = 4*L - 3 \tag{2-1}$$

where L is the number of times the loop must be traversed. We want the

* Those who are not familiar with BASIC may skip the paragraphs that explain the BASIC program. They should, however, still be able to understand the material following this explanation.

loop to terminate after $N = 20001$. Substituting 20001 in Equation (2-1) gives an L of 5001. Thus the loop should be traversed 5001 times.

The correct results are shown in the printout. Do not worry about the fact that $N = 20005$. That is because N was increased in the last loop but was not yet added to the result.

Now that the problem has been solved using a higher level language, BASIC, we will consider how to solve it using our rudimentary computer with only LOAD, ADD, STORE, and BRANCH instructions available.

Actually the assembly language program for this problem is very similar to the BASIC program. We must keep track of three numbers: S, the sum; N, the number to be added; and L, the loop counter. An elementary flowchart, shown in Figure 2-13, can be drawn. This flowchart shows the action

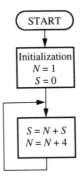

Figure 2-13 A partial flowchart for the problem in Section 2-7.4.

of the loop, but there is a problem—the loop never terminates. A complete flowchart that takes termination into account is shown in Figure 2-14.

Now that the flowchart is complete, it must be translated into assembly language instructions. Perhaps the first step is to assign memory locations for all the variables. Arbitrarily, let us select C100 for S, C101 for N, and C102 for L. Note that the number 4 is used. This number is called a *constant* and can also be placed in memory, at C103, for example.*

Before starting the program, the proper numbers must be written into these locations (0 into C100, 1 into C101, 5001 into C102, and 4 into C103. Now the coding can be written almost directly from the flowchart as shown in Figure 2-15. The comments correspond to the flowchart. Note that not every instruction requires 3 bytes, as we will see when we study the **68HC11**.

---

* In this example it is certainly true that a single byte is not large enough to accommodate the variables. However this complication is best ignored at this time. The student may imagine that each memory location is large enough to hold the required data.

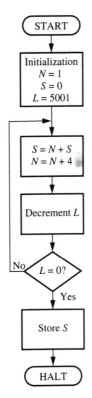

Figure 2-14 A complete flowchart for the problem in Section 2-7.4.

2-7.5 Event Detection

In the previous section a loop counter was used; the loop ended when the counter decremented to 0. *Event detection* is an alternate way to terminate a loop. This means waiting for an event to occur that will cause the loop to terminate. In this problem the loop should continue until $N = 20001$. This means that the loop can terminate when N is greater than 20001. This is the event. The code of Figure 2-15 can be modified by replacing the DECRE-MENT ACCUMULATOR instructions with instructions such as

LOAD C101; LOAD N

SUBTRACT 20002

If the results of the subtraction are negative, the loop should continue; otherwise it should stop. Thus a BMI (Branch on Minus) instruction can be used to branch back to the start of the loop.

Locations	Instructions	Comments
0020, 0021, 0022	LOAD C101	S = S + N
0023, 0024, 0025	ADD C100	
0026, 0027, 0028	STORE C100	
0029, 002A, 002B	LOAD C101	N = N + 4
002C, 002D, 002E	ADD C103	
002F, 0030, 0031	STORE C101	
0032, 0033, 0034	LOAD C102	L = L − 1
0035	DECREMENT ACCUMULATOR	(Decrement loop counter)
0036, 0037, 0038	STORE C102	If loop counter ≠ 0
0039, 003A	BNE TO 0020	go back to the start
		of the loop and repeat.
003B, 003C, 003D	LOAD C100	When the program arrives
		here the loop is finished.
003E, 003F, 0040	STORE	Store the results.
0041, etc.		Halt or continue.

Figure 2-15   The code for the flowchart in Figure 2-14.

Summary

This chapter introduced the operation of a computer. First, all the hardware components were listed and described. Particular emphasis was placed on the various forms of memories available. Then the operation of the LOAD, ADD, and STORE instructions was discussed in detail to show how the hardware registers and instructions work together.

Distinctions were drawn among machine, assembly, and high level languages. Program documentation, a very important part of programming, was introduced, and flowcharts and comments were discussed. Finally, JUMP and BRANCH instructions were explained, and a program using them was written.

Glossary

Access time   The time required for a memory to provide valid data after the address lines or control lines are stable.

Accumulator   A register in a computer that contains an operand used in an arithmetic operation.

ALU—Arithmetic-Logic Unit   The component of a computer that performs the arithmetic and logic operations.

Assembly Language   A language for writing instructions that uses mnemonics instead of numbers.

Buffer   A designated area of memory.

Bus   A group of wires that carries a set of signals between digital devices, such as between a μP and the memory.

Clock   The oscillator that controls the timing in a μP. Typically, it is crystal-controlled.

Cycle time   The minimum time the address, data, and control lines in a memory must be stable in order to write data.

Dynamic RAM   A RAM that stores its information on capacitors and has to be refreshed periodically.

EEPROM   An Electrically Erasable Programmable ROM. It can be altered while the IC remains in its socket. EPROMs must be removed and changed using a programmer.

EPROM   An Erasable Programmable ROM.

Flip-Flop   An electronic circuit that functions as a 1-bit memory.

Flowchart   A graphic method used to outline or show the action of a program.

Initialization   Setting initial values into the proper registers and memory locations at the beginning of the program.

Instruction Set   The set of instructions available in a computer or μP.

Load   To read data from memory into a μP.

Looping   Executing the same sequence of instructions several times.

Machine language   Instructions expressed as a set of numbers that the computer can understand.

Monitor   A program in the computer that takes control when it is turned on. In another context, the term monitor also refers to the CRT screen used with a computer.

Motherboard   The large board in a PC that contains most ICs.

Nondestructive readout   Reading data from memory where the process of reading does not change the data in memory.

Op code—Operation Code   The part of an instruction that specifies the operation to be performed by that instruction.

PC—Program Counter   The register that tells the computer where to find the next instruction. In other contexts, PC stands for Personal Computer.

Program   A set of instructions that cause the computer to perform a particular operation.

Programmer   The person who writes the instructions for a computer. It is also the name for the device that places information into an EPROM.

RAM—Random Access Memory   A memory that can be both read and written to.

Refresh   The process of restoring or rewriting data in dynamic RAM, so that it does not decay and become lost.

Register   A part of a μP that holds a group of related bits usually containing a character, command, or number.

ROM—Read Only Memory   Memory that can only be read during normal computer operation.

Software   The totality of the programs written for a particular computer or μP.

Static RAM   A RAM that stores its information in flip-flops.

Store   To write data from a computer's accumulator to its memory.

Volatility   Volatile memories lose their data when power is disconnected.

Word   The number of bits read from or written to the memory during each memory access.

Problems

Section 2-2

2-1 The size of a memory is 256K bytes.
   a. How many bits in the address bus?
   b. How many bits in the data bus?
   c. What other lines are required between the memory and the μP?

2-2 For Problem 2-1, sketch the buses between the memory and the μP
   a. If the memory is RAM.
   b. If the memory is ROM.

2-3 An Intel **8086** or **8088** has a 20-bit address bus. How many locations can such a μP address?

Section 2-6

2-4 Write the assembly language instructions corresponding to the following statement:

$$C = A - B + C + 2 * D$$

2-5 Assume a MULTIPLICATION instruction is available. Write the assembly language instructions for the equation:

$$E = (A + B) * (C + D)$$

Section 2-7

2-6 You are to multiply the numbers A, B, C . . . together and to stop when the product is greater than 1500. Draw a flowchart for this problem.

2-7 Your alarm clock goes off at 8 o'clock every day. If it's Saturday or Sunday, you shut it off and go back to bed. If it's Wednesday, you can sleep an extra hour. You then eat breakfast. If it's raining, you take an umbrella. If the car starts, you go to work; otherwise you go back to bed. Draw the flowchart for these activities.

3

Introduction to the 68HC11

INTRODUCTION

The material presented in the two previous chapters was general; it applied to any computer. The remainder of this book considers only the **68HC11**. In this chapter the register set, modes of operation, and some of the instructions, basically the LOAD, STORE, and ADD instructions, are covered.

INSTRUCTIONAL OBJECTIVES

After reading the chapter, the student should be able to

- List the registers in the **68HC11** and explain their function.
- Determine exactly what an instruction does by referring to its description in Appendix A.
- Determine which modes are appropriate for each instruction and be able to use each instruction in its proper modes.
- Load an index register, either in the immediate mode or from memory.
- Determine how the condition codes will be set as the result of an operation.
- Calculate the offsets for BRANCH instructions.
- Write the code for programs using the instructions covered in this chapter.

SELF-EVALUATION QUESTIONS

Watch for the answers to the following questions as you read the chapter. They should help you to understand the material presented:

1. How is the D accumulator related to the A and B accumulators?
2. What is the difference between the direct and the extended modes?
3. What is a prebyte? What is its function?
4. What conditions cause overflow?
5. What is the difference between a SUBTRACT instruction and a COMPARE instruction?

6. Why does a COMPARE instruction often precede a BRANCH instruction?

7. What are the differences between a JUMP and a BRANCH?

8. What are the range limitations of BRANCH instructions?

3-1 A First Look at the 68HC11

The **68HC11** microprocessor (μP) comes as a single Integrated Circuit (IC) packaged in either a 52-pin Plastic Leadless Chip Carrier (PLCC) or a 48-

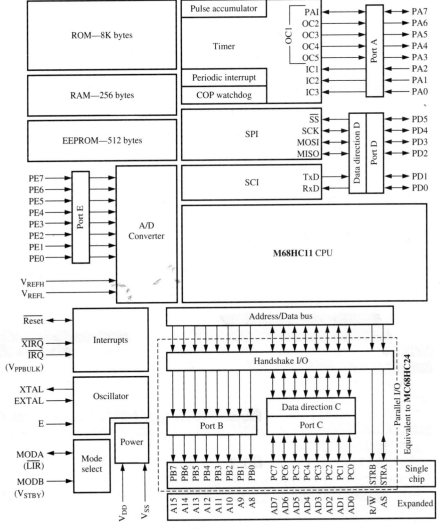

Figure 3-1   A block diagram of the 68HC11. *(Redrawn Courtesy of Motorola, Inc.)*

pin Dual In-line Package (DIP). Photographs of these packages have been shown in Chapter 1 (see Figure 1-1).

Figure 3-1 is a block diagram of the **68HC11**. In addition to the Central Processing Unit (CPU), the box marked **M68HC11CPU**, the IC chip contains ROM, RAM, A/D converters, and so on. This chapter and the three following chapters are concerned with the *programming* of the CPU, and discussion of the other components of the μP will be deferred until Chapter 8. Now the reader should visualize the simplified computer system shown in Figure

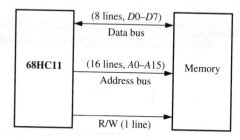

Figure 3-2   A simplified computer system.

3-2, which consists only of the **68HC11** CPU and a RAM. The figure confirms what we have stated in Chapter 2:

1. All memory addresses require 16 bits.

2. All data transfers use an 8-bit word.

3-1.1   The Register Set of the 68HC11

The register set of the **68HC11** is shown in Figure 3-3. Observe that some of the registers are 8 bits long and some are 16. *Do not attempt to place an 8-bit number in a 16-bit register, or vice versa.* This is a mistake that is commonly made by beginning students.

The registers in the **68HC11** are

1. Two 8-bit accumulators, A and B   Motorola μPs contain two accumulators. Instructions can work with either one of them. The 16-bit D accumulator shown in Figure 3-3 actually consists of the A and B registers joined together (*concatenated* is the long word that impresses people). It is also used as a 16-bit accumulator. But remember, there are *not* three different accumulators in the **68HC11**. When the **D** (for **Double**) **accumulator** is used, it is the concatenation of A and B with A as the most significant byte of the double accumulator.

2. The Program Counter (PC)   As discussed in Chapter 2, this 16-bit register provides the address of the next instruction to be executed.

3. Two index registers, X and Y   The index registers are generally associated with addresses and are 16 bits long. They are discussed in Section 3-4.

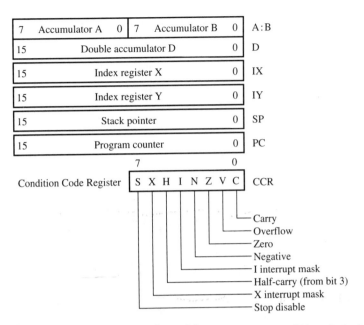

Figure 3-3   The 68HC11 programmer's model. *(Redrawn Courtesy of Motorola, Inc.)*

4. The Stack Pointer (SP)   This 16-bit register is also associated with addresses, but it is not used in this chapter. A discussion of it is deferred until Section 6-3.

5. The Condition Code Register (CCR)   The CCR is an 8-bit register that contains information about the results of the last arithmetic or logic operation. It is discussed in Section 3-5.

E X A M P L E   3 - 1

If the A accumulator contains the number AA and the B accumulator contains BB, what does the instruction ADD 1234* to D accomplish?

Solution

The D accumulator must contain the number AABB. When 1234 is added to it, the results are BCEF. This sum is written to the D accumulator. Thus the A accumulator, which is the top half of D, contains BC and the B accumulator contains EF.

3-2   The Instruction Set of the 68HC11

Each instruction of the **68HC11** is thoroughly described in Appendix A. Unfortunately, this appendix is 112 pages long, but the instructions can be understood if they are discussed one at a time.

* In this chapter all numbers will be considered as hexadecimal. This simplifies the chapter.

Load Accumulator LDA

Operation: ACCX ◆ (M)

Description: Loads the contents of memory into the 8-bit accumulator. The condition codes are set according to the data.

Condition codes and Boolean formulae:

S	X	H	I	N	Z	V	C
—	—	—	—	↕	↕	0	—

N R7
Set if MSB of result is set; cleared otherwise

Z $\overline{R7} \cdot \overline{R6} \cdot \overline{R5} \cdot \overline{R4} \cdot \overline{R3} \cdot \overline{R2} \cdot \overline{R1} \cdot \overline{R0}$
Set if result is $00; cleared otherwise

V 0
Cleared

Source form: LDAA (opr); LDAB (opr)

Addressing modes, machine code, and cycle-by-cycle execution:

Cycle	LDAA (IMM)			LDAA (DIR)			LDAA (EXT)			LDAA (IND, X)			LDAA (IND, Y)		
	Addr	Data	R/W̄	Addr	Data	R/W̄	Addr	Data	R/W̄	Addr	Data	R/W̄	Addr	Data	R/W̄
1	Op	86	1	Op	96	1	Op	B6	1	Op	A6	1	Op	18	1
2	Op+1	ii	1	Op+1	dd	1	Op+1	hh	1	Op+1	ff	1	Op+1	A6	1
3				00dd	(00dd)	1	Op+2	ll	1	FFFF	—	1	Op+2	ff	1
4							hhll	(hhll)	1	X+ff	(X+ff)	1	FFFF	—	1
5													Y+ff	(Y+ff)	1

Cycle	LDAB (IMM)			LDAB (DIR)			LDAB (EXT)			LDAB (IND, X)			LDAB (IND, Y)		
	Addr	Data	R/W̄	Addr	Data	R/W̄	Addr	Data	R/W̄	Addr	Data	R/W̄	Addr	Data	R/W̄
1	Op	C6	1	Op	D6	1	Op	F6	1	Op	E6	1	Op	18	1
2	Op+1	ii	1	Op+1	dd	1	Op+1	hh	1	Op+1	ff	1	Op+1	E6	1
3				00dd	(00dd)	1	Op+2	ll	1	FFFF	—	1	Op+2	ff	1
4							hhll	(hhll)	1	X+ff	(X+ff)	1	FFFF	—	1
5													Y+ff	(Y+ff)	1

Figure 3-4   The LDA instruction. (*Redrawn Courtesy of Motorola, Inc.*)

The most commonly used instruction is probably the LOAD instruction. The page of Appendix A that describes the LOAD instruction is reproduced here as Figure 3-4 for convenience. At the top of the figure is the Motorola *mnemonic* for the LOAD instruction. It is LDA, which will be used throughout the remainder of the book. It stands for "Load Accumulator," as also shown on the top line of the figure. The three-letter mnemonic LDA is insufficient, however, because there are two possible accumulators that can be loaded. Therefore an additional letter, A or B, must be added to distinguish between the accumulators. This is shown as the *source form* at about the middle of the figure and is the form the EVB (see Chap. 4) will recognize. Many other instructions, such as the ADD instruction, will also take an extra letter to specify which accumulator is involved.

The second line of the figure is the *operation*. It is shown as ACCX←(M). In standard programming notation the parentheses mean "the contents of." This graphical statement means that the instruction causes the contents of memory to be copied into ACCX, the specified accumulator (accumulator A or B).

The *description* is simply a written description of the instruction's operation. This is followed by the *condition codes*. These lines and equations tell how the condition codes (sometimes called **flags**) are affected by the instruction. Condition codes are discussed in Section 3-5.

There are two tables, labeled *addressing modes, machine code*, and *cycle-by-cycle execution* at the bottom of the figure. One table is for the LDAA (Load Accumulator A) instruction and the other is for LDAB. The meaning of these tables is discussed in the next section.

3-3 Instruction Modes

For most instructions in a computer there are *variations* in the way a particular instruction can be executed. Each variation is called a *mode*. Table 3-1 lists the modes that the **68HC11** can use to execute its instructions.

Table 3-1 The Instruction Modes of the 68HC11

Mode	Designation
Immediate	IMM
Direct	DIR
Extended	EXT
Indexed by X	IND,X
Indexed by Y	IND,Y
Inherent	INH
Relative	REL

Not all modes apply to all instructions. The LDAA and LDAB instructions of Figure 3-4 can be executed in any of the first five modes listed in Table 3-1.

3-3.1 The Immediate Mode

The immediate mode is used when the number to be used in the instruction is *known* at the start of the program and will not change. The *pound (#)* symbol is used to denote immediate instructions. For example, the instruction

LDAA   #7B

is an immediate instruction (indicated by the # sign) that will load the number 7B into the A accumulator.

The execution of the LDAA # instruction is described by the box in the upper left-hand corner of the tables of Figure 3-4. It is reproduced here as

Cycle	LDAA (IMM)		
	Addr	Data	R/$\overline{W}$
1	Op	86	1
2	Op+1	ii	1
3			
4			
5			

Cycle	LDAB (IMM)		
	Addr	Data	R/$\overline{W}$
1	Op	C6	1
2	Op+1	ii	1
3			
4			
5			

Figure 3-5   **The LDA instruction in the immediate mode.** (*Redrawn Courtesy of Motorola, Inc.*)

Figure 3-5 for convenience. From it we can obtain the following information:

1. The instruction is indeed a load immediate. It is titled LDAA (IMM).

2. The instruction takes two cycles to complete. A **cycle** is the *time of one clock cycle* as determined by the *crystal clock* connected to the **68HC11**. It is precisely 500 ns for most **68HC11s**. Thus the LDAA would require exactly 1 μs to execute.

3. The instruction requires 2 bytes of memory. The first byte is the Op code for this instruction. The table tells us that the Op code is 86 for a load A immediate. The second byte of the instruction is labeled "ii." Here ii stands for the *immediate operand* to be used, 7B in the previous example. Observe that the immediate operand is 1 byte long, as it must be placed in an 8-bit register.

4. The R/W line is high or 1, indicating that both memory cycles are read cycles. (A low or 0 in the column would indicate a write cycle.)

5. The instruction executes as follows: First, a memory read is required to find the Op code. The memory reads out 86. Then a second memory read is used to find the immediate data. The memory reads out ii. This data is then placed in accumulator A and the computer goes on to the next instruction.

3-3.2 The Direct and Extended Modes

Both the extended and direct addressing modes use a byte from the memory. The *extended mode* means that the 16-bit memory address of the byte is found in the *two* locations of the program following the Op code. The sample programs presented in Chapter 2 used extended mode instructions.

From Figure 3-6, which is also part of Figure 3-4, repeated here for convenience, we can see that the LDAA (EXT) or the LDAB (EXT) each

LDAA (EXT)			LDAB (EXT)		
Addr	Data	R/$\overline{W}$	Addr	Data	R/$\overline{W}$
Op	B6	1	Op	F6	1
Op+1	hh	1	Op+1	hh	1
Op+2	ll	1	Op+2	ll	1
hhll	(hhll)	1	hhll	(hhll)	1

Figure 3-6 The LDA instruction in the extended mode. (*Redrawn Courtesy of Motorola, Inc.*)

takes four cycles, but they only occupy 3 bytes of memory. The Op code (B6 or F6) is fetched during the first cycle. Then the address is fetched during the next two cycles. The address is symbolized by hhll in Figure 3-6. During the fourth cycle the address hhll is placed on the address bus and the *contents* of this address (hhll) are read from memory and written to the selected accumulator.

Direct instructions save one memory cycle and 1 byte of program space by only using 1 byte to specify the memory location. But how can 1 byte (8 bits) specify a 16-bit address? *In direct instructions the eight MSBs of the address are all 0.* Thus direct addresses can only specify one of the first 256 locations, from 00 to FF. If the address is higher than FF (such as 0123 or C101), extended addressing must be used. Direct addresses save both time and space, but they cannot be used in all cases.

E X A M P L E   3 - 2

If hex location 007E contains the number CC, what number is placed in accumulator A as a result of the following instructions?

a. LDAA   #7E

b. LDAA $7E

Solution

The first instruction is an immediate instruction. It simply places 7E in the accumulator.

The second instruction is a direct instruction (the $ indicates a hex number or address). It places the *contents* of 7E, the number CC, in the accumulator. This example should clarify the difference between immediate and direct instructions.

3-3.3 The Indexed Modes

Figure 3-3 shows that there are two 16-bit index registers in the **68HC11**. *Indexed instructions use these index registers.* For the LDA instruction, which is typical of most instructions, there are two forms of the indexed instructions:

LDAA (or LDAB) offset,X   Instruction indexed by X

LDAA (or LDAB) offset,Y   Instruction indexed by Y

The **offset** is an 8-bit number that follows the Op code in an indexed instruction. The *memory address* that the instruction references is given by *the sum of the 8-bit offset and the 16-bit contents* of the designated index register. If the programmer decides to use an index register directly (with no offset) a byte of 00 must follow the Op code because the **68HC11** *always considers the byte following the Op code as an offset.*

E X A M P L E   3 - 3

If the X index register (hereafter simply called X) contains the number C121, what does the instruction LDAB 33,X do?

Solution

LDAB means load a byte into the B accumulator. The address of the byte is given by the sum of the contents of X, C121, and the offset, 33. Consequently, the *contents* of C154 are loaded into B.

Figure 3-7, which is also taken from Figure 3-4, shows the Op codes for loading accumulator A or B in the indexed mode. If X is the index register to be used, the instruction requires 2 bytes. The first byte is the Op code, A6, and the second byte is the offset. The code is identical with respect to

| LDAA (IND, X) | | | LDAA (IND, Y) | | |
Addr	Data	R/$\overline{\text{W}}$	Addr	Data	R/$\overline{\text{W}}$
Op	A6	1	Op	18	1
Op+1	ff	1	Op+1	A6	1
FFFF	—	1	Op+2	ff	1
X+ff	(X+ff)	1	FFFF	—	1
			Y+ff	(Y+ff)	1

| LDAB (IND, X) | | | LDAB (IND, Y) | | |
Addr	Data	R/$\overline{\text{W}}$	Addr	Data	R/$\overline{\text{W}}$
Op	E6	1	Op	18	1
Op+1	ff	1	Op+1	E6	1
FFFF	—	1	Op+2	ff	1
X+ff	(X+ff)	1	FFFF	—	1
			Y+ff	(Y+ff)	1

Figure 3-7   The LDA instruction in the indexed mode. *(Redrawn Courtesy of Motorola, Inc.)*

Y, except that instructions using Y for an index register have a *prebyte* (which is usually 18). The **prebyte** indicates that the instruction uses Y instead of X. The rest of the code is identical to the code used for X.

E X A M P L E   3 - 4

What is the code for LDAA 4A,Y?

Solution

This instruction requires 3 bytes—the prebyte, the Op code, and the offset, as Figure 3-7 shows. The code is

    18 A6 4A

3-3.4   Inherent Instructions

Inherent instructions are instructions that do not require any further information from memory. After the Op code is fetched the computer knows exactly what to do and does not need to make any additional memory reference. Examples of inherent instructions are CLEAR A, INCREMENT A, SHIFT B, and so on. INCREMENT A, for example, simply adds 1 to the contents of A without having to obtain any further information from memory. Most inherent instructions are only 1 byte long and take two cycles.

The LOAD instructions are not inherent. After the Op code is fetched and the μP determines that a LOAD instruction is to be executed, the μP always has to go back to memory for more information (the data in immediate mode, the address in direct or extended modes, and the offset in indexed modes).

3-3.5   The Relative Mode

The relative mode applies to BRANCH instructions. They are discussed in Section 3-7.2.

3-4   Index Register Instructions

In the discussion of registers we often say, "assume the contents of the index register are . . . ." The reader may wonder how to set a particular value into an index register. In this section we consider instructions that *control* the index registers.

There are two types of instructions that concern an index register and you must be able to distinguish between them. They are

1. Instructions that *use* the contents of the index register. These are the indexed instructions discussed in Section 3-3.3. The index register is used with its offset to calculate the memory address of the operand or data.

Table 3-2   Instructions that Affect Index Registers

INX INY	Increment
DEX DEY	Decrement
CPX CPY	Compare
LDX LDY	Load
STX STY	Store

2. Instructions that *affect* or *change* the contents of the index register. A short list of the most commonly used of these instructions is given in Table 3-2.

The INCREMENT and DECREMENT instructions simply add or subtract 1 to X or Y. They are often used as loop counters or to step a program through a memory buffer.

The COMPARE instructions are discussed in Section 3-6.

3-4.1   The LOAD INDEX REGISTER Instructions

The LOAD INDEX REGISTER instructions (LDX and LDY) are used to place a number in one of the index registers. They are most frequently used in the immediate mode, where the *2 bytes* following the Op code specify the address. Note that 2 bytes are needed for this instruction (as opposed to 1 byte for a LDA #) because index registers are 16 bits long. For example, the instruction

```
LDX    #ABCD
```

codes as CE AB CD, where CE is the Op code, as determined from Appendix A. This instruction requires 3 bytes in memory and loads the number ABCD into X. The similar instruction to load Y would code as 18 CE AB CD. Here 18 CE is the Op code, and the instruction takes 4 bytes. *Warning*: Do not forget the # sign in this instruction. Students often cause disasters by making this mistake.

Index registers can also be loaded from memory, using the direct, extended, or indexed modes. Each mode specifies one address. But because an index register needs 16 bits, the contents of the specified address *and the next sequential address* are used. For example, the instruction LDX $ABCD loads the contents of ABCD and ABCE into the index register.

E X A M P L E   3 - 5

What does the following program segment do?

    LDY   #1000
    LDX   33,Y

Solution

The first instruction is in the immediate mode and loads 1000 into Y. The second instruction is in the indexed mode; it calculates an address of 1033. The contents of 1033 and 1034 are loaded into X.

3-4.2 STORE INDEX REGISTER Instructions

STORE INDEX REGISTER instructions are used when the programmer must preserve the contents of an index register. They write the contents of the index register into memory. These contents can be retrieved when needed by LOAD INDEX REGISTER instructions.

The STORE INDEX REGISTER instructions (STX and STY) can be executed in the direct, extended, and indexed modes. The index register will be stored in the addressed location *and the next location*. Be careful; a STORE INDEX REGISTER instruction changes two locations.

E X A M P L E   3 - 6

A programmer needs to preserve X and Y. The following segment was written:

    STX $1000
    STY $1001

What is wrong with the program?

Solution

The first instruction stores X in 1000 and 1001. The second instruction stores Y in 1001 and 1002. When the most significant byte of Y is written to 1001, it overwrites the least significant byte of X; therefore X is not preserved. The programmer forgot to allocate 2 bytes of memory for a STORE INDEX REGISTER instruction. (See Prob. 3-2.)

3-5 The Condition Code Register

The CCR, shown in Figure 3-8, consists of 8 bits. The lower 4 bits retain the *characteristics* of the last arithmetic or logic operation and are discussed in this section. Discussion of the upper 4 bits is deferred until the conditions that control them are considered.

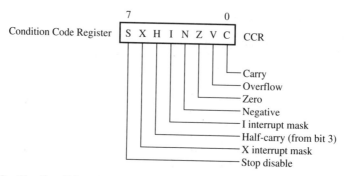

Figure 3-8   The Condition Code Register (CCR). (*Redrawn Courtesy of Motorola, Inc.*)

3-5.1   The N Bit

Bit 3 of the CCR is the N (Negative) bit. It is set whenever the result of the last operation is negative and is cleared whenever the result is positive. Its logic equation is

$$N = R7$$

where R7 means bit 7 of the *result*. If the result of an arithmetic or logical operation is negative, the N bit is set; otherwise it is cleared.

3-5.2   The Z Bit

The Z (zero) bit is set whenever the result of the last operation is exactly 0. Its logic equation is

$$Z = \overline{R7} \cdot \overline{R6} \cdot \overline{R5} \cdot \overline{R4} \cdot \overline{R3} \cdot \overline{R2} \cdot \overline{R1} \cdot \overline{R0}$$

which is another way of saying that all 8 bits of the result must be 0. Most operations clear the Z bit because most results are not 0. When the Z bit is set, the N bit is cleared because 0 is a positive number.

3-5.3   The Carry Bit

Bit 0 of the CCR is the carry bit or carry flag. It is generally set under the following conditions:

1. An addition that produces a carry.

2. A subtraction that produces a borrow.

3. A SHIFT or ROTATE instruction that shifts out a 1.

Remember that the converse will *clear* the carry flag (an addition that doesn't produce a carry will clear the flag, etc.)

3-5.4   The Overflow Bit

The V bit (for oVerflow) is set when there *may* be an error in a mathematical operation. High schools teach that adding two positive numbers will never give a negative result. With a computer, however, it's easy. For example, adding 64 and 64 gives C8. Here two positive numbers add to give a negative result, but this will set the V bit to warn the user that this result has occurred. Usually this is unimportant, but occasionally it indicates that an error has occurred. The V bit is like an insurance policy: you generally don't need it, but it should be there when you do.

When adding numbers, the V bit will be set if two positive numbers are added and give a negative result, or if two negative numbers are added and give a positive result. If numbers of *opposite* signs are added, there will *never be overflow*. The logic equation for overflow in addition is

$$V = X_7{\cdot}M_7{\cdot}\overline{R_7} + \overline{X_7}{\cdot}\overline{M_7}{\cdot}R_7 \tag{3-1}$$

where X_7 is the MSB of the first operand (augend), M_7 is the MSB of the second operand, and R_7 is the MSB of the result. Observe that the X_7, M_7, and R_7 all determine the signs of their respective numbers.

The first term in the equation ($X_7{\cdot}M_7{\cdot}\overline{R_7}$) is true only if $X_7 = M_7 = 1$ and $R_7 = 0$. But this implies that two negative numbers have been added (X and M) and have produced a positive result. The second term of the equation ($\overline{X_7}{\cdot}\overline{M_7}{\cdot}R_7$) implies that the sum of two positive numbers gives a negative result. The overflow equation takes care of both possibilities.

E X A M P L E   3 - 7

Consider the program segment

    LDAA   #8A

    ADDA   #AA

What are the states of the N, Z, V, and C flags after the second instruction?

Solution

The program first loads 8A into accumulator A and then adds AA to obtain a result of 34, with a carry. At the end the N flag is 0 because 34 is a positive number and Z is also 0 because the result is not 0. The C flag is a 1 because the addition produced a carry-out of the MSB, and V is also a 1 because two negative numbers were added and the result was positive. The addition corresponds to the first term in the overflow equation.

Overflow can also occur during subtraction. The operation of the subtraction instruction is described on page A-97 of Appendix A as ACCX (ACCX) − (M). It means that the contents of memory (M) or an immediate operand form the subtrahend. They are subtracted from the contents of an accumulator (the minuend) and the results go to that accumulator. The numbers can be represented by

$$X - M = R$$

where X, M, and R are all 8-bit numbers whose signs are given by their MSBs (X_7, M_7, and R_7).

For subtraction, the equation for overflow is

$$V = X_7 \cdot \overline{M_7} \cdot \overline{R_7} + \overline{X_7} \cdot M_7 \cdot R_7 \tag{3-2}$$

Consider the first term. It indicates that there was a negative number in the accumulator ($X_7 = 1$), that a positive number was subtracted from it ($M_7 = 0$), and that the result was positive ($R_7 = 0$). This is impossible mathematically but not in a computer, so the V flag will be set to warn the user. The second term indicates that a negative number was subtracted from a positive number and the results were negative. This also sets the V flag.

E X A M P L E   3 - 8

How are the flags set after each of the following operations?

a. 75 − D3

b. A3 − D0

Solution

a. The results of this subtraction are A2. The N flag is set because A2 is a negative number and the Z flag is cleared. The C flag is set because the subtrahend is larger (in absolute magnitude) than the minuend, or this subtraction generates a borrow. The V flag is also set because we have subtracted a negative number from a positive number and obtained a negative result. The second term of the overflow equation for subtraction is satisfied.

b. The results of this subtraction are D3. As in part (a), $Z = 0$ and $N = C = 1$. Now, however, there is no overflow, so $V = 0$. As Equation (3-2) implies, there will never be overflow from a subtraction operation if both the minuend and the subtrahend have the same sign.

To further clarify overflow, let us reconsider part (a) of Example 3-8. If the numbers 75 and D3 are converted to decimal, they become 117 and −45. The decimal answer is then 117 − (−45) = +162. But the number 162 is *out of bounds* because 8-bit numbers are restricted to the range from −128 to +127. This is why overflow occurred. Any decimal result that produces a number that is out of bounds also sets the V flag.

3-5.5 Determining the Condition Codes

Each instruction in Appendix A has a small chart that shows the way it affects the condition codes. This chart consists of four symbols:

↕ This symbol indicates that the condition code is set or cleared in accordance with the result.

− This symbol indicates that the condition code is *unchanged* by the instruction. After the instruction the condition code is the same as it was before the instruction.

0 This symbol indicates that the condition code was cleared by the instruction.

1 This symbol indicates that the condition code was set by the instruction.

E X A M P L E   3 - 9

Consider the instruction CLRA (CLEAR the A accumulator). What should the condition codes be after it is executed?

Solution

This instruction writes a 0 into A. After it is executed we would expect N, V, and C to be 0, but Z to be 1 because A = 0. If the chart on page A-41 is examined, these are indeed the results.

The careful student should examine the condition code chart for every instruction, but some general guidelines are listed for the instructions considered in this chapter:

- ADD and SUBTRACT instructions affect all four condition codes considered here.
- LOAD and STORE instructions set N and Z. Thus the *characteristics* of the number (Is it negative? Is it zero?) are retained. They clear the V bit on the assumption that the user is not interested in overflow if he or she is loading or storing. They do not affect the carry flag. This is important in multibyte arithmetic. The reader can refer to Figure 3-4 for confirmation.
- Branches never affect any condition codes.

3-6   COMPARE Instructions

To help the programmer, the **68HC11** has a rich set of COMPARE instructions. A COMPARE instruction operates as a *pseudo-subtraction*. Its operation is listed in Appendix A as

$$(ACCX) - (M)$$

which means the contents of M are subtracted from an accumulator. The *difference* between COMPARE and SUBTRACT instructions is that COMPARE instructions *discard* the results, whereas SUBTRACT instructions place their results in an accumulator. COMPARE instructions, however, set the four condition codes we have discussed in accordance with the result of the pseudo-subtraction. They generally precede conditional branches that use these condition codes to determine whether or not to branch. Thus *the only register that is ever altered by a COMPARE instruction is the CCR.*

3-6.1 Compare Accumulators

Compare accumulator instructions, labeled CMP in Appendix A, are relatively simple. They compare an 8-bit number to an 8-bit accumulator, A or B. If the immediate mode is used, the number to be compared follows the Op code. If one of the addressing modes is used, the number at the selected address is compared to the accumulator.

E X A M P L E   3 - 10

Consider the following program segment:

    CLRB
    CMPB   #75

a. What is the machine code for this segment?

b. What are the contents of B and the CCR after the segment is executed?

Solution

a. CLRB (Clear B) is an inherent instruction. It is found on page A-41 to be 5F.

CMP is found on page A-43. The Op code for CMPB in the immediate mode is C1. Thus the machine code becomes

    5F   Clear B
    C1   Compare B
    75   Immediate operand

b. The instruction causes the subtraction $00 - 75 = 8B$. The condition codes are set as follows.

The number 8B is negative and not 0. Therefore $N = 1$ and $Z = 0$. There is no overflow [Eq. (3-2) is not satisfied]. The C flag is set because the subtrahend is larger than the minuend or because a borrow out was generated.

The contents of B after the CMP instruction is still 00 because COMPARE instructions do not change the accumulator. If the CMPB were changed to a SUBB, the B accumulator would contain 8B.

3-6.2   Compare Index Registers

The COMPARE X (CPX) and COMPARE Y (CPY) instructions compare a number with the contents of the respective index register. Because the index registers are 16 bits long the number compared must also be 16 bits long. If the comparison uses the immediate mode, the contents of the *next two* memory locations are used to give the 16-bit immediate operand. If the comparison uses a memory mode, a single 8-bit memory address is selected, but the 16 bits required for the comparison come from that memory location and the *next* memory location. The operation of these instructions is listed in Appendix A as $(IX) - (M:M + 1)$, indicating that the number subtracted from the index register requires two locations, M and M + 1.

E X A M P L E   3 - 11

If location 0009 contains CC and 000A contains DD, what do X and the condition codes contain after the following program segment?

    LDX   #CCDD
    CPX   09

Solution

The first instruction is an immediate instruction that loads CCDD into X. The second instruction is a COMPARE using the direct mode because there are only 8 bits specified for the address. The other 8 bits are therefore 00. But this means that the contents of locations 9 and A are both used to form the 16-bit number to be compared. As we are comparing two identical numbers, the result will be 00, which will set the Z flag and clear the other three flags. At the end of the instruction the number in X will still be CCDD.

E X A M P L E   3 - 12

What numbers are compared by the following program segment?

    LDY   #1000
    CPY   33,Y

Solution

The first instruction loads 1000 into Y. The second instruction is in the IND,Y mode. The address is the offset, 33, plus the contents of Y, 1000, or 1033. Therefore the contents of 1033 and 1034 are compared to the contents of Y (they are subtracted from 1000) and the condition codes are set accordingly.

3-6.3   Compare Double Register

The last COMPARE instruction in the **68HC11** is the CPD. It compares the contents of the double register to a 16-bit number.

E X A M P L E   3 - 13

If A contains 77 and B contains 66, what does the instruction CPD #ABCD do?

Solution

This is an immediate compare of the double register. It subtracts ABCD from 7766 to give CB99, which is discarded. The operation sets the N, V, and C flags.

3-7   JUMPs and BRANCHes

JUMPs and BRANCHes alter the linear flow of a program by changing the number in the PC. This causes the next instruction to be executed at the new address in the PC rather than at the next sequential location. We have already used these instructions in the looping programs discussed in Chapter 2.

3-7.1   JUMP Instructions

JUMP instructions simply place the *target address* in the PC. The **target address** is *the address we want the program to jump to*. There is only one JUMP instruction in the **68HC11** and it is *unconditional*.

The JUMP instruction can be executed in one of two modes: extended or indexed. It most commonly uses the extended mode. For example, the instruction JMP $C100 simply causes the next instruction to be executed at $C100.

E X A M P L E   3 - 14

What is wrong with the following program?

```
    0025    JMP     $C100
              .
              .
              .
    C100    ADDA    #55
    C102    JMP     $0025
```

Solution

The first instruction causes the program to jump to C100, where it performs an addition, and then jumps back to 25, which causes it to jump back to C100 and repeat indefinitely. This is a simplified version of the *endless loop* program. Endless loops are programs that loop forever and do nothing. They are not recommended, but the programmer who hasn't inadvertently written at least one endless loop is very rare.

Actually the program of this example can be useful if an engineer is conducting a hardware test of the operation of the ADD instruction. Then the endless loop causes the μP to run repetitively, allowing the engineer to monitor its operation with an oscilloscope.

JUMP instructions can also be used in the indexed mode, using X or Y. Here the result of the address calculation is placed in the PC instead of used as a memory address. For example, if Y contains 1000, the instruction JMP 33,Y will cause the program to jump to 1033.

3-7.2 BRANCH Instructions

BRANCH instructions differ from JUMP instructions in that they do not supply a direct address. When a BRANCH instruction is used, the target address must be calculated, and it is always *relative* to the current PC location. Target addresses must be within about 128 locations, forwards or backwards, from the current PC.

The simpler BRANCH instructions in the **68HC11**, which are needed for the problems of this chapter, are listed in Table 3-3. Other, more complex BRANCH instructions are discussed in Chapter 5.

Table 3-3 Branch Instructions in the 68HC11

Instruction	Condition	Operation
BRA	None	BRanch Always
BCC	C = 0	Branch if Carry Clear
BCS	C = 1	Branch if Carry Set
BVC	V = 0	Branch if oVerflow Clear
BVS	V = 1	Branch if oVerflow Set
BNE	Z = 0	Branch on Not Equal to 0
BEQ	Z = 1	Branch on EQual to 0
BPL	N = 0	Branch on PLus
BMI	N = 1	Branch on MInus

The first instruction in the table, the BRA for BRanch Always, is the *unconditional* branch. It *always* branches. The other branches are all *conditional*. Each of these instructions first tests one of the four condition codes we have discussed. If the test is satisfied, the program branches; otherwise it continues by executing the next sequential instruction. For example, suppose the program is doing an addition and must branch to an error routine if overflow occurs. This can be done simply by following the ADD instruction by a BVS, which will branch only if the results of the addition set the V bit.

The BRANCH instructions in Table 3-3 are all 2-byte instructions. The first byte is the Op code, and it is followed by a byte of *offset* which is added to the PC to tell the program where to branch to. Unlike offsets used with index registers, *branch offsets can be positive or negative*, depending on the MSB of the offset. The branch location or target address is given by Equation (3-3).

Target address

= PC location of the Op code of the instruction *following* the branch

+ the offset

(3-3)

The offset can also be considered as the number of locations to be skipped, forwards or backwards. For example, if the offset is FB, −5, the program will go back five locations from the address of the Op code of the next instruction.

E X A M P L E   3 - 15

If the instruction in C100 is BRA 2A (which codes as 20 2A), where does the program branch to?

Solution

Because all BRANCH instructions discussed in this section are 2-byte instructions, the *address of the next instruction* is at C102. The program will branch to

C102 + 2A or C12C.

The instruction at C12C is the next instruction the program will execute.

The branch offsets can be positive or negative. A *positive offset* means that the program will branch *forwards* (the next instruction will be at a higher location). Conversely, a *negative offset* means that the program will branch *backwards*, to a lower location. Knowing the direction of the branch allows the user to check his or her calculations.

Offsets are 8-bit numbers that are added to the contents of the PC, a 16-bit number. To avoid problems, the offset should be converted to a 16-bit number by *extending* it. An 8-bit number is extended to a 16-bit number by the following rules:

1. If the MSB of the number is 0 (the number, or offset in this case, is positive), extend the number adding eight 0s; for example, the number 76 extended becomes the 16-bit number, 0076.

2. If the MSB of the number is 1 (the number is negative), extend the number by adding eight 1s; for example, AB extended becomes FFAB.

E X A M P L E   3 - 16

The instruction 25 xx is in location 10B. The 25 is the Op code for BCS (Branch if Carry Set). Where do you branch to if

a. The carry flag is clear.

b. The carry flag is set and xx equals
   (1) CC
   (2) 55

Solution

a. If the carry flag is clear, the program will not branch. The Op code, 25, occupies 10B and the offset must be in 10C. Therefore the address of the Op code of the next instruction must be in 10D. Because the program does not take the branch, it *falls through* and executes the instruction at 10D.

b. If the offset is CC, a negative number, the program must branch backwards. Then Equation (3-3) can be used to find the result:

$$
\begin{array}{ll}
 010D & \text{(the address of the Op code of the next instruction)} \\
+\ \underline{\text{FFCC}} & \text{(the given offset, CC, extended)} \\
 00D9 &
\end{array}
$$

   Therefore the program will branch to D9, which is indeed a backward branch.
   If the offset is a positive number, it must cause a forward branch (a branch to a higher location). The branch location is found by adding 010D and 0055 to give 0162. This is where the program will branch to.

E X A M P L E   3 - 17

What offset must be used to branch from 10B to

a. F0
b. 150

Solution

In this example the target (the location to be branched to) is known, and the offset is the unknown number. It can be found by algebraically transposing Equation (3-3).

   Offset = target − next instruction's Op code location

As in Example 3-16, the Op code of the next instruction is 10D. For part (a) the answer is

$$
\begin{array}{r}
00F0 \\
-\ 010D \\
\hline
FFE3
\end{array}
$$

Of course, FFE3 is a 16-bit number, but it is really an 8-bit offset, E3, extended. The FF is discarded and the code will look like the following:

010B	25	Branch Op code
010C	E3	Offset (8 bits)
010D	Op code of the next instruction	

This problem specifies a backward branch and the offset, E3, is indeed negative.

For part (b) the problem becomes simply 0150 − 010D = 0043. Again, the 00 is discarded and the 8-bit offset is 43. This is a positive number and the branch is forward.

E X A M P L E   3 - 18

What does the instruction 20 FE (BRA FE) do?

Solution

Because FE = −2, this instruction causes the program to branch back two locations from the Op code of the next instruction. But because the instruction itself takes two locations, it will always branch back to itself. It will keep executing itself repeatedly. This is a very tight loop that does nothing, but 20 FE is sometimes used as a HALT. Some programmers put it at the end of a program so that the program will loop after it is finished until the programmer can stop the loop and examine the results.

3-7.3   Branches out of Range

Suppose there is a BRANCH instruction in C21E and the target location is C140. To find the offset, C220 must be subtracted from C140, giving FF20. Many students will write 20 as the offset, but wait a bit. This implies a *positive offset* (20), but a *backward* branch. Our suspicion is further heightened by the fact that there is a negative extension, FF, but a positive number. Something is wrong!

The difficulty is that the program attempts to branch too far. Using hex numbers, a program *can only branch 7F locations forward* (from the Op code of the next instruction) and *80 locations backward*. This is the limitation of an 8-bit offset.* What has been attempted here is a *branch out of range*.

---

* Most 16-bit μPs allow for 16-bit branch offsets and can branch much farther (about 32000 decimal locations).

The target was beyond the limit of the offset. Here the distance between the target and the instruction was E0, which is too far.

3-7.4 BRANCHing Around the JUMP

When a branch is out of range, as in the previous paragraph, the branch can be accomplished by a combination of BRANCH and JUMP instructions. If the instruction is unconditional (i.e., we would like to write BRA to C140 in C11E but cannot because it is too far for the branch), the BRA can simply be replaced by a JUMP instruction. Here JMP C140 will do. But remember, the JMP takes 3 bytes whereas the BRA only requires 2.

If the branch is conditional, it can be done by

1. Replacing the BRANCH instruction by a BRANCH of the *opposite* condition.

2. Following this instruction by a JUMP to the target location.

3. Branching around the JUMP instruction to the next instruction.

E X A M P L E   3 - 19

When the program reaches C21E, it must go to C140 if the carry flag is set. Write the code to accomplish this.

Solution

The code will look like the following:

C21E	24	Op code for BCC
C21F	03	Offset
C220	6E	Op code for JMP (extended)
C221	C1	Address
C222	40	
C223		Op code of next instruction

Observe that a BRANCH IF CARRY SET was required. We used the opposite branch (BRANCH IF CARRY CLEAR) to branch around the JUMP. If the carry flag is set, the program will *not* take the branch at C21E and will execute the JMP at C220. If the carry is clear when the program reaches C21E, it will branch around the JMP and execute the next instruction at C223.

3-7.5 Good News About Branches

BRANCH instructions are tedious to calculate. The good news is that assemblers relieve the programmer of the task; they calculate the offsets. Even in the one-line assembler on the EVB (see Chap. 4), the user enters the target address directly and the assembler enters the correct offset. The assembler will also generate an "OUT OF RANGE" error message if the branch is too far.

Still, users must understand branch calculations if they are to understand the machine code. Besides, instructors love to put offset calculations on examinations.

3-8   Additional Examples

In this chapter only a few of the **68HC11** instructions have been studied. The remaining instructions will be covered in Chapters 5 and 6. Nevertheless, some worthwhile programs can be written using the material in this chapter. This section presents some examples that you may use as guides for writing your own programs.

E X A M P L E   3 - 20
_____

There is a set of 8-bit numbers in a buffer (a *buffer* is a set of memory locations reserved for data) from C100 to C150. Write a program to add these numbers.

Solution

Because this is the first program presented, it will be discussed in great detail in order for you to understand every step in the solution. First, the general approach should be considered. This program does repeated additions so a loop should be written. Also, the program must step through memory, adding C100, then C101, and so on. This can be done by setting an index register to point to the start of the buffer and then incrementing the index register each time through the loop.

Now specific functions can be assigned to specific registers. We designate accumulator A to hold the sum, accumulator B to hold the loop counter, and X to act as the index register. These must be initialized as follows:

- Accumulator A, the sum, must be 0 at the start.
- Accumulator B is the loop counter. At the start it should be set equal to the number of times around the loop. It can then be decremented during each pass of the loop. By testing the Z flag, the looping can be stopped when the counter decrements to 0.
- X should be set to point to the start of the buffer, C100.

The next step could be to draw a flowchart, as shown in Figure 3-9. Notice the difference between the initialization and the action command boxes. The initialization commands are only executed at the start of the program, whereas the action commands are executed each time through the loop.

We are now ready to write the program. It must be placed somewhere in memory and *not* in the buffer area. Location C020 will arbitrarily be

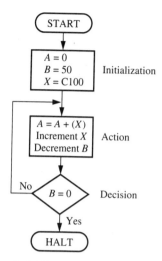

Figure 3-9 Flowchart for Example 3-20.

assigned as the first instruction in the program. With the flowchart to help, the program can be written very simply as shown in Figure 3-10.

The program functions by performing an ADD to A in the indexed mode (step 4), incrementing X to point to the next buffer location and decrementing B. As long as B is not 0, the program will continue to loop. The location of each instruction shows that the program must loop back from C02D to C026. The correct offset was calculated using the methods of Section 3-7 and found to be F9. The program was halted by a BRA*, which means branch to itself, 20 FE, as discussed in Section 3-7.2.

For simplicity this program did not worry about any overflows that might have occurred during the additions. They will be discussed further in Chapter 5.

Step	Label	Locations	Instructions	Code	Comments
1	START	C020	CLRA	4F	Initialization.
2		C021	LDAB #$50	C6 50	
3		C023	LDX #C100	CE C1 00	
4	LOOP	C026	ADDA 0,X	AB 00	Add buffer to A.
5		C028	INX	08	Increment X.
6		C029	DECB	5A	Decrement B.
7		C02D	BNE LOOP	26 F9	Branch to loop as long as B ≠ 0.
8		C02D	BRA*	20 FE	Halt.

Figure 3-10 The program for Example 3-20.

The next example is the **block transfer**. This is a very commonly used computer program to *move or relocate* a buffer containing data or code from one area of memory to another. The **68HC11** contains two index registers that greatly simplify this problem.

E X A M P L E   3 - 21
_____

Write a program that will transfer a data buffer from C100 to C3FF to D200 to D4FF.

Solution

In this problem the buffer from C100 to C3FF will be called the *source* buffer and the other memory area will be called the *destination* buffer. After a little thought we realize that the program can operate by reading a byte from the source and sending it to the destination. This must be done in a loop until all the data is transferred. The X index register can be initialized to point to the source buffer and Y can point to the destination. A single 8-bit register is not sufficient to act as a loop counter because the loop requires more than 256 iterations. One way to use a loop counter would be to place it in memory, but it is probably easier to use a COMPARE instruction, as discussed in Section 3-6.

The program is shown in Figure 3-11. First, X and Y are initialized to point to the source and destination buffers. The LDAA and STAA instruc-

```
         Instructions              Comments

         LDX   #$C100      Point X to source buffer.
         LDY   #$D200      Point Y to destination buffer.
    LOOP LDAA 0,X     ⎫
         STAA 0,Y     ⎬    Transfer a byte.
                      ⎭
         INX          ⎫
                      ⎬    Increment X and Y.
         INY          ⎭
         CPX   #C400   ⎫
         BNE   LOOP    ⎬    Loop finished?
         HALT              Yes. Halt.
```

Figure 3-11 The program for Example 3-21.

tions transfer a byte from the source to the destination. The index registers are then incremented for the next iteration of the loop. Finally, X is compared to C400, the first location *after* the loop. When it reaches C400, the loop terminates and the program halts.

In this problem several steps used in the previous example were omitted:

1. The flowchart was not drawn.

2. The exact coding and its locations were not specified. Because of this,

Line No.		Instruction	Operand	Comments
1	START	LDX	#C100	Point X to the start of the buffer.
2		LDY	#D000	Put the location of
3		STY	COA0	the positive buffer in COA0.
4		LDY	#D400	Put the location of the
5		STY	COA2	negative buffer in COA2.
---	---	End of initialization	---	---
6	LOOP1	LDAA	0,X	Load the number into A.
7		BMI	LOOP2	Go to LOOP2 if it is negative.
8		LDY	COA0	Put the positive number
9		STAA	0,Y	in the positive list.
10		INY		Increment Y to prepare for next positive number.
11		STY	COA0	Store it away.
12	LOOP3	INX		
13		CPX	#C400	Is the list finished?
14		BNE	LOOP1	No. Continue.
15		HALT		Yes. Stop.
16	LOOP2	LDY	COA2	Put the address of the negative list in Y.
17		STAA	0,Y	Store the negative number.
18		INY		Increment Y to prepare for the next negative number
19		STY	COA2	and store it away.
20		BRA	LOOP3	Reenter the main loop.

Figure 3-12 The program for Example 3-22.

the branch offset cannot be calculated. The target of the branch is identified by the label LOOP in the program.

The program is actually written in assembly language, but the one-line assembler on the EVB can handle it. If you are having trouble understanding the problem, perhaps you should go back and write out the steps that were omitted (see Prob. 3-10).

E X A M P L E   3 - 22

There are a set of numbers in locations C100–C3FF. Write a program to copy all the positive numbers into a list starting at D000 and copy all the negative numbers into a list starting at D400.

Solution

This program requires three index registers: one to hold the original list, one to hold the list of positive numbers, and one to hold the list of negative numbers. Because only two index registers are available, Y will be used to point to both the positive and negative lists. This can be done by swapping the contents of Y in and out of memory during each loop.

The program is shown in Figure 3-12. It starts by initializing X to point to the original list at C100. It also places the address of the positive list in C0A0 and the address of the negative list in C0A2.

After initialization it reads the first number. If it is positive, it loads the address of the positive list into Y and then places the number in memory. It then increments Y and stores it to await the next positive number. The program then increments X and tests to determine whether the list is finished.

If the number in the list is negative, it will set the N condition code at line 6 and branch at line 7. The LOOP2 program then copies the negative number into the negative list and returns.

Summary

This is the first chapter that pertains specifically to the **68HC11**. The register set was described and the instruction set was introduced. The operation of some of the most basic instructions, such as LOAD and STORE, was explained. The various modes used by the **68HC11** were discussed. Special emphasis was placed on the index register and indexed instructions because these are very important.

The CCR was described. BRANCH and JUMP instructions were considered and the method of calculating branch offsets was explained. Finally, some examples were presented to illustrate the concepts.

Glossary

Block transfer   Copying or transferring a program or data from one area of a computer's memory to another.

Condition Code Register (CCR)   A register that holds the flags that are set to indicate the characteristics of the results of the last instruction.

Cycle   One period of the clock.

D—Double Accumulator   A 16-bit accumulator comprised of 8 bits of A and 8 bits of B.

Flags   Parts of the condition codes. The N, Z, V, and C flags were considered in this chapter.

Inherent instructions   Instructions that are totally specified by their Op codes, and that do not require any additional information from memory to execute.

Offset   **a.** A positive 8-bit number added to an index register to compute an address.

   b. A positive or negative number added to a branch instruction to compute the target location of the branch.

Prebyte   A byte of the Op code indicating that the instruction pertains to the Y register instead of X.

Target Address   The address where a program will branch to if the branch is taken.

Problems

3-1 The contents of memory are given as

C100	45
C101	56
C102	65
C103	76
C104	87
C105	AB

For each part of this problem, assume that X contains C100 at the start.

What data is loaded into what registers (or memory) in response to the following instructions?

a.	LDAB	C102	**e.**	LDX	4,X
b.	LDX	#C102	**f.**	STX	C104
c.	LDX	C102	**g.**	STX	3,X
d.	LDAA	3,X			

3-2 Write the contents of both index registers and the contents of locations C100–C103 after each instruction of the following program is executed:

		X	Y	C100	C101	C102	C103
LDX	#1023						
LDY	#2345						
STX	$C100						
STY	$C101						
LDX	$C101						

3-3 What gets loaded in response to the following sequence of instructions?

    LDX   #1033
    LDX   33,X

3-4 What are the results of the following arithmetic operations? List the N, Z, V, and C bits of the condition codes. All numbers are in hex.
   a. 66 + 66
   b. 66 − 66
   c. 66 − 67
   d. 66 + 9A
   e. 90 + AA

3-5 The hex equivalents of the decimal numbers 75 and 64 are to be added. Without converting these numbers to hex, explain whether there will be overflow.

3-6 If B contains $77 and X contains C100, what are the results of the following instructions? Where necessary, use the memory contents given in Problem 3-1.
   a. CMPB   #AA
   b. CMPB   C102
   c. CPX   C102
   d. CPX   #C102
   e. CPX   4,X

3-7 The opcode for a BEQ instruction is in C014 followed by the offset xx in C015. Where do you execute the next instruction?
   a. If Z = 0.
   b. If Z = 1 and xx equals
      (1) F9.
      (2) B8.
      (3) 4F.

3-8 There is a BRANCH instruction in C014. What offset do you use if you want to branch to
   a. C001.
   b. BFE2.
   c. C077.

3-9 When the program reaches C014 it must branch to C300 if the results are negative (the N bit is set). What instructions should be written in C014 to accomplish this?

3-10 Draw the flowchart and write the exact code for Example 3-21. Be sure to calculate the branch offset.

For Problems 3-11, 3-12, and 3-13 the memory buffer is assumed to extend from C100 to C4FF.

3-11 Some locations in the buffer contain the number AB. Assume there are less than 256 (decimal) of them. Write a program to count the number of ABs and put the total in C600.

3-12 Write a program to place the highest number in the buffer (in absolute value) in C600.

3-13 Write a program to place all the numbers in the buffer that are higher than A0 in a list starting at C600.

4

*The Evaluation Board**

INTRODUCTION

The study of computers is not as effective as it could be if the computer is considered solely as an abstract device. The study is much more meaningful if students can get some "hands-on" experience. This allows them to try out the programs that have already been written. When writing programs, there are many ways to make mistakes and checking the performance of a program on an actual computer is the best means of determining whether the program has been written correctly.

For the **68HC11**, Motorola has developed a small, inexpensive EValuation Board (EVB) for use in the laboratory. This board is intended to reinforce the discussion of the **68HC11** instructions and features by allowing the students to examine them by actually using the μP.

INSTRUCTIONAL OBJECTIVES

This chapter studies the characteristics and features of the EVB. After reading the chapter, the student should be able to

- Connect the EVB and invoke the Buffalo monitor.
- Read and write data to memory.
- Write small programs and execute them.
- Insert and remove breakpoints in a program.

SELF-EVALUATION QUESTIONS

Watch for the answers to the following questions as you read the chapter. They should help you to understand the material presented:

1. What areas of memory are available to the user?

2. How is the bit rate for the communications channel set?

3. Does a RESET destroy the user's program or data?

4. How does a user write a BRANCH instruction using ASM?

5. What good are NOPs inside a program?

* Refer to *M68HC11 EVB Evaluation Board User's Manual.* Motorola, Phoenix, AZ, 1986.

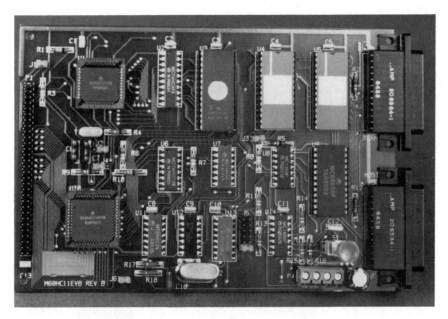

Figure 4-1 The M68HC11 evaluation board. (*Reprinted Courtesy of Motorola, Inc.*)

4-1 The Physical Construction of the Evaluation Board

Figure 4-1 is a photograph of the EVB. The board accommodates both small- and large-scale ICs. The major ICs on the board are shown in the block diagram of Figure 4-2.

4-1.1 The Hardware of the Evaluation Board

Figure 4-2 shows that there are six major components on the EVB:

1. The MCU (block 1) This is the **68HC11** μP. The rest of the board consists of support chips for the μP.

2. The PRU (block 2) This is a Port Replacement Unit (PRU) that enhances the I/O capabilities of the **68HC11** on the board. It is discussed in Chapters 8 and 12, where some of the **68HC11**'s I/O capabilities are explored.

3. The latch (block 3) The latch ICs latch the memory address (see Sec. 12-3).

4. The monitor EPROM (block 4) This erasable programmable Read Only Memory contains the **Buffalo monitor**. This is the program that controls the **68HC11** when power is applied. It enables the user to read and write memory and to examine the programs he or she has written to the μP.

5. The user RAM (block 5) This is an 8K-byte RAM where the user can write and test programs and data.

6. The ACIA (block 6) The ACIA (Asynchronous Communications Interface Adapter) is a special purpose IC used to control communications between the EVB and a terminal or host computer (see Sec. 4-2).

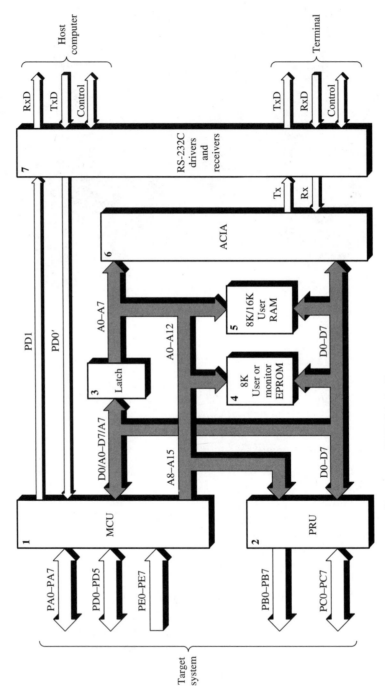

Figure 4-2   EVM block diagram. *(Redrawn Courtesy of Motorola, Inc.)*

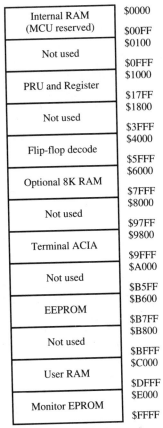

$0000–$0035 User RAM
$0036–$004A User stack pointer
$004B–$00C3 Monitor variables
$00C4–$00FF Vector jump table

Internal RAM (MCU reserved)	$0000 / $00FF
Not used	$0100 / $0FFF
PRU and Register	$1000 / $17FF
Not used	$1800 / $3FFF
Flip-flop decode	$4000 / $5FFF
Optional 8K RAM	$6000 / $7FFF
Not used	$8000 / $97FF
Terminal ACIA	$9800 / $9FFF
Not used	$A000 / $B5FF
EEPROM	$B600 / $B7FF
Not used	$B800 / $BFFF
User RAM	$C000 / $DFFF
Monitor EPROM	$E000 / $FFFF

Figure 4-3   EVB memory map diagram. *(Redrawn Courtesy of Motorola, Inc.)*

7. The RS-232 drivers and receivers (block 7)   These ICs change the voltage levels so that the μP and the ACIA are compatible with the modem or terminal voltages.*

The ICs on the EVB require four voltages, +5, +12, −12, and ground. The connections to accommodate these voltages are on the lower right-hand side of the board as shown in Figure 4-1. There is also an 8-MHz crystal oscillator on the board that supplies the basic 2-MHz clock for the **68HC11**.

4-1.2   The Memory Map of the Evaluation Board

Figure 4-3 is the memory map of the 64K-byte memory used by the **68HC11** on the EVB. It shows that certain areas of memory are reserved for specific functions. The five major dedicated areas of memory are

1. Internal RAM   Locations 0000–00FF—This area of RAM is internal—the RAM is inside the **68HC11**. It is reserved for interrupt vectors (see Chap. 8) and for some monitor variables. The monitor program, although

---

* A complete discussion of RS-232 voltage levels is given in J. D. Greenfield, *Practical Digital Design Using ICs*, latest edition, Prentice-Hall, Englewood Cliffs, N.J.

in EPROM, occasionally has to store some data into read/write memory and uses this area. Although locations 0000–0035 are designated as user RAM, we recommend that the user stay away from this area entirely.

2. PRU and Register decode   Locations 1000–103F—There are many registers in the **68HC11** that control its various I/O functions. This area is reserved for them. It is primarily RAM, but some bits of some registers are EPROM. It is shown as a larger area in Figure 4-3, but the registers only use locations 1000–103F.

3. EEPROM   Locations B600–B7FF—This is a 512-byte area of EE-PROM (Electrically Erasable PROM) that is also within the **68HC11**. The user can write some small programs in this area that should be retained when power is turned off. The EEPROM is discussed in Section 12-2.3.

4. User RAM   Locations C000–DFFF—This 8K-byte area consists of RAM ICs that are external to the **68HC11**. It is where the user can write and test his or her programs. We recommend that students use this area exclusively. The programs in Chapter 3 tried to set a precedent by using this area of memory.

5. Monitor EPROM   Locations E000–FFFF—This area is reserved for the EPROM that contains the Buffalo monitor (see Sec. 12-2.3).

Other areas of the memory map, such as the flip-flop decode and terminal ACIA, are reserved for special functions. They do not need to concern you at this time, but they are not available for general use.

4-1.3   Terminal Connections

A terminal consists of a keyboard and a display, such as a video monitor, and is used to communicate with a computer. This communication uses a serial bit stream and data is often transmitted over phone lines using modems. A typical terminal is shown in Figure 4-4.

The EVB comes with two type DB-25 female connectors that are on the right-hand side of the board as shown in Figure 4-1. The connector farther from the power pins is used to connect the EVB to a terminal via a cable. The near connector is used for communications with a host computer, which can download programs. Most often, the EVB communicates with a terminal instead of a host computer.

Personal computers can be made to function as terminals. The IBM-PC, for example, works very well. It requires

1. A communications card or channel   This card is required for asynchronous communication between the PC and any external communication device, such as a modem.

2. A communications software package   This is a program that contains the software necessary for the control of communications. Several communications packages are available. Typical examples are PROCOMM and KERMIT.

3. A cable   The cable used to connect the EVB and the IBM-PC is shown in Figure 4-5. It consists simply of a ribbon cable, a male DB-25 connector to mate with the EVB, and a female to connect with the PC.

Figure 4-4   A terminal connected to a μP.

The communications channel will not work unless both the PC and the EVB run at the same speed or baud rate. On the PC side communications packages allow the user to choose among the various common baud rates using the software to select the rate. The EVB has several rates available. There is a jumper on the EVB (labeled J5) that sets it for a particular rate. The most commonly used rate is 9600 bits per second (bps).

4-1.4   The Input/Output Connector

The EVB contains a 60-pin header shown on the left-hand side of the board. It is primarily for Input-Output (I/O). The pinout of the header is given in Figure 4-6. Power (+5 V) and ground should be applied to the input pins, but they can be monitored at pins 26 and 1, respectively.

Pins 5, 7, and 8, labeled E, EXTAL, and XTAL, are concerned with the clock driving the **68HC11**. The output of the internal crystal oscillator appears at XTAL. If an external clock is to drive the system, it should be placed on the EXTAL pin, and jumper J2 on the board must be changed. The E clock is one-fourth the frequency of the crystal oscillator. This determines the cycle time of the μP and can be monitored by applying an oscilloscope to the E pin. In most systems the internal 8-MHz crystal oscillator is used, the E clock is 2 MHz, and each **68HC11** cycle takes 500 ns.

The other pins on the connector are used for input and output with the **68HC11**. These are discussed in Chapters 9, 10, and 11.

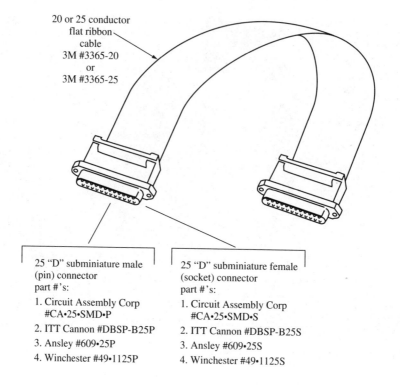

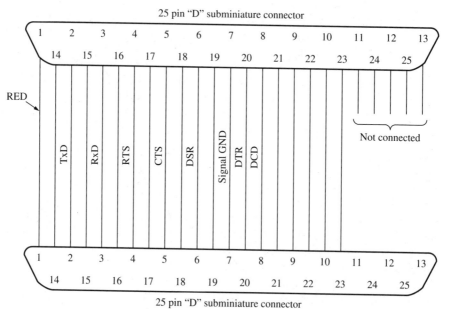

Figure 4-5   The cable connecting the EVB and a PC. (*Redrawn Courtesy of Motorola, Inc.*)

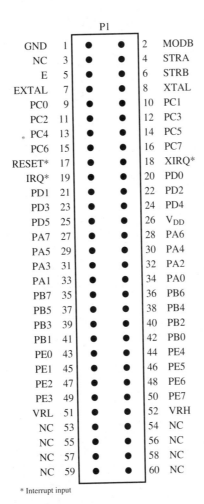

Figure 4-6 The I/O connector on the EVB. (*Redrawn Courtesy of Motorola, Inc.*)

4-2 Starting the Evaluation Board

The following steps must be followed to get the EVB "up and running" with a terminal or PC:

1. Apply power to the EVB on all four pins.

2. Connect the cable between the terminal and the EVB.

3. Set the terminal's baud rate to match that of the EVB. If a PC is used, the communications software must be started. Most communications software packages allow the user to select one of several commonly used bit rates.

4. Push the red RESET button on the EVB. This will cause the **68HC11** to jump to the monitor program starting at E000. One of the first things the monitor program does is to print the Buffalo message on the screen. It should look like this:

BUFFALO 2.5 (ext)—Bit User Fast Friendly Aid to Logical Operation

5. Push the RETURN key on the terminal or PC. This should cause the EVB to display a > symbol. This is the prompt, which indicates that the system is up and running.

4-2.1 Effects of RESET

Pushing the **RESET** button will always restart the monitor at E000. This may change the contents of lower memory, between 0000 and 00FF, and the contents of some of the registers at 1000. It should never change the information in the user RAM. Therefore, if the computer must be stopped and returned to monitor control, user programs will not be destroyed if they are where they should be—between C000 and DFFF.

4-3 Reading and Writing Memory

After the system is up and the > symbol is displayed, there are commands that allow the user to read and write memory and the registers. These will be discussed in this section.

4-3.1 The Register Modify Command

The register modify command can be invoked by typing **RM** (Register Modify) at the terminal. The terminal will then display the current contents of all the registers. The contents of these registers can be changed by typing the register letter and then the data to be set into the register.

Figure 4-7 shows the action of the RM command. After RM was entered the registers were displayed. It can be seen that the PC was at C007, the Y index register contained 7982, and so on. Just typing RM displays the registers and the contents of the PC on the next line. The PC can then be changed as shown in Figure 4-7a. If RM is followed by a letter that designates a register, then that register can be changed. In Figure 4-7b the X register was changed to C020.

4-3.2 The Memory Display Command

To show the contents of an area of memory on the screen, the user can type **MD** (Memory Display). MD must be followed by at least one address. If it is followed by a single address, nine lines of data will be displayed. Each line consists of 16 bytes and starts on a 16-byte boundary. Figure 4-8a shows such a display.

Instructions	Comments
>RM	
P-C007 Y-7982 X-FF00 A-44 B-70 C-C0 S-0054	Display P register contents.
P-C007 C020	Modify P register contents.
>	

A

Instructions	Comments
>RM X	
P-C007 Y-7982 X-FF00 A-44 B-70 C-C0 S-0054	Display X register contents.
X-FF00 C020	Modify X register contents.
>	

B

Figure 4-7 The RM (Register Modify) command. (*Courtesy of Motorola, Inc.*)

```
>MD F7D0
F7D0  AA AA AA AA AA AA AA AA AA AA AA AA AA AA AA AA
F7E0  AA AA AA AA AA AA AA AA AA AA AA AA AA AA AA AA
F7F0  AA AA AA AA AA AA AA AA AA AA AA AA AA AA AA AA
F800  AA AA AA AA AA AA AA AA AA AA AA AA AA AA AA AA
F810  AA AA AA AA AA AA AA AA AA AA AA 12 CD 34 56 EE
F820  AA AA AA AA AA AA AA AA AA AA AA AA AA AA AA AA
F830  AA AA AA AA AA AA AA AA AA AA AA AA AA AA AA AA
F840  AA AA AA AA AA AA AA AA AA AA AA AA AA AA AA AA
F850  AA AA AA AA AA AA AA AA AA AA AA AA AA AA AA AA
>
```

A

```
>MD C000 C020
C000  FF FF FF FF FF FF FF FF FF FF FF FF FF FF FF FF
C010  FF FF FF FF FF FF FF FF FF FF FF FF FF FF FF FF
C020  FF FF FF FF FF FF FF FF FF FF FF FF FF FF FF FF
>
```

B

Figure 4-8 The MD (Memory Display) command. *(Courtesy of Motorola, Inc.)*

E X A M P L E   4 - 1

In Figure 4-8a, what are the contents of location F81C?

Solution

Location F81C is the twelfth entry on the line beginning at F810 (remember to start counting at 0). It is circled on the figure. F81C contains CD.

If two addresses are displayed, as in Figure 4-8b, the contents of all memory that start on or between those addresses are shown.

4-3.3 The Memory Modify Command

The memory modify command is used to write data into memory. It is invoked by typing **MM** (Memory Modify) on the terminal followed by an address. The contents of that address will then be displayed. The user now has three main possibilities:

1. Type a byte of data followed by a space. This changes the contents of memory and displays the next location, which can now be changed.

2. Simply type a space. This leaves the location unchanged, but presents the next location for modification.

3. Type a return. This causes the monitor to exit MM.

   The action of the MM command is shown in Figure 4-9. First, the contents of C100–C10F were displayed. Then MM was invoked. The underlines show where data was inserted. The up arrows indicate where spaces were typed. Finally, the memory contents were reexamined, and we can see that the first three locations were modified.

4-3.4 The Block Fill Command

The Block Fill (**BF**) command is used to fill an area of memory (a buffer) with the same data byte. The command is typed as

BF   <Address 1> <Address 2> <Data>

where Address 1 and Address 2 are the start and end addresses of the buffer that is being filled with the data.

   Figure 4-10 shows the action of the BF command. First, the contents of memory from C100 to C14F were displayed using the MD command. Then the BF command was issued to fill locations C112–C12C with the byte EE. Finally, the buffer was displayed again so that the user could see that locations C112–C12C were filled with EE.

```
>MD C100 C100

C100 12 34 56 78 BF 00 FF 00 7F 00 7F 00 FF 00 4Vx

>MM C100

C100 12 AA 34 BB 56 CC 78 BF 00
        ←      ←      ←      ←      ↓

>MD C100 C100

C100 AA BB CC 78 BF 00 FF 00 7F 00 7F 00 FF 00 x
^
```

Figure 4-9 The action of the MM (Memory Modify) command.

```
>MD C100 C140

C100  AA BB CC 78 BF 00 FF 00 FF 00 7F 00 7F 00 FF 00
C110  96 00 05 00 03 00 30 00 83 00 11 00 04 00 11 04
C120  FF 00 FF 00 BF 00 FF 00 BF 00 7F 00 7F 00 FF 00
C130  05 00 07 00 03 00 00 00 03 00 15 00 05 00 11 00
C140  DF 00 DF 00 9F 00 FF 00 BF 00 7F 00 7F 00 FF 00
>BF C112 C12C EE

>MD C100 C140

C100  AA BB CC 78 BF 00 FF 00 FF 00 7F 00 7F 00 FF 00
C110  96 00 EE EE EE EE EE EE EE EE EE EE EE EE EE EE
C120  EE EE EE EE EE EE EE EE EE EE EE EE EE 00 FF 00
C130  05 00 07 00 03 00 00 00 03 00 15 00 05 00 11 00
C140  DF 00 DF 00 9F 00 FF 00 BF 00 7F 00 7F 00 FF 00
>
```

Figure 4-10   Action of the BF (Block Fill) command.

4-3.5 The Block Move Command

The Block Move (**MOVE**) command actually copies a block of data from one area of memory, called the *source*, to another area of memory, called the *Destination*. The command is typed as

MOVE   <Address 1> <Address 2> <Destination>

As in the BF command, Address 1 and Address 2 are the start and end addresses of the source buffer, and Destination is the starting address of the destination buffer. For example, the command

MOVE   C100 C14F C290

will copy the $(50)_{16}$ bytes that are in locations C100–C14F into a destination buffer from C290 to C2DF.

4-4 Writing Commands into Memory

The previous section discussed the various methods of writing and reading data. This section explains the **ASM** (Assemble) command, which is used to write commands into memory. ASM is a *one-line assembler* that allows the user to enter commands in assembly language format and converts them into machine language. This simplifies the task of entering programs into the **68HC11**.

Typing ASM <address> causes the prompt to indent and displays the command at the specified address in assembly format. The user can then type a command. If the **68HC11** cannot understand the command, it types "MNEMONIC NOT FOUND" and does not advance the PC. If the command is acceptable, ASM types the code for it and proceeds to the next location. ASM can be exited by typing (CTRL)A, which returns the user to the main command line of the Buffalo monitor.

Figure 4-11 is an example using ASM. The commands LDAA #55 and STAA C0 were entered at locations C000 and C002, respectively.

Examples	Descriptions
>ASM C000	
C000 STOP $FFFF	Immediate mode addressing, requires #
>LDAA #55	before operand.
86 55	
C002 STOP $FFFF	Direct mode addressing.
>STAA C0	
97 C0	
C004 STOP $FFFF	

Figure 4-11 The ASM (Assemble) command. (*Courtesy of Motorola, Inc.*)

4-4.1 Branches Using ASM

BRANCH instructions are written in the general form BXX <address>, where BXX is the mnenonic Op code for the branch and address is the actual address to be branched to. The user does not have to calculate offsets when using branch instructions with ASM; the assembler does it. Figure 4-12 shows a typical branch instruction entered under ASM.

Example	Description
C006 STOP $FFFF	Branch offsets calculated automatically;
>BRA C030	address required as conditional branch
20 28	operand.

Figure 4-12 A BRANCH instruction under ASM. (*Courtesy of Motorola, Inc.*)

When using backward branches, such as for loops, the user generally knows the address to be branched back to. Sometimes the user will not know the address for a forward branch because the code has not yet been entered. In that case we recommend that the user insert a dummy branch, BRA FE, for example, and continue entering code until the place to be branched to is reached. Once the proper target address is found, the user can replace the dummy branch instruction with the correct instruction.

Sometimes the user inadvertently specifies a target location that is too far away from the instruction. ASM will respond with a "BRANCH OUT OF RANGE" message to show that the user is trying to branch farther than 128 locations from the instruction.

4-4.2 Entering Programs Using ASM

The beginning student typically enters programs under ASM and utilizes the simplifications it provides. Here are two suggestions:

1. Do not start programs at C000. C000 is the lowest address in user RAM and the tendency is to select it as the address of the first instruction. The problem is that sometimes students find that an instruction that should occur at the beginning of the program (a location they may have forgotten to initialize) has been omitted. There is no room below C000 to write the instruction and the entire program may have to be reentered. If the student starts the programs at C020, for example, there is some space at the start of the program for these instructions.

2. When writing long programs, add NOP instructions throughout the program. NOP (No Operation) instructions do nothing except waste time and memory bytes, but these are generally not crucial. If a student finds that an instruction has been omitted, however, the NOPs that have been entered can now be replaced with the new instruction. This will often save the trouble of reentering the entire program.

An alternate method of inserting instructions into a long program that is already entered is to JUMP or BRANCH to a free area of memory, insert the new instruction plus the instruction that has been overwritten by the JUMP or BRANCH, and then JUMP or BRANCH back to the main program.

E X A M P L E 4 - 2

A section of code is given here:

C115	INX
C116	INX
C117	LDAA #33
C119	NEXT INSTRUCTION

The user finds that an LDX #C234 must be added in at C115. How can this be done? Assume that the space at C500 is available.

Solution

The user can put in a JMP C500 at C115. But this is a 3-byte instruction that eliminates the instructions at C115, C116, and C117. These will have to be replaced. The code at C115 will be

C115	JMP C500
C118	This byte is now irrelevant or wasted.
C119	NEXT INSTRUCTION

while the code at C500 is

C500	LDX #C234
C503	INX
C504	INX
C505	LDAA #33
C507	JMP C119

The instructions in the original program that were overwritten have been reentered in the C500 area along with the LDX instruction. The main program was reentered by the JUMP instruction.

4-5 Running Programs

Once the program has been entered using ASM, it can be run at full speed or run in *single-step* or *trace* mode.

4-5.1   Trace Mode

The program can be run a step at a time (single-step) by

1. Setting the PC to the first location in the program.

2. Typing T (Trace) on the keyboard.

The program will then execute one instruction, print out the register contents at the end of the instruction, and halt. The user can now examine the contents of the various registers to determine if the instruction has behaved properly. If the user types T followed by a number, the program will trace that number of instructions before halting. Figure 4-13 shows the use of the trace command.

Warning: The trace command modifies some of the bits in the registers at 101E, 101F, 1020, and 1024. The reader does not need to be concerned about it at this time.

Examples	**Descriptions**

>T   Single trace.

```
Op-86
P-C002 Y-DEFE X-FFFF A-44 B-00 C-00 S-004B
>
```

>T 2   Multiple trace (2).

```
Op-B7
P-C005 Y-DEFE X-FFFF A-44 B-00 C-00 S-004B

Op-01
P-C006 Y-DEFE X-FFFF A-44 B-00 C-00 S-004B
>
```

Figure 4-13   Use of the T (Trace) command. (*Courtesy of Motorola, Inc.*)

4-5.2   The Go Command

The user's program can be executed at full speed by typing **G** (Go) followed by an address. The program will start to execute at that address. If the user simply types G, the program will start to execute at the contents of the PC.

Before running a program, the user must provide a method for stopping it. There are three reasonable methods of stopping a program:

1. Using a branch-to-itself instruction   This instruction, BRANCH to the current PC, codes as 20 FE and puts the program in an endless loop. The user can only regain control of the μP by pressing the RESET button.

2. Typing an SWI   SWI, a **Software Interrupt** instruction, prints out the registers and returns control to the monitor. Its code is 3F, and it is discussed in Section 8-8.4. SWI is a superior way to halt a program, and we recommend that it be the last instruction in every program.

3. Using breakpoints   Breakpoints are discussed in the next section.

4-5.3   Breakpoints

A **breakpoint** causes a program to stop at a specific address. The user must set the breakpoint by typing **BR** <address> before running the program. The address must be the Op code of an instruction—it cannot be the address or immediate part of the instruction.

To run a program, the user sets breakpoints where the program should pause. Then the PC is set to the start of the program and the G command is given. The program runs until it encounters the first breakpoint, where it stops and prints out the registers. The user can monitor the program's progress thus far by examining the registers as well as tracing the next few instructions or issuing a **P** (Proceed) command. The P command allows the program to resume execution at the current location. It will then advance to the next breakpoint or the SWI at the end of the program. Figure 4-14 illustrates the options available with breakpoints and gives examples of their use.

Again, a word of warning is appropriate. The **68HC11** executes breakpoints by inserting an SWI (Op code 3F) in the code at the breakpoint lo-

Command Formats	Description
BR	Display all current breakpoints.
BR ⟨address⟩	Set breakpoint.
BR ⟨addr1⟩⟨addr2⟩ . . .	Set several breakpoints.
BR -	Remove all breakpoints.
BR -⟨addr1⟩⟨addr2⟩ . . .	Remove ⟨addr1⟩ and add ⟨addr2⟩.
BR ⟨addr1⟩-⟨addr2⟩ . . .	Add ⟨addr1⟩, clear all entries, then add ⟨addr2⟩.
BR ⟨addr1⟩-⟨addr2⟩ . . .	Add ⟨addr1⟩, then remove ⟨addr2⟩.

Figure 4-14   Breakpoint formats and examples. (*Courtesy of Motorola, Inc.*)

	Examples	Description

>BR C003   Set breakpoint at address
                  location C003.

C003 0000 0000 0000
>

>BR C003 C005 C007 C009   Sets four breakpoints.
C003 C005 C007 C009   Breakpoints at same
>   address will result in
                                 only one breakpoint
                                 being set.

>BR   Display all current
                  breakpoints.

C003 C005 C007 C009
>

>BR - C009   Remove breakpoint at
                  address location C009.

C003 C005 C007 0000
>

>BR -C009   Clear breakpoint table and
                  add C009.

C009 0000 0000 0000
>

>BR -   Remove all breakpoints.
0000 0000 0000 0000
>

>BR E000   Only RAM locations can be
                  breakpointed.

rom-E000   Invalid address message.
0000 0000 0000 0000
>

>BR C005 C007 C009 C011 C013   Maximum of four
                              breakpoints can be set.

Full   Buffer full message.
C005 C007 C009 C011
>

Figure 4-14   (*continued*)

cation. When the breakpoint is reached, it replaces the SWI with the proper instruction. Sometimes a student's program hangs before the breakpoint is reached. If the student exits by hitting RESET, the SWI will remain in the program and must be removed by an MM command.

4-6   Other Capabilities of the Evaluation Board

The EVB has several other capabilities that have not been mentioned. It can be loaded with a program in an S record format, it can communicate with a host computer, and the monitor has several subroutines that might be useful. These capabilities will be discussed in the later chapters of this book.

4-7   The Universal Evaluation Board

In 1991, Motorola unveiled a new board for evaluation purposes, the **M68HC11EVBU** Universal Evaluation Board. The board is shown in Figure 4-15. It contains only the **68HC11** (the large IC in the PLCC), a timer IC, the **MC68HC68T1**, and a communications IC, the **MC14507**. The I/O connector is at the center of the board and there is a large work area on the

Figure 4-15   The M68HC11 Universal Evaluation Board (EVBU). (*Reprinted Courtesy of Motorola, Inc.*)

right side where the user may install ICs to connect to the **68HC11**. The EVBU uses the **MC68HC11E9** version of the **68HC11***. The EVBU has the following features:

1. It is designed to operate in single-chip mode (modes are discussed in Section 8-2.4). This restricts the user to 325 bytes of RAM.

2. The EVBU can be used in other modes, but this requires additional hardware, which can be added in the ample work area provided. There is no PRU.

3. The Buffalo monitor is in ROM. It is very similar to the monitor on the EVB.

4. The board requires only a single +5 V power supply.

5. The EVBU is intended to be used in colleges and universities, where small programs and experimental work using I/O are common. The large work area facilitates experimentation.

Summary

This chapter introduced the student to the EVB, the board supplied by Motorola for experiments with the **68HC11.** First, the hardware components on the board and the memory map were discussed. Then methods of reading and writing memory were explained. Finally, methods of entering programs and running them were discussed. At this time you should be able to use the board to examine and debug your programs.

Glossary

ACIA—Asynchronous Communications Interface Adapter   A Motorola IC **(MC6850)** used for communications. It is used in the EVB to communicate with a host computer.

ASM—(Assembly)   The mode of operation that allows a user of the EVB or EVBU to enter instructions in mnemonic form. The ASM operation will then generate the correct codes and assemble these instructions.

Breakpoint   A stopping point in a program. Setting a breakpoint at an address causes the program to stop when reaching that address so that the program's progress can be examined.

Buffalo monitor   The monitor or program, in ROM, that controls the EVB or EVBU, and allows it to perform its various commands, such as ASM.

---

* More information on the **'E9** can be found in the Technical Data Manual on this μP (Motorola no. **MC68HC11E9/D**). More information on the EVBU can be found in the User's Manual for the board **(MC68HC11EVBU/AD1)**, published by Motorola, 1990.

PRU—Port Replacement Unit   An IC that allows the EVB to make I/O ports B and C available to the user.

RESET   The act of resetting or restarting the EVB or EVBU. It is invoked by pushing the button on the EVB or EVBU. Programs in endless loops can only be stopped by resetting.

SWI—Software Interrupt   A **68HC11** instruction, discussed further in Chapter 8. On the EVB or EVBU it should be the last instruction in a program, and will return control of the board to the Buffalo monitor.

Transfer, Arithmetic, and Logic Instructions

5

INTRODUCTION

This chapter continues the study of the **68HC11** instruction set by considering the basic transfer, arithmetic, and logic instructions. Testing the instructions on the EVB will clarify and solidify the explanations, and the student should do this especially if he or she is confused at any point.

INSTRUCTIONAL OBJECTIVES

After reading the chapter, the student should be able to

- CLEAR, INCREMENT, DECREMENT, COMPLEMENT, and NEGATE a memory location using only one instruction.
- Transfer data between memory and the accumulators.
- Add, subtract, multiply, and divide hexadecimal numbers.
- Shift and rotate numbers in the accumulators or in memory.
- Perform logic operations.
- Set or clear bits in memory or in an accumulator.
- Add numbers decimally.

SELF-EVALUATION QUESTIONS

Watch for the answers to the following questions. They should help you to understand the material presented.

1. In what modes can't the CLEAR instruction be used? Explain if you can.
2. How can the user determine whether one or two memory locations are involved in a data transfer?
3. What is the use of a TST instruction?
4. What is the difference between an LSR and an ASR? What is the difference between a SHIFT and a ROTATE?
5. What is a mask? What instructions use it?
6. What is the H bit? Where is it used?

5-1   Transfer Instructions

Table 5-1 shows the instructions that are available on the **68HC11** to transfer data between memory and the registers or to transfer data between the registers. Transfers that involve the stack or stack pointer are deferred until the stack concept is discussed in Chapter 6.

5-1.1   CLEAR Instructions

A CLEAR instruction sets all the bits of a location to 0. It also sets the Z flag (indicating the contents are 0) and clears the N, V, and C flags.

There are three CLEAR instructions available: CLRA, CLRB, and CLR. The first two are inherent instructions that clear the A and B accu-

Table 5-1   The Transfer Instructions in the 68HC11

Function	Mnemonic	IMM	DIR	EXT	INDX	INDY	INH
Clear memory byte	CLR			X	X	X	
Clear accumulator A	CLRA						X
Clear accumulator B	CLRB						X
Load accumulator A	LDAA	X	X	X	X	X	
Load accumulator B	LDAB	X	X	X	X	X	
Load double accumulator D	LDD	X	X	X	X	X	
Push B onto stack	PSHB						X
Store accumulator A	STAA		X	X	X	X	
Store accumulator B	STAB		X	X	X	X	
Store double accumulator D	STD		X	X	X	X	
Load index register X	LDX	X	X	X	X	X	
Load index register Y	LDY	X	X	X	X	X	
Store index register X	STX	X	X	X	X	X	
Store index register Y	STY	X	X	X	X	X	
Transfer A to B	TAB						X
Transfer A to CCR	TAP						X
Transfer B to A	TBA						X
Transfer CCR to A	TPA						X
Exchange D with X	XGDX						X
Exchange D with Y	XGDY						X

mulators, respectively. The CLR instruction clears a memory location. Table 5-1 shows that the CLR instruction can be executed in the extended, indexed by X, and indexed by Y modes. This is true of many other **68HC11** instructions, such as INCREMENT, DECREMENT, NEGATE, and so on that access memory; they can only be executed in these modes.

E X A M P L E   5 - 1
_____

The contents of the registers, taken from an EVB printout, are shown in Figure 5-1. What does the instruction

    CLR   25,Y

do?

Solution

The instruction adds 25 to the contents of Y (7982) to obtain the address, 79A7. Then it clears (zeros out) this location.

```
P-C007 Y-7982 X-FF00 A-44 B-70 C-C0 S-004A
>
```

Figure 5-1   Sample register contents. (*Courtesy of Motorola, Inc.*)

_____

5-1.2   LOAD and STORE Instructions

The LOAD and STORE instructions were introduced in Chapter 3. Table 5-1 shows that the LOAD instructions can be executed in all modes except inherent. Table 5-2 provides a review of these modes.

Table 5-2   The Mode of a LOAD Instruction

Mode	Example	Operation
Immediate	LDAA #xx	$xx \rightarrow A$
Direct	LDAA $xx	$(00xx) \rightarrow A$
Extended	LDAA $xxxx	$(xxxx) \rightarrow A$
Indexed X	LDAA xx, X	$(X + xx) \rightarrow A$
Indexed Y	LDAA xx, Y	$(Y + xx) \rightarrow A$

The STORE instructions operate similarly except that a STORE IMMEDIATE is impossible.

The *double* instructions work as above except that they affect *two* memory locations.

E X A M P L E   5 - 2

Using the register set of Figure 5-1, what does the instruction

    LDD   17,X

do?

Solution

First, the instruction calculates the memory address as FF17 (17 plus FF00, the contents of X). It then loads the contents of FF17 into A and the contents of FF18 into B.

The index register instructions, LDX, LDY, STX, and STY load and store the index registers. Of course, these require two memory locations because the index registers contain 16 bits.

E X A M P L E   5 - 3

Using the register set of Figure 5-1, what does the instruction

    STY   44,Y

do?

Solution

First, the instruction calculates the address as 79C6 (Y + 44). Then it stores the most significant byte of Y, 79, into 79C6 and the least significant byte, 82, into 79C7. Note that it uses two memory locations to accommodate the 16 bits of Y.

5-1.3   Register Transfer Instructions

The rest of the instructions in Table 5-1 transfer data between registers. The contents of the A accumulator can be copied into B, or vice versa, by the TAB and TBA instructions. The CCR can be sent to A by using a TPA, and A can be used to set the condition codes by using a TAP. This can be useful if the programmer needs to preserve the condition codes at one point in a program and to restore them later. The two *exchange* instructions, XGDX and XGDY, allow the user to exchange the X and Y index registers with the 16-bit double accumulator.

E X A M P L E   5 - 4

Starting with the register set of Figure 5-1, what will happen if the instruction XGDX is executed?

Solution

This instruction exchanges the contents of the double register (A and B) with X. After it is executed the register contents will be

$$A = FF \qquad B = 00 \qquad X = 4470$$

5-2   Addition Instructions

The addition instructions in the **68HC11** are given in Table 5-3. The ADD instructions add an 8-bit number to either A or B. They are available in all memory modes. There is also an ABA instruction that adds A and B, with the results going to A.

Table 5-3   The Addition, Increment, and Decrement Instructions in the 68HC11

Function	Mnemonic	IMM	DIR	EXT	INDX	INDY	INH
Add memory to A	ADDA	X	X	X	X	X	
Add memory to B	ADDB	X	X	X	X	X	
Add accumulators	ABA						X
Add with carry to A	ADCA	X	X	X	X	X	
Add with carry to B	ADCB	X	X	X	X	X	
Add memory to D (16 bit)	ADDD	X	X	X	X	X	
Increment memory byte	INC			X	X	X	
Increment accumulator A	INCA						X
Increment accumulator B	INCB						X
Increment index register X	INX						X
Increment index register Y	INY						X
Decrement memory byte	DEC			X	X	X	
Decrement accumulator A	DECA						X
Decrement accumulator B	DECB						X
Decrement index register X	DEX						X
Decrement index register Y	DEY						X
Add accumulator B to X	ABX						X
Add accumulator B to Y	ABY						X

E X A M P L E   5 - 5

The partial contents of memory are given in Table 5-4. For *each part* the initial contents of the registers are

$$A = CD \quad B = 8A \quad X = C110$$

What are the results of each of the following instructions?

a. ADDA #35

b. ADDB $35

c. ABA

d. ADDA 0A,X

Solution

a. This instruction simply adds 35 to the A, giving 35 + CD = 02. It will clear N, V, and Z but set the C (carry) flag.

b. This instruction adds the contents of 35, found to be 2B from Table 5-4, to B. The results in B are B5. This addition sets N but leaves the other flags clear.

c. ABA adds A and B. The sum, 57, is placed in A; B remains 8A. This addition sets the C flag because it generates a carry and sets the V flag because two negative numbers were added to give a positive result.

d. The address of this instruction is C11A, which contains 33. After it is executed A will contain CD + 33 or 00. The Z and C flags will be set.

5-2.1 ADD WITH CARRY Instructions

The ADD WITH CARRY instructions (ADCA and ADCB) perform the required addition and then add *one more if the carry flag is set at the start of the addition*. The carry flag at the end of the addition is set in accordance with the result. ADC instructions can be used for multiple byte instructions, as Example 5-6 illustrates.

E X A M P L E   5 - 6

There is a 16-bit number in locations 30 and 31 of Table 5-4. It is 53FA. Write a program to add the number in locations 30 and 31 to the 16-bit number in 32 and 33, E395, and put the sum in 34 and 35. Graphically, the problem looks like

$$
\begin{array}{cc}
(30) & (31) \\
+ \ \underline{(32)} & \underline{(33)} \\
(34) & (35)
\end{array}
$$

where the parentheses indicate the contents of the location.

Table 5-4 Initial Memory and Register Contents for Some Examples in Chapter 5

Memory Contents		Initial Register Content
ADDR	**CODE**	
002A	01	A = CD
2B	A0	B = 8A
2C	14	
2D	6C	X = C110
2E	AC	
002F	8D	
30	53	
31	FA	
32	E3	
33	95	
34	9E	
0035	2B	
36	EB	
37	CE	
38	00	
39	01	
003A	4C	
C11A	33	
1B	26	
1C	0A	
1D	39	
1E	25	
C11F	25	
20	20	
21	D2	
22	BF	
23	FE	
24	FA	
C125	4F	
26	BD	
27	2A	
28	2A	
29	2D	
C12A	0E	

Solution

There is never a carry *into* the least significant bytes, so they can simply be added. Their sum, however, may produce a *carry-out*. If so, it must be added to the more significant bytes. This requires an ADCA. The program is

LDAA 31	(FA)
ADDA 33	(8F)
STAA 35	(8F)
LDAA 30	(53)
ADCA 32	(37)
STAA 34	(37)

The program will work for any numbers. The numbers in parentheses are the contents of A after each instruction for the given numbers. In this case the ADCA added 53 + E3 and then one more because the C bit was set by the previous addition. The C bit is also set at the end of the program because the second addition also produced a carry.

E X A M P L E   5 - 7

Write a program to add all the bytes between C100 and C1FF. They are all to be considered as positive numbers.

Solution

It is easy to add the numbers in the buffer. The problem is that they will probably overflow an 8-bit accumulator, and a 16-bit space should be provided for the answer. Perhaps the simplest solution is to use one accumulator, A, for example, to hold the lower byte of the sum and to use A to hold the more significant byte.

The upper byte can be obtained by incrementing A every time there is a carry-out of the lower byte. This can be done by using a BRANCH ON CARRY SET (BCS) instruction to branch to a program that increments B each time the C bit is set, but there is a more elegant way. The instruction ADCB #00 will increment A only if the carry flag is set.

The program is as follows:

	CLRA		
	CLRB		Clear both bytes of the sum.
	LDX	#C100	Set the index register to the start of the buffer.
LOOP	ADDB	0,X	Add the least significant byte.
	ADCA	#00	Increment A when the carry flag is set.
	INX		

```
CPX      #C200    Done?
BNE      LOOP     No. Execute the loop again.
SWI               Yes. Halt.
```

The index register was compared to one location after the buffer to determine when to stop. But as soon as it pointed beyond the buffer, the program terminated without executing the loop.

5-2.2   The DOUBLE ADD Instruction

The DOUBLE ADD instruction, ADDD, adds a 16-bit number to the contents of the double accumulator formed by A and B. If a memory location is specified, it is the contents of that memory location and the next that are added.

E X A M P L E   5 - 8

Using Table 5-4, what are the contents of A and B after the following instructions:

a. ADDD #C11C

b. ADDD $C11C

Solution

a. This immediate instruction adds the number C11C to D, giving C11C + CD8A = 8EA6. Thus A contains 8E and B contains A6.

b. This is an extended instruction whose memory address is C11C. It adds what is in C11C and C11D to the double accumulator. The results are 0A39 + CD8A = D7C3.

Example 5-6 could have been done very simply by using the double accumulators. The coding is

```
LDD      30    Load the contents of 30 and 31 into D.
ADDD     32    Add the contents of 32 and 33.
STD      34    Put the results in 34 and 35.
```

E X A M P L E   5 - 9*

Write a program to add all the bytes between C100 and C1FF. These are signed numbers and must be added correctly.

* This example contains more advanced concepts and may be omitted on first reading.

Solution

Because some numbers in the buffer are positive and others are negative, trying to add them in an 8-bit register can result in overflow problems. One solution is to make 16-bit numbers out of the 8-bit numbers by *extending* them. Extension is a valuable concept in computers and the modern 16-bit computers have an EXTEND instruction. Once the numbers have been extended, they can be added directly and without overflow problems. The sign of the answer will also be correct.

An 8-bit number can be extended to a 16-bit number by adding 00 if it is positive and FF if it is negative. This preserves the sign of the original number. Problem 5-7 should clarify the concept. The following program segment will extend an 8-bit number in B to a 16-bit number in the D register:

```
          CLRA
          LDAB    number
          BPL     RESUME
          LDA     #FF
RESUME    . . . .
```

If the number loaded into B is positive, the program will branch to RESUME, leaving 00 in A, the most significant byte of D. If the number is negative, the program loads FF into A.

Now we can attack the problem. Assume locations C200 and C201 are set aside for the result. The code would look like the following:

```
          CLR     C200
          CLR     C201          Clear both bits of the sum.
          LDX     #C100         Point X to the start of the
                                buffer.
LOOP      CLRA
          LDAB    0,X           Place the first byte in B and ex-
                                tend it to form the D register.
          BPL     RESUME
          LDAA    #FF
RESUME    ADDD    C200          Add the sum.
          STD     C200          Store the sum.
          INX
          CMPX    #C200         Loop finished?
          BNE     LOOP          No, do it again.
          . . . . . . . . . . . Yes, go on.
```

At the end of this program the proper number with the proper sign will be found in C200 and C201.

5-2.3 INCREMENT and DECREMENT Instructions

INCREMENT is an instruction that adds 1 to A or B (INCA or INCB) or to a memory location (INC). Like the CLR instruction, INC can only reference memory using the extended or indexed modes. For example, the instruction INC 25,Y increments the contents of the location that is 25 beyond the value in Y. Two other INCREMENT instructions, INX and INY, increment the X and Y index registers, respectively.

The **DECREMENT** instructions deduct 1 from the registers or memory locations they specify. They operate in the same modes as the INCREMENT instructions.

5-2.4 The ABX and ABY Instructions

The ABX and ABY instructions add the B accumulator to the specified index register. This is very useful for constructing jump tables or for code conversion (see Example 5-10).

Table 5-5 A Hex-to-ASCII Conversion Table

Hex Digit	ASCII Equivalent
0	30
1	31
2	32
3	33
4	34
5	35
6	36
7	37
8	38
9	39
A	41
B	42
C	43
D	44
E	45
F	46

Table 5-5 is a simple code conversion table that converts each of the 16 hex digits into its ASCII* equivalent.

* ASCII (The American Standard Code for Information Interchange) is the code recognized by keyboards, printers, and video terminals. The complete ASCII chart is given in Appendix B.

E X A M P L E   5 - 10

Assume that the ASCII table is in memory starting at C100. Given a hex digit in C050, write a program to place its ASCII equivalent in A.

Solution

The program is shown below. It loads the digit into B, adds B to X, and then loads A from X:

LDX	#C100	Point X to the start of the table.
LDAB	$C050	Load digit into B.
ABX		
LDAA	0,X	Load ASCII equivalent into A.

If the number in C050 is 5, for example, X will contain C105 after the ABX. But the contents of C105 are 35—the ASCII equivalent of 5—and it is this number that is loaded into A.

5-3   Subtraction Instructions

The SUBTRACT instructions in the **68HC11** are listed in Table 5-6. In general, they subtract the contents of memory (or an immediate operand) from an accumulator. If the subtrahend (usually the contents of memory) is greater than the contents of the accumulator, the result is negative and the C flag is set to indicate this condition. In some cases the N flag will not be set; this is shown in Example 5-11.

E X A M P L E   5 - 11

If accumulator B contains 30, what is the result of the instruction

    SUBB   #F0

Solution

The result of the subtraction is 40. The only condition code that will be set is the C flag because the minuend is larger (in absolute value) than the subtrahend. The decimal equivalent of this subtraction is

$$48 - (-16) = 64$$

Notice that the N and V bits were not set.

The V bit is set under subtraction when an impossible situation occurs, such as subtracting a negative number from a positive number and obtaining a negative result. The action of the V bit was discussed in Section 3-5.4.

Table 5-6 Subtracts, Compare, Negate, and Test Instructions in the 68HC11

Function	Mnemonic	IMM	DIR	EXT	INDX	INDY	INH
Subtract memory from A	SUBA	X	X	X	X	X	
Subtract memory from B	SUBB	X	X	X	X	X	
Subtract with carry from A	SBCA	X	X	X	X	X	
Subtract with carry from B	SBCB	X	X	X	X	X	
Subtract memory from D (16 bit)	SUBD	X	X	X	X	X	
Compare A to B	CBA						X
Compare A to memory	CMPA	X	X	X	X	X	
Compare B to memory	CMPB	X	X	X	X	X	
Compare D to memory (16 bit)	CPD	X	X	X	X	X	
Twos complement memory byte	NEG			X	X	X	
Twos complement accumulator A	NEGA						X
Twos complement accumulator B	NEGB						X
Test for zero or minus	TST			X	X	X	
Test for zero or minus A	TSTA						X
Test for zero or minus B	TSTB						X

5-3.1 SUBTRACT-WITH-CARRY and SUBTRACT-DOUBLE Instructions

The **SUBTRACT-WITH-CARRY** (SBCA and SBCB) instructions perform the subtraction and also subtract 1 from the result if the carry flag at the *start* of the instruction is high. Thus the carry flag acts as a borrow in subtraction. After the instruction is executed the SBC instructions set the carry flag in accordance with the result. The SBC instructions are useful when performing multibyte subtractions, as Example 5-12 shows.

E X A M P L E   5 - 12

a. Write a program to perform the following subtraction:

$$
\begin{array}{cc}
(C030) & (C031) \\
-(C032) & (C033) \\
\hline
(C034) & (C035)
\end{array}
$$

b. If C030–C031 contains AB07 and C032–C033 contains ABAA, show the results and the carry flag after each subtraction.

Solution

The program is similar to multibyte addition (refer to Example 5-6). This time the index register is used:

LDX	#C030	Point the index register at the buffer.
LDAA	1,X	Place the least significant (LS) byte of the minuend in A.
SUBA	3,X	Subtract the LS byte.
STAA	5,X	Store it.
LDAA	0,X	Place the most significant (MS) byte of the minuend in A.
SBCA	2,X	Subtract the MS byte (with carry).
STAA	4,X	Store it.

The subtraction on the LS bytes is simple because *there is never a borrow into the LS bytes*. A subtraction with carry is required on the more significant bytes, however, because the subtraction on the LS bytes might have generated a borrow.

b. For the given numbers the first subtraction is 07 − AA, which equals 5D. Because the subtrahend is larger than the minuend, the carry flag will be set by this instruction. The second subtract instruction takes AB from AB, giving 00. If this were a simple subtraction, the carry flag would be clear at the end of the instruction. Here, however, it is an SBC and the carry flag is set. This causes the 00 to decrement to FF, and the carry flag will be set when the instruction finishes.

The SUBTRACT-DOUBLE (SUBD) subtracts the contents of two memory locations from the D accumulator. Its use would have simplified the program of Example 5-12. The program could have been

LDD	C030
SUBD	C032
STD	C034

The SUBD instruction is not recommended for larger additions or subtractions, however, because there is no ADDD with carry or SUBD with carry.

5-3.2 COMPARE Instructions

The COMPARE instructions were discussed in Section 3-6. To review, a COMPARE instruction is identical to a subtraction except that the results are discarded. Therefore *a COMPARE instruction will not change any registers* except the CCR.

COMPARE instructions are used to set the flags prior to a conditional branch.

5-3.3   NEGATE Instructions

The NEGATE instructions find the 2s complement negation of a number by subtracting it from 00. The **68HC11** can negate A, B, or memory in the extended or indexed modes. For example, if A contains FF, the instruction NEGA causes the computer to perform the operation $00 - FF = 01$. Thus A will contain 01 or $+1$, which is indeed the negative of FF (-1).

The NEGATE instruction sets the flags in the same manner as other subtraction instructions. Thus it will set the C flag in all cases *except when the number to be negated is 00* because it is subtracting a number from 00. If the number to be negated is 00, it will set the Z flag. There is no instruction to negate the D register. It can be negated, however, by a small program (see Prob. 5-11).

5-3.4   The TEST Instruction

The TEST (TST) instruction subtracts 00 from an accumulator or memory. This does not change the destination, but it sets the N and Z flags in accordance with *characteristics* of the number (Is it negative? Is it 0?) TST also clears the V and C flags. For example, if A contains FF, TSTA will set the N flag so that the user can determine that A contains a negative number.

5-4   Multiplication and Division Instructions

The **68HC11** has multiplication and division instructions. This is an improvement over earlier 8-bit μPs. These instructions are listed in Table 5-7.

5-4.1   Multiplication

Multiplication tends to produce larger numbers in the result. For binary multiplication the number of bits in the product must equal the sum of the bits in the multiplier and multiplicand. In the **68HC11** there is a single

Table 5-7   Multiplication and Division Instructions in the 68HC11

Function	Mnemonic	INH
Multiply (A × B ♦ D)	MUL	X
Fractional Divide (D ÷ X ♦ X; r ♦ D)	FDIV	X
Integer Divide (D ÷ X ♦ X; r ♦ D)	IDIV	X

1-byte multiply instruction, MUL. It multiplies the contents of A by the contents of B and puts the result in the 16-bit D register. Thus MUL changes the contents of A and B when it is executed.

E X A M P L E   5 - 13

Write a program to multiply the byte in C100 by the byte in C200 and place the results in C300 and C301.

Solution

The program is simply

LDAA	C100	Put multiplier in A.
LDAB	C200	Put multiplicand in B.
MUL		Multiply.
STD	C300	Store results.

The MUL instruction is *unsigned*; all numbers are assumed to be positive. We tried multiplying -2 by -3 (FE times FD) and obtained FB06 instead of 0006. The computer interpreted FE as decimal 254 and FD as 253. The result, FB06, converts to decimal 64,262, which is indeed the product of 253 and 254.

E X A M P L E   5 - 14*

Write a program to do signed multiplication. This means that the result should appear in the D register with the proper sign.

Solution

The solution will be described conceptually.

1. Reserve a dummy location in memory, say C050, and clear it.

2. Load the multiplier into A. If it is negative (as determined by the N bit), do the following:
   a. Increment C050.
   b. Negate A so that it is positive.

3. Load the multiplicand into B. If it is negative
   a. Decrement C050.
   b. Negate B.

4. Multiply. The result in D is the product of the positive numbers.

5. Test C050. If it is zero, the product is correct. Otherwise negate the D register (see Prob. 5-11).

* This example is a little more complex and may be skipped on first reading.

5-4.2   Rounding in Multiplication*

The MUL instruction multiplies two 8-bit numbers to give a 16-bit product. Sometimes only the eight MSBs (Most Significant Bits) of the product are of interest and the user may decide to discard the eight LSBs (Least Significant Bits). Rounding allows the user to increment the value of the MSBs if the value of the LSBs being discarded is greater than 0.5, where the MSBs are valued from 0 to 255. The MUL instruction simplifies rounding by setting the C flag to 1 if the MSB of accumulator B is 1. This indicates that the value of the lower half of the number is greater than 0.5. If the user wishes to round, he or she can then perform an ADCA #00 to increment the result in A if the carry flag is set.

5-4.3   Integer Division

The INTEGER DIVISION instruction (IDIV) divides the 16-bit number in X into the 16-bit number in D. After the division the quotient is placed in X and the remainder is placed in D.

E X A M P L E   5 - 15

If A contains 00, B contains E3, and X contains 000A, what will they hold after an IDIV instruction?

Solution

This problem is really 00E3/A. The result is 16 with a remainder of 7. After it is run X will contain 16 and D will hold 7 (A = 00, B = 07).

   This problem can be checked using decimal numbers. In decimal it converts to 227/10, which equals 22 with a remainder of 7. But, of course, $(22)_{10} = (16)_{16}$.

5-4.4   Fractional Division

The FRACTIONAL DIVISION instruction (FDIV) also divides the number in D by the number in X. Here, however, the divisor is assumed to be greater than the dividend so that a fractional answer (a number less than 1) results. The quotient is expressed as a binary fraction in X and the remainder goes to D.

   This instruction was tested with A = 0, B = B, and X = 16. In decimal numbers this is dividing 11 by 22. The result was 8000 in X and 0 in D; 8000 is the equivalent of 0.5 expressed as a binary fraction.

* This section may be omitted on first reading.

5-5   SHIFT and ROTATE Instructions

A SHIFT instruction shifts the operand 1 bit. Both left and right shifts are available. A SHIFT instruction vacates a bit at one end of the register, which must be replaced, and shifts the bit at the other end of the register out of the register. The bit that is shifted out always goes into the carry FF. A left shift vacates b0 and shifts b7 into the carry FF, whereas a right shift vacates b7 and shifts b0 into the carry FF.

Table 5-8 shows the SHIFT and ROTATE instructions available on the **68HC11**. With some exceptions they all operate on the A, B, or D accumulators (e.g., ASLA, ASLB, and ASLD instructions are available) or di-

Table 5-8   SHIFT and ROTATE Instructions in the 68HC11

Function	Mnemonic	IMM	DIR	EXT	INDX	INDY	INH
Arithmetic shift left memory	ASL			X	X	X	
Arithmetic shift left A	ASLA						X
Arithmetic shift left B	ASLB						X
Arithmetic shift left double	ASLD						X
Arithmetic shift right memory	ASR			X	X	X	
Arithmetic shift right A	ASRA						X
Arithmetic shift right B	ASRB						X
(Logical shift left memory)	LSL			X	X	X	
(Logical shift left A)	LSLA						X
(Logical shift left B)	LSLB						X
(Logical shift left double)	LSLD						X
Logical shift right memory	LSR			X	X	X	
Logical shift right A	LSRA						X
Logical shift right B	LSRB						X
Logical shift right D	LSRD						X
Rotate left memory	ROL			X	X	X	
Rotate left A	ROLA						X
Rotate left B	ROLB						X
Rotate right memory	ROR			X	X	X	
Rotate right A	RORA						X
Rotate right B	RORB						X

rectly on operands in memory (ASL). Memory operands must be addressed in the extended or indexed modes.

5-5.1 LOGICAL SHIFT Instructions

A LOGICAL SHIFT always brings a 0 into the vacated bit position. The action of the logical shifts is shown in Figure 5-2, which shows the 0 being shifted in. It also shows that the bit shifted out goes into the carry flag.

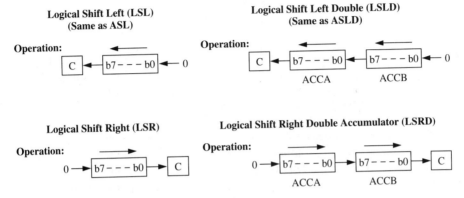

Figure 5-2 Logical shifts in the 68HC11. (*Redrawn Courtesy of Motorola, Inc.*)

E X A M P L E   5 - 16

If A contains A6 and B contains E3, what are the results of the following instructions?

a. LSLA

b. LSRB

c. LSLD

d. LSRD

Solution

The most precise way to attack shift problems is to write the numbers out bit-by-bit and then shift them. Figure 5-3 shows the solution for parts (a) and (b) using this method. The answers for these parts are seen to be 4C and 71, respectively.

The double shifts require shifting the 16-bit number A6E3. The student might make a drawing similar to Figure 5-3 to find the results. The LSLD gives 4DC6. Note that the 1 in the MSB of B was shifted into the LSB of A by the double shift. The LSRD instruction gives 5371. The carry flag was set in both cases.

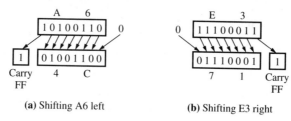

(a) Shifting A6 left   **(b)** Shifting E3 right

Figure 5-3   Logical shifts for Example 5-16.

5-5.2   ARITHMETIC SHIFT Instructions

Although ARITHMETIC SHIFT LEFT (ASL) and LOGIC SHIFT LEFT (LSL) are described separately in Table 5-8 and even listed separately in Appendix A, they are actually identical (having the same Op code), and they have been described in the previous paragraph.

The effect of a shift left is to multiply the number in a register by 2. This happens precisely unless a 1 is shifted out of the register. In Example 5-16 we saw that when the number A6 was shifted left it became 4C, which is definitely not twice A6. This is because a 1 was shifted out of the A accumulator. But if the 1 in the carry FF is considered, the number is 14C, which is indeed twice A6.

ARITHMETIC SHIFT RIGHT (ASR) keeps *the most significant bit the same* while it shifts right, as shown in Figure 5-4. Thus if the MSB were a

Arithmetic Shift Right (ASR)

Operation: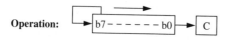

Description: Shifts all of ACCX or M one place to the right. Bit 7 is held constant. Bit 0 is loaded into the C bit of the CCR. This operation effectively divides a 2s complement value by 2 without changing its sign. The carry bit can be used to round the result.

Figure 5-4   The Arithmetic Shift Right. (*Redrawn Courtesy of Motorola, Inc.*)

0, it would remain a 0, and if it were a 1, it would remain a 1. This *preserves the sign of the number* and makes the ASR the equivalent of dividing the number by 2.

E X A M P L E   5 - 17

In Example 5-16 we saw that when the number E3 was logically shifted right it became 71. What if it is arithmetically shifted right? Is this equivalent to dividing it by 2?

Solution

The ASR preserves the 1 in the MSB, so the results are F1 instead of 71. The number E3 corresponds to -29 decimal, and F1 corresponds to -15. So the number has indeed been halved. If we started with an even number, it would have been halved precisely.

5-5.3   ROTATE Instructions

The ROTATE instructions are shown in Figure 5-5. A ROTATE is like a SHIFT except that the contents of the carry FF at the start of the ROTATE

Rotate Left (ROL)   **Rotate Right (ROR)**

Figure 5-5   The Rotate instructions in the 68HC11. *(Redrawn Courtesy of Motorola, Inc.)*

are brought into the vacant bit, as shown in the figure. Thus it is really a 9-bit rotate; the 8 bits of the operand and the carry FF are rotated. Nine ROTATE instructions in a row will restore the register to its original condition.

There are no instructions to rotate the D register, nor to ASR the D register. The following code segment effectively performs an ASR on the D register.

        ASRA
        RORB

The ASRA put the LSB of A in the carry FF. The ROR was necessary to take it out of the carry FF and to transfer it into the MSB of B.

E X A M P L E   5 - 18

There is a 16-bit number in C100 and C101. Write a program to shift it left.

Solution

There are two possible solutions. The first program uses the double registers:

        LDD   C100
        LSLD
        STD   C100

An alternate solution is to shift the numbers in memory:

LSL   C101   Shift the lower byte.
ROL   C100   Rotate the upper byte.

Again, a ROTATE was necessary to get the carry from the lower byte to the upper byte.

5-6 Logic Instructions

Table 5-9 lists the logic instructions available on the **68HC11**. The major logic instructions are **AND**, **OR**, and **EOR** (Exclusive Or). These instructions use one of the two accumulators for one of the operands and memory or an immediate value for the other operand. The D register is not used for logic instructions. In general, these instructions clear the V flag, leave C alone, and set N and Z in accordance with the results.

The AND instructions calculate the logical AND of the two operands. If the corresponding bits of both operands are 1 the corresponding output bit is 1; otherwise it is 0. If Y is to be the logical AND of A and B, it is written as $Y = AB$.

Table 5-9 Logic Instructions in the 68HC11

Function	Mnemonic	IMM	DIR	EXT	INDX	INDY	INH
AND A with memory	ANDA	X	X	X	X	X	
AND B with memory	ANDB	X	X	X	X	X	
Bit(s) test A with memory	BITA	X	X	X	X	X	
Bit(s) test B with memory	BITB	X	X	X	X	X	
Ones complement memory byte	COM			X	X	X	
Ones complement A	COMA						X
Ones complement B	COMB						X
OR A with memory (exclusive)	EORA	X	X	X	X	X	
OR B with memory (exclusive)	EORB	X	X	X	X	X	
OR A with memory (inclusive)	ORAA	X	X	X	X	X	
OR B with memory (inclusive)	ORAB	X	X	X	X	X	
Clear bit(s) in memory	BCLR		X		X	X	
Set bit(s) in memory	BSET		X		X	X	

The OR instruction finds the logical OR of the operands. If either bit of the operands is 1, the output will be 1. If Y is the logical OR of A and B, it is written as $Y = A + B$.

The EOR instruction finds the Exclusive-Or of the operands. The output of the corresponding bits will be 1 only if exactly one of the two input bits is 1. If Y is the Exclusive-Or of A and B, it is written as $Y = A \oplus B$. All of these instructions act basically like the digital logic gates of the same name.

E X A M P L E   5 - 19

If C100 contains C9 and A contains 63, what are the results of each of the following instructions?

a. ANDA   C100

b. ORAA   C100

c. EORA   C100

Solution

a. The AND operation is

	(C100)	1 1 0 0 1 0 0 1	C9
AND	(A)	0 1 1 0 0 0 1 1	63
Result	(AB)	0 1 0 0 0 0 0 1	41

   The result, found in A, is 41. It has a 1 wherever both operands are 1.

b. The OR operation is

	(C100)	1 1 0 0 1 0 0 1	C9
OR	(A)	0 1 1 0 0 0 1 1	63
Result	(A + B)	1 1 1 0 1 0 1 1	EB

   The result in A, EB, has a 1 in all bit positions where either A or the contents of C100 has a 1.

c. The Exclusive-Or (EORA) operation is

	(C100)	1 1 0 0 1 0 0 1	C9
EOR	(A)	0 1 1 0 0 0 1 1	63
Result	(A $\oplus$ B)	1 0 1 0 1 0 1 0	AA

   The result of the EORA instruction, AA, has a 1 where exactly one of the operands has a 1.

5-6.1 The BIT Instructions

The **68HC11** provides two BIT instructions, BITA and BITB. The BIT instructions perform an AND of the designated accumulator and memory or an immediate operand but discard the results. Like the COMPARE instructions, the BIT instructions are used only to set the condition codes. BIT instructions set N and Z in accordance with the result, clear V, and leave C unchanged. They are most often used to determine whether a particular bit in a word is a 1 or a 0. This is often used during I/O operations.

E X A M P L E   5 - 20

Write a code segment to determine whether bit 4 of location C100 is a 1 or 0.

Solution

One solution is to use a BIT instruction to test the bit in C100. The code might be

    LDAA   C100

    BITA   #10

Note that a hexadecimal 10 only has a 1 in bit position 4. The results of these instructions will be zero ($Z = 1$) if bit 4 is 0, and nonzero (actually 10) if bit 4 is 1. In that case the Z flag will clear. After the BITA instruction the Z flag can be used to determine whether bit 4 of C100 was 0 or 1.

5-6.2 Setting Bits

Sometimes it is required to set one or more bits in an accumulator or memory location without changing any of the other bits in that location. This often occurs when doing I/O where each bit controls an external device.

If the byte is already in an accumulator, perhaps the simplest way to do it is to OR the accumulator with a **mask**. A mask is the term given to a byte that contains 1s in the positions where the bits are to be set. For example, if we must set bits 3 and 5 of accumulator B, the mask will contain 1s in these bit positions; it will then be 28. The instruction ORAB #28 or ORAB #mask will set bits 3 and 5 in the selected accumulator.

A special instruction, BSET, has been provided to set bits in memory. BSET operates in one of three modes: direct, indexed by X, and indexed by Y.

BSET in the *direct mode* requires 3 bytes consisting of Op code, address, and mask. The Op code is 14, so a direct BSET would look like 14 23 05, for example. Here the address is 0023 and the mask is 5. This instruction sets bits 0 and 2 of location 0023. Like any direct instruction, it can only be used to reference locations 0000–00FF.

BSET in the indexed by X instruction also requires 3 bytes: the Op code (1C), the offset, and the mask. It sets the bits specified by the mask at the referenced location. For example, the instruction 1C 05 01 sets bit 0 of the word at the contents of X plus 5.

BSET in the indexed by Y mode operates similarly. It is a 4-byte instruction, however, because it requires a 2-byte Op code—18 1C.

The following example is somewhat more complex but uses several of the logic instructions.

E X A M P L E   5 - 21

There is a 7-bit number in bits b6–b0 of C100. Write a program to set bit 7 of C100 so that it contains *odd parity*. That is, the number of bits that are 1 in C100 must always be odd at the end of the program. For instance, if C100 contains 0F originally, which means it has an even number of 1s (four), it must contain 8F after the program is run. Bits b6–b0 are unchanged but b7 is a 1 so that location C100 now contains five 1s.

Solution

The following program works by assuming at first that bits b6–b0 are all 0, and therefore it sets b7 to 1. It then checks each bit. If the bit is a 1, it inverts b7:

	LDX	#C100	
	LDAA	#01	Set up A to check the first bit.
	BSET	0,X 80	Set b7 of C100. Note that X was previously loaded with C100 so that this instruction could be used.
L1	BITA	0,X	Is the bit a 1?
	BEQ	L2	This branch will be taken if the bit is 0.
	LDAB	0,X	If we are here, the bit was 1. These instructions invert b7.
	EORB	#80	
	STAB	0,X	
L2	LSLA		Set up for the next bit.
	CMPA	#80	Done?
	BNE	L1	No. Examine next bit.
	SWI		Yes. Stop.

See also Problem 5-22.

5-6.3   Clearing Bits

Clearing bits is very similar to setting bits. Bits can be cleared in a location by ANDing it with a number that contains 0s in the positions to be cleared and 1s elsewhere. The BIT CLEAR (BCLR) instruction can be used to clear bits in a memory location. It is similar to the BSET instruction, using the same modes and mask. The bits that are cleared correspond to the 1s in the mask.

E X A M P L E   5 - 22

Write a code segment to clear bits 0 and 1 of

a. Accumulator B.

b. Location CC.

Solution

a. The instruction ANDB #FC will clear bits 0 and 1. Note that FC contains 1s in all bit positions except 0 and 1.

b. Here we can take advantage of the fact that CC is in lower memory and the direct mode can be used. The instruction

BCLR CC 03

will clear the bits. The mask, 03, has 1s in positions 0 and 1.

5-6.4   Complement Instructions

These instructions will complement an accumulator or memory location, which means they will invert all the bits of the location. Complementation is performed by subtracting the byte from FF. For example, if A contains CC, it will contain 33 after COMA is executed.

5-7   Decimal Arithmetic

Some computer applications require decimal arithmetic. They assume decimal inputs and require decimal outputs. For example, if 26 and 35 are added, the user might like the result to be 61, but the computer sum will be 5B. Decimal arithmetic allows the user to convert these answers to their correct decimal value. The **68HC11** can only correct addition operations; other operations cannot be conveniently performed decimally.*

---

* A discussion of decimal subtraction can be found in J. D. Greenfield and W. C. Wray, *Using Microprocessors and Microcontrollers: The Motorola Family*, 2nd ed. This book was originally published by John Wiley, 1988. It may now be obtained from Prentice-Hall, Englewood Cliffs, N.J.

5-7.1 The H Bit

The **H Bit** is bit 5 of the Condition Code Register (CCR). Because it is used only for decimal arithmetic, it has not been discussed up to now. The H bit is only set or cleared by 8-bit addition instructions; all other instructions, including ADDD, ADX, and ADY, do not affect the H bit.

The H bit is set during an addition instruction when there is a carry-out of bit 3 and into bit 4. This is a carry from the lower nibble to the upper nibble.

E X A M P L E   5 - 23

What is the result of the following instructions? Include the H bit.

a. Adding 37 and 38

b. Adding 37 and 39

Solution

a. Adding 37 + 38 gives 6F. The H bit is not set because there was no carry-out of the lower nibble.

b. Adding 37 + 39 gives 70. Now there is a carry-out of the lower nibble and the H bit is set. As we will see, this is an indication that 70 is *not* the sum of 37 and 39 if decimal arithmetic is being used.

5-7.2 Decimal Addition

Decimal addition presupposes that both numbers to be added are decimal. The result of the addition, however, will be in hex as we have seen. The **DECIMAL ADJUST ACCUMULATOR (DAA)** instruction should be used immediately following the addition if a decimal result is required. It modifies the hex sum to give the proper decimal sum. The DAA, however, only applies to accumulator A, so A must hold the sum of the addition.

The DAA depends on the sum and the H bit. It adds a number to the hex result in accordance with Table 5-10 to correct the result.

E X A M P L E   5 - 24

Add 56 + 46, and follow with a DAA. Show that the decimal result is correct.

Solution

After the ADD instruction the result is 9C. The carry bit and the half-carry bit are both clear. Because the lower half-byte is between A and F, the only

Table 5-10   The Action at a DAA Instruction

State of C Bit Before DAA (Column 1)	Upper Half-Byte of ACCA (Bits 7–4) (Column 2)	Initial Half-Carry H Bit from CCR (Column 3)	Lower Half-Byte of ACCA (Bits 3–0) (Column 4)	Number Added of ACCA by DAA (Column 5)	State of C Bit After DAA (Column 6)
0	0–9	0	0–9	00	0
0	0–8	0	A–F	06	0
0	0–9	1	0–3	06	0
0	A–F	0	0–9	60	1
0	9–F	0	A–F	66	1
0	A–F	1	0–3	66	1
1	0–2	0	0–9	60	1
1	0–2	0	A–F	66	1
1	0–3	1	0–3	66	1

NOTE: Columns (1) through (4) of the table represent all possible cases that can result from any of the operations ABA, ADD, or ADC, with initial carry either set or clear, applied to two binary-coded-decimal operands. The table shows hexadecimal values.

applicable lines in Table 5-10 are lines 2, 5, and 8. A further examination of the table reveals that only line 5 applies in this case. Line 5 states that the DAA will add 66 to the 9C to give 02. It will also set the carry flag. If a carry-out is considered as 100 in decimal, the sum is 102, the correct decimal sum.

E X A M P L E   5 - 25

Add 48 + 59, and follow with a DAA.

Solution

The sum is A1 and the H bit is set. This matches line 6 of the table. Again, 66 is added, giving 07 and setting the carry bit. The final result is therefore 107, which is the correct decimal answer.

5-7.3   Decimal Incrementing

Some applications require that a number be incremented decimally. You might want a number to progress from 00 to 99 and then produce a carry-out when it overflows. Again, the DAA helps, but the number to be incremented must be in A. The instruction sequence INCA–DAA might cause a problem if the C or H bit is set. The INCA does not affect the C or H bit. Instead, we recommend

ADDA #01

DAA

Summary

This chapter introduced and explained many of the basic **68HC11** instructions. It started with LOAD and STORE, the instructions that move data between accumulators or between the accumulators and memory. It then progressed to the instructions that perform arithmetic operations, emphasizing addition, subtraction, multiplication, and division. SHIFT and ROTATE instructions were discussed next, followed by logic operations, including AND, OR, and EOR. Methods of testing, setting, and clearing bits were considered. Finally, techniques for performing decimal addition were discussed.

Glossary

AND   An instruction that calculates the logical AND of the two operands. If the corresponding bits of both operands are 1, the bit in the result is 1.

DECIMAL ADJUST ACCUMULATOR (DAA)   An instruction that adjusts the A accumulator to present the result of an addition in decimal.

Decrement   To subtract a number from the contents of a register or memory location. Generally these are reduced by 1.

EOR   An instruction that finds the EXCLUSIVE-ORs of two operands. If the corresponding bits of both operands are different, the bit in the result is 1.

H bit   A bit in the CCR that is set when a a carry from bit position 3 to bit position 4 occurs during an addition instruction. It is used for decimal arithmetic operations by the DAA instruction.

Increment   To add a number to the contents of a register or memory location. Generally these are increased by 1.

Mask   Used when only some bits of a byte are to be affected by an instruction. The mask byte indicates which bits are to be affected and which bits are to be left unchanged.

OR   An instruction that finds the logical OR of two operands. If the corresponding bit in either operand is 1, the bit of the results will be 1.

SUBTRACT-WITH-CARRY   A SUBTRACT instruction that will reduce the result by 1 if the C bit is set.

Problems

Section 5-1

5-1 Using the register set of Figure 5-1, what happens when the following instructions are executed *in sequence*:

CLRB		INX	
CLR	33,X	STAB	33,X
INC	33,X	LDD	32,X

5-2 Write a program to clear all memory locations between C100 and C2FF.

5-3 Using the register set of Figure 5-1, what happens when the following instructions are executed:
   a. STX   A,Y
   b. ABY
   c. XGDX

5-4 In C100–C2EF there is a set of 16-bit numbers.
   a. Copy all the numbers whose most significant nibble is A into a list starting at C300.
   b. Copy the addresses of all these numbers into a list starting at C400.

Section 5-2

5-5 Write a program to add the 24-bit number in 2A, 2B, and 2C of Table 5-4 to the number in 2D, 2E, and 2F. In other words, do the following:

$$\begin{array}{r} (2A)\ (2B)\ (2C) \\ +\ \underline{(2D)\ (2E)\ (2F)} \\ (3A)\ (3B)\ (3C) \end{array}$$

5-6 Show the contents of the sum and the condition codes after each addition in Problem 5-5.

5-7 For each set of the following decimal numbers
   a. Convert them into 8-bit hex numbers.
   b. Extend them into 16-bit numbers.
   c. Add them and show the result is correct in both magnitude and sign.
      (1) $+34, +86, -45$
      (2) $+34, +45, -113$
      (3) $-2, -4, -55$

Section 5-3

5-8 Repeat Problems 5-5 and 5-6 but subtract instead of add.

5-9 There is an 8-byte number in locations C040–C047 and another in C050–C057.
   a. Write a program to add them. Put the results in C060–C067.

b. Write a program to subtract them.

(*Hint*: Use an index register and a looping program.)

5-10 Using Table 5-4, what are the results of the following instructions?

a. CLR	B,X	**d.** COMA	
b. NEGB		**e.** TST	13,X
c. NEG	12,X	**f.** TST	0038

5-11 A student had to negate the double register. He tried the following three programs:

a. NEGB	**b.** NEGB	**c.** NEGB	
NEGA	COMA	ADCA	#00
		NEGA	

Try these programs if the number in the double register is

(1) 1465 **(2)** 1400.

Which program works for all cases?

5-12 There is a 24-bit number in C100, C101, and C102. Write a program to negate it.

Show the step-by-step results of the program if the number is

a. 465700 **b.** AB00CD.

Section 5-4

5-13 Using the memory contents of Table 5-4, what are the results of the following program?

```
LDAA    2B
LDAB    2C
MUL
```

5-14 Write the code for Example 5-14.

5-15 Divide 4A3 by 1B using IDIV. Show your quotient and remainder and where they are. Verify, if possible, using the EVB.

5-16 Divide 1B by 4A3 using the EVB. Show the results using IDIV and FDIV.

Section 5-5

5-17 If A contains BB and B contains CD and the carry FF is a 1, what are the results of the following instructions?

a. ASLA		**e.** LSRB
b. ASRB		**f.** RORA
c. LSLD		**g.** ROLB
d. LSRD		

5-18 There is a 24-bit number in locations C100–C102. Write a program to
   a. Shift it to the left.
   b. Arithmetic shift it to the right.
      If the number is 7653AB, show the contents of memory and the registers used for each step of the program.
      Check your program's answer by doing a hand calculation (in BASE 2) on the full 24-bit number.

Section 5-6

5-19 If A contains 56, what is the result of each of the following instructions?
   a. ANDA   #33
   b. ORA   #33
   c. EORA   #33
   d. BITA   #80

5-20 Write a code segment to determine if
   a. Bit 6 of B is a 0.
   b. Bits 5 and 6 of B are both 0.

5-21 If C100 contains 65, follow Example 5-21 and show what the registers and C100 contain after each step.

5-22 Write a program to check the parity of the entire byte in C100. Go to EVEN or ODD depending on the parity of the byte.

5-23 If BUFFER is defined as the memory locations between C100 and C2FF, write a program segment to
   a. Clear BUFFER.
   b. Complement BUFFER.
   c. Set bit 7 of every byte in BUFFER.
   d. Clear bits 2 and 3 of every byte in BUFFER.

5-24 Consider the following program:

```
LDD     #F00D
LDX     #C100
STD     0,X
BSET    0,X 44
BCLR    1,X 11
```

What numbers are in C100 and C101 at the end?

5-25 Determine the contents of A and the condition codes after each of the following instructions has been executed. Use Table 5-4 for the initial contents of memory. Assume the carry 8-bit is set.
      Each part is independent of the others. The initial conditions apply for each case.

Fill in the following table:

		A	B	H	N	Z	V	C
a. 47	ASRA							
b. A6 15	LDAA							
c. 85 20	BITA #							
d. A9 0D	ADCA							
e. 92 36	SBCA							
f. D3 39	ADDD							
g. 88 77	EORA							

Section 5-7

5-26 Write a program to add an 8-digit decimal number in locations C100–C103 to an 8-digit decimal number in C110–C113. Show the results including the condition codes if 45 87 29 86 is in C100 and 22 99 69 15 is in C110.

5-27 Given the following program:

		A	H	N	V	C	Z
LDAA	$60						
ORAA	#08						
ADDA	#57						
DAA							
CMPA	#80						

If the number in 60 is 41, fill in the table. Use X for any unknown values.

5-28 There is a set of numbers in the buffer from C0F0 to C2F0. Some of them are not BCDs (Binary Coded Decimal) (they contain As, Bs, etc.).
   a. Write a routine to determine whether a number is BCD.
   (*Hint*: One way is to add 66 to the number. If the C bit or H bit is set, the number is not BCD. You may want to use a TPA to transfer the condition codes into A.)
   b. Write a routine to count the number of non-BCD numbers in the field.

6 | Branches, Stacks, and Subroutines

INTRODUCTION

This chapter concludes the study of the **68HC11** instruction set by considering signed branches, stacks, and subroutines. Subroutines are extremely important and are frequently used by programmers.

INSTRUCTIONAL OBJECTIVES

After reading the chapter, the student should be able to

- Explain the difference between signed and unsigned numbers and use the proper branches for his or her particular application.
- Use the Boolean equations for a BRANCH instruction to determine whether that instruction will branch.
- Use BRSET and BRCLR instructions.
- Use the stack for pushing and pulling data.
- Write subroutines.
- Use time delay subroutines and nested subroutines.
- Use subroutines in the EVB monitor.

SELF-EVALUATION QUESTIONS

Watch for the answers to the following questions as you read the chapter. They should help you to understand the material presented:

1. What are some uses for NOPs? What is the difference between an NOP instruction and a BRN instruction?

2. How can the contents of the condition code register be preserved?

3. Why must the stack always be in RAM?

4. Why don't two consecutive PUSH instructions write to the same memory location?

5. Why does a TSX add 1 to the SP before writing it in X? Does it affect the SP?

6. Why must the stack be balanced before an RTS is executed?

7. Why do some programmers place a "Destroy" comment at the beginning of a subroutine?

6-1 Branch Instructions

BRANCH instructions were introduced in Section 3-7. Methods of calculating branch offsets and target addresses were also presented there. Table 6-1 is a complete table of the BRANCH instructions available on the **68HC11**.

BRANCH instructions can be divided into three types—simple, unsigned, and signed. The simple branches, presented in Table 3-3, are simple because they depend only on a *single* condition code. For example, BCC will branch if the carry flag is clear.

The BRANCH NEVER (BRN) instruction is also a simple branch. As its name implies, it never branches. BRN does absolutely nothing but it takes three clock cycles to do nothing; that is why BRN is listed as a three-cycle NOP in Table 6-1. The **NOP** stands for No Operation, an instruction that does nothing. The BRN takes an offset, but it is irrelevant because it will never branch.

While branch instructions use the condition codes to decide whether to branch, *they never change or affect the condition codes*. The BRA and BRN instructions do not use the condition codes.

6-1.1 Signed Branches

Signed numbers are numbers that are treated by the computer as 2s complement numbers. They contrast with unsigned or absolute numbers.

E X A M P L E   6 - 1

Consider the numbers A0 and 60. What are their decimal equivalents and which is greater if they are treated as

a. Absolute numbers?

b. Signed numbers?

Solution

a. As absolute numbers, A0 (10100000 in binary) is greater than 60 (01100000). Their decimal equivalents are 160 and 96, respectively. Note that negative numbers are not used when expressing **absolute numbers**.

b. As signed numbers 60 is definitely greater because it is positive and A0 is negative. Here 60 corresponds to $+96$ and A0 corresponds to -96.

Table 6-1 Branch Instructions in the 68HC11

Function	Mnemonic	REL	DIR	INDX	INDY	Comments
Branch if carry clear	BCC	X				$C = 0$?
Branch if carry set	BCS	X				$C = 1$?
Branch if equal zero	BEQ	X				$Z = 1$?
Branch if greater than or equal	BGE	X				Signed $\geq$
Branch if greater than	BGT	X				Signed $>$
Branch if higher	BHI	X				Unsigned $>$
Branch if higher or same (same as BCC)	BHS	X				Unsigned $\geq$
Branch if less than or equal	BLE	X				Signed $\leq$
Branch if lower (same as BCS)	BLO	X				Unsigned $<$
Branch if lower or same	BLS	X				Unsigned $\leq$
Branch if less than	BLT	X				Signed $<$
Branch if minus	BMI	X				$N = 1$?
Branch if not equal	BNE	X				$Z = 0$?
Branch if plus	BPL	X				$N = 0$?
Branch if bit(s) clear in memory byte	BRCLR		X	X	X	Bit manipulation
Branch never	BRN	X				3-cycle NOP
Branch if bit(s) set in memory byte	BRSET		X	X	X	Bit manipulation
Branch if overflow clear	BVC	X				$V = 0$?
Branch if overflow set	BVS	X				$V = 1$?

The application often determines whether signed numbers or absolute numbers should be used. If the input to a computer is a pressure or some other quantity that can never be negative, absolute numbers are preferable; they allow for 256 different values. Here FF is certainly greater than 01.

If quantities that can be both positive and negative are being measured, signed numbers are preferred. Temperature might be an example. Because temperatures below zero are certainly possible, probably the best idea is to assign FF for a temperature of $-1°$, and so on, in accordance with 2s complement numbering. Now 01 is greater than FF ($+1°$ is warmer than $-1°$).

COMPARE instructions, discussed in Section 3-6, are often used to *set* the condition codes. They usually precede BRANCH instructions. The BRANCH instructions determine whether or not to branch depending on the condition codes. Signed and unsigned branches react differently depending on the condition codes.

E X A M P L E   6 - 2

If E0 is subtracted from 70, what are the results and what condition codes are set?

Solution

The results of the subtraction are 90.

The N flag is set because 90 is a negative number.

The C flag is set because E0 is greater than 70 (the subtrahend is greater than the minuend, considered as absolute numbers).

The V flag is set because there is overflow.

The Z flag is clear because the result is not zero, and the H flag is irrelevant.

Table 6-2 lists the four signed branch instructions in the **68HC11**. The Boolean equations are also listed. These equations depend on the condition codes; if they are satisfied, the branch will be taken.

Table 6-2   Signed Branches in the 68HC11

Operations	Mnemonic	Relative			Index			Extend			Implied			Branch Test
		OP	~	#	OP	~	#	OP	~	#	OP	~	#	
Branch if $\geq$ Zero	BGE	2C	4	2										$N \oplus V = 0$
Branch if $>$ Zero	BGT	2E	4	2										$Z + (N \oplus V) = 0$
Branch if $<$ Zero	BLT	2D	4	2										$N \oplus V = 1$
Branch if $\leq$ Zero	BLE	2F	4	2										$Z + (N \oplus V) = 1$

The signed branch instructions are

BGE—Branch if Greater than or Equal   This instruction will branch if the minuend is greater than or equal to the subtrahend in the COMPARE instruction. The COMPARE instruction sets the condition codes that must satisfy the equation $N \oplus V = 0$. This most often happens with a normal subtraction that leaves both N and V equal to zero. Example 6-2 showed a case where both N and V were 1, satisfying the equation. This is correct because 70 is certainly greater than E0, if the numbers are considered as *signed*.

BGT—Branch if Greater Than   This instruction will branch if the minuend is greater than but not equal to zero. The equation is

$$Z + N \oplus V = 0$$

E X A M P L E   6 - 3

Consider the code segment

        CMPA   #30
        BGT

Will the branch be taken if A contains

a. 2F?

b. 30?

c. 31?

Solution

a. The COMPARE instruction will calculate 2F − 30. The results will set N but not V. The equation for a BGT, $Z + N \oplus V = 0$, is not satisfied because $N \oplus V = 1$. Therefore the branch is not taken. This is correct because 30 is greater than 2F.

b. The COMPARE instruction will calculate 30 − 30. The results will clear N and V but set Z. Because $Z = 1$, the equation is not satisfied and the branch will not be taken. Another way of looking at this is to say that A is equal to, not greater than, 30.

c. Now the results are positive and the COMPARE instruction will clear N, Z, and V. The equation is satisfied and the branch will be taken.

BLT—Branch Less Than   This branch will be taken if the minuend is less than, but not equal to, the subtrahend in the COMPARE instruction. Its equation is $N \oplus V = 1$. In Example 6-3 this equation is satisfied only when A = 2F or only when the minuend is less than the subtrahend.

BLE—Branch Less than or Equal This branch will be taken when the minuend is less than or equal to the subtrahend. Its equation is $Z + N \oplus V = 1$. In Example 6-3 it is satisfied when A = 2F and when A = 30.

6-1.2 Unsigned Branches

Some problems use **unsigned numbers**. These can also be described as absolute or always positive numbers. A set of branch instructions also exists for absolute numbers. They use the words *lower* or *higher* instead of *less than* or *equal to*, and their equations depend on Z and C instead of N or V.

The unsigned branches are given in Table 6-3. They correspond to the signed branches. They are

Table 6-3   Branch Instructions in the 68HC11

Instruction	Mnemonic	Equation
Branch if higher	BHI	$C + Z = 0$
Branch if carry clear	BCC	$C = 0$
Branch if lower or the same	BLS	$C + Z = 1$
Branch if carry set	BCS	$C = 1$

BHI—Branch if Higher This instruction will branch if the minuend is higher in absolute value than the subtrahend. The equation is $C + Z = 0$. Previously we saw that if E0 were subtracted from 70, the corresponding unsigned branch would be taken because $N \oplus V = 0$. Here, however, the branch will not be taken because the subtraction sets the C flag and the equation $C + Z = 0$ is not satisfied. This is correct, of course. As an absolute number E0 is greater than 70 and the branch should not be taken.

When looking for an instruction that corresponds to BGE, we would like to find a branch if higher or the same. After a little thought, we realize that this function is served by an instruction that already exists, BCC. If the minuend is larger than or equal to the subtrahend the carry flag will be clear. Note that if the minuend and subtrahend are equal, the C flag will be clear but Z will be set. Under these conditions BCC will branch but BHI will not.

BRANCH LOWER does not exist. If the subtrahend is greater than the minuend, the carry flag will be set. Thus BRANCH LOWER is the same as BCS (Branch on Carry Set).

BLS—Branch if Lower or the Same This instruction completes the signed branches. It branches if the minuend is lower than or equal to the subtrahend. Its equation is $C + Z = 1$. If the minuend is lower, C will be set, and if the minuend is the same, the Z bit will be set.

6-1.3   BRCLR and BRSET

The **68HC11** also contains two more sophisticated BRANCH instructions: BRCLR, BRANCH IF BITS CLEAR, and BRSET, BRANCH IF BITS SET. They are similar to BSET and BCLR in that they take a mask and only operate in the direct and indexed modes.

The BRCLR instruction consists of four parts:

1. The Op code

2. The direct address, for the direct mode, or the offset if an indexed mode is being used

3. The mask

4. The relative branch, rr

It will branch to rr plus the address of the Op code of the next instruction if all the bits specified in the mask are 0.

E X A M P L E   6 - 4

Write instructions, starting at C100, to branch to C0E0 if bits 0, 1, and 2 of D050 are all clear. Use the Y index register.

Solution

The following two instructions are one way to do it:

```
C100    LDY     #D000           18 CE D0 00
C104    BRCLR   50,Y 7   C0E0   18 1F 50 07 D7
```

Notice the coding. The LDY immediate takes 4 bytes, two for the Op code and two for the immediate operand. The BRCLR takes 5 bytes. It requires two for the Op code (18 1F), one for the offset (50), one for the mask that specifies that bits 0, 1, and 2 are to be clear (07), and one for the displacement that will branch from C109, the address of the next instruction, to C0E0. This is D7 (see Sec. 3-7).

The BRSET operates in the same manner and the same modes as the BRCLR. It branches only if all the bits corresponding to the mask are set.

E X A M P L E   6 - 5

There is a buffer between C200 and C3FF. Write a program to count the number of odd numbers in the buffer and put the result in C050.

Solution

Odd numbers all have an LSB of 1. They can be detected by a BRSET with
a mask equal to 01. The following program should solve the problem:

```
          LDX      #C200              Point X to the start of the buffer.
          CLR      C050               Clear the count.
LOOP      BRSET    0,X 01    ADD1     Branch to ADD1 if the number is odd.
BACK      INX                         Point to the next number in the buffer.
          CPX      #C400              Done?
          BNE      LOOP               No, do it again.
          SWI                         Yes. Halt.
ADD1      INC      C050               Increment count.
          BRA      BACK               Go back.
```

6-2 Condition Code Register Instructions

Table 6-4 lists a group of **68HC11** instructions that allow the user to set or
clear various condition codes. Special instructions exist to set or clear C,
V, and I. The I bit is discussed in Section 8-7.4.

Table 6-4 Condition Code Instructions in the 68HC11

Function	Mnemonic	INH
Clear carry bit	CLC	X
Clear interrupt mask bit	CLI	X
Clear overflow bit	CLV	X
Set carry bit	SEC	X
Set interrupt mask bit	SEI	X
Set overflow bit	SEV	X
Transfer A to CCR	TAP	X
Transfer CCR to A	TPA	X

The TPA transfers the 8 bits of the condition code register into A. The
TAP is the reverse; it transfers data from A into the condition codes. Thus,
if the condition codes must be preserved at one point in a program and then
retrieved later, this can be done by issuing a TPA and then storing the codes
at a vacant location in memory. They can be retrieved later by loading A
from the location and doing a TAP.

6-3 The Stack

In many of the previous programs memory has been divided into two areas:
an area to hold programs and an area to hold data. We often call the latter
a buffer area. Data is often variable and usually written into RAM so that

it can be changed. Programs that are under development and changed often must also be written in RAM. Conversely, programs that may not change are usually written into ROM or EPROM. Monitors are an example of these programs.

The stack is a third area of memory that is allocated for special functions; an area of memory is set aside and called the **stack**. Because it must be read and written to, *the stack is always in RAM.* The **stack pointer (SP)** is the register in the **68HC11** that we have not discussed previously. It is a 16-bit register that holds the *address* of the stack area. *The SP points to the highest vacant location in the stack.** As the stack is used, the SP decrements.

The PUSH and PULL instructions available for the **68HC11** are given in Table 6-5. A **PUSH** instruction writes data from the source to the stack

Table 6-5   PUSH and PULL Instructions in the 68HC11

Function	Mnemonic
Push A onto stack	PSHA
Push B onto stack	PSHB
Push X onto stack	PSHX
Push Y onto stack	PSHY
Pull A from stack	PULA
Pull B from stack	PULB
Pull X from stack	PULX
Pull Y from stack	PULY

and then decrements the SP. There are four PUSH instructions: PSHA, PSHB, PSHX, and PSHY.

Figure 6-1 illustrates the action of a PUSH instruction. The PSHA copies the contents of A into memory at the SP location and then decrements the SP. The PSHX copies X into two successive memory locations and decrements the SP twice. Note that for each PUSH the SP is decremented. The

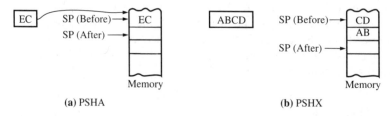

(a) PSHA   (b) PSHX

Figure 6-1   The action of PSH instructions in the 68HC11.

* In most 16-bit microprocessors the stack pointer points to the highest previously used location in the stack.

memory location that is pushed into is now used, but the SP is decremented so that it points to the new highest location in memory that is available. Obviously we cannot keep pushing indefinitely or the stack will move down and overwrite the program area and anything else in RAM.

PULL instructions are the inverse of PUSHes. First, they increment the SP. Then they copy from memory, at the SP location, into the designated register. The PULL instructions are also shown in Table 6-5. Once a memory location has been PULLed, which means it is read into a register, that location is assumed to be vacant, as indicated by the fact that the SP is now pointing at it.

E X A M P L E   6 - 6

If the SP is initially at C100, what do each of the following instructions do, and what are the contents of the stack after each instruction?

PSHA	PULX
PSHB	PULA
PSHY	PULB

Solution

The first instruction pushes, or writes, the contents of A into C100. It then decrements the SP to C0FF.

The second instruction writes the contents of B into C0FF, and again decrements SP to C0FE.

The third instruction writes the contents of Y into C0FE and C0FD. SP now contains C0FC.

The fourth instruction is a 2-byte PULL. It first increments the SP and reads from C0FD, then increments again and reads from C0FE. To two bytes read in are stored in X and the SP ends up at C0FC.

The fifth instruction increments the SP to C0FF and writes its contents to A.

The sixth instruction increments the SP to C100 and writes its contents to B.

This program copies Y into X and exchanges the contents of A and B. Notice that the stack finished where it started, at C100. This always happens when each PUSH is balanced by a PULL.

6-3.2 Stack Pointer Instructions

The instructions that *affect* or change the SP are listed in Table 6-6. They are similar to instructions that affect the index registers.

The LOAD STACK POINTER (LDS) instruction can be executed in the immediate or memory modes. It loads the SP with an immediate value,

Table 6-6   Stack Pointer Instructions in the 68HC11

Function	Mnemonic	IMM	DIR	EXT	INDX	INDY	INH
Load stack pointer	LDS	X	X	X	X	X	
Store stack pointer	STS		X	X	X	X	
Increment stack pointer	INS						X
Decrement stack pointer	DES						X
Transfer SP to X	TSX						X
Transfer SP to Y	TSY						X
Transfer X to SP	TXS						X
Transfer Y to SP	TYS						X

in which case it is followed by 2 bytes of data, or by a memory location. The contents of the specified memory location and the next one are then loaded into the SP.

Because the SP must point to a RAM area of memory, it cannot be allowed to be random. Most programs that use the SP start with an LDS in the immediate mode to position the stack properly. If, for example, the user decides to place the stack area at CFFF and below, the first instruction in the program should be an LDS #CFFF.

The STORE STACK POINTER (STS) stores the contents of the SP in two consecutive memory locations. It can be used to preserve the SP in memory if it will be needed later in the program.

The INCREMENT and DECREMENT SP (INS and DES) instructions operate exactly as their name implies. Generally, they should be *balanced*; each increment should have a compensating decrement to keep the SP in the same place.

E X A M P L E   6 - 7

The initial contents of the registers are given in Figure 6-2.

```
P-C007 Y-7982 X-FF00 A-44 B-70 C-C0 S-C04A
>
```

Figure 6-2   The register set for Example 6-7. *(Courtesy of Motorola, Inc.)*

a. Where is the SP and exactly what is in the stack after the following sequence of instructions is executed?

PSHA

PSHB

PSHY

b. At some later point in the program it is necessary to restore B to its original value. How can this be done?

Solution

a. The stack and its contents are shown in Figure 6-3. The first instruction, PSHA, wrote the contents of A (44) into the stack at CO4A and then decremented the SP to CO49. This prepared it for the PSHB instruction. After the last instruction the SP will have been decremented four times and contain CO46.

	Location	Contents
Initial SP →	C04A	44
	C049	70
	C048	82
	C047	79
New SP →	C046	

Figure 6-3 The stack and its contents for Example 6-7.

b. The original value of B can be restored by the following sequence of instructions:

INS

INS

PULB

The first two instructions increment the SP to C048. Then the PULB copies the 70 at C049 into B. Note that at the end of the program segment the SP contains C049.

The last four instructions in Table 6-6 are transfers between the SP and the index registers. When an address is transferred from the SP to an index register, the address is incremented. This is done deliberately. The SP always points to a vacant location. It is assumed that the reason for transferring the SP to an index register is to have the index register point to data on the stack. By incrementing, the index register now points to the last byte of data that was actually stored on the stack.

E X A M P L E   6 - 8

a. If the program of Example 6-7, part (a), was followed by a TSX instruction, what data is placed in X?

b. At a later point in the program, how can B be restored without changing the SP?

Solution

a. As shown in Figure 6-3, the SP at the end of the program was C046. The TSX will increment this value and place it in X, so X will contain C047 after the TSX instruction.

b. The following two instructions can be used to restore B:

TSX

LDAB   2,X

The first instruction puts C047 in X and the second instruction loads the contents of C049, 70, into B. Neither of these instructions affects the SP.

Transfers from an index register to the SP (TXS and TYS) do not change the value in the index register, but they decrement the number being transferred before writing it to the SP. This compensates for the increment when going the other way and positions the SP properly.

6-3.3   Subroutines

Consider a long main program as shown in Figure 6-4. Assume that when the program reaches point A it must calculate the sine of the angle in register A. Also, assume that a program exists for calculating the sine of a number

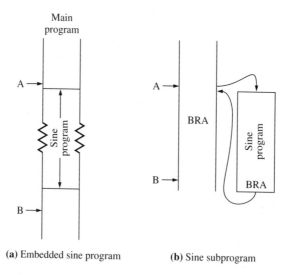

(a) Embedded sine program   (b) Sine subprogram

Figure 6-4   Use of subprograms.

Table 6-7 Subroutine Calls and Returns

Function	Mnemonic	REL	DIR	EXT	INDX	INDY	INH
Branch to subroutine	BSR	X					
Jump to subroutine	JSR		X	X	X	X	
Return from subroutine	RTS						X

and that it takes abut 50 instructions. The sine program can simply be *embedded* in the main program as shown in Figure 6-4a. The program then simply picks up the value in the A register and calculates its sine. Indeed, this may be the most economical way to do it if the sine only needs to be calculated once.

Now, however, suppose the sine must also be calculated when the program reaches point B. We could again embed the sine program at this point, but then we would be duplicating code that is already written.

We could make the sine routine a subprogram as shown in Figure 6-4b. When the main program reaches point A, it could JUMP or BRANCH to the sine program and then branch back when it is finished. When it reaches point B, it could also jump to the sine program. The problem, however, occurs when it is finished. The BRA that is the last instruction in the sine program would take it back to point A. In almost all cases if the sine program started at point B, it should return to point B when it is finished.

The problem posed in the previous paragraph can be solved by making the sine program a *subroutine*. A **subroutine** is a subprogram that is usually executed more than once during the execution of the main program.* It is reached by a JUMP TO SUBROUTINE (JSR) or BRANCH TO SUBROUTINE (BSR) instruction. Once entered, the subroutine is executed and then the program returns to its original point.

The instructions that concern subroutines are listed in Table 6-7. The action of these instructions is shown in Figure 6-5. Subroutines are entered by a BRANCH or JUMP to their starting address using a BSR or JSR. These instructions are the same as a BRA (there are no conditional branches to subroutines) or a JMP except that they *stack the address of the next instruction*. This means that they write the address of the instruction following the BSR or JSR onto the stack. In this way the program remembers where to return when the subroutine finishes.

---

* Some programmers prefer to have almost their entire program consist of subroutines. Their main programs then look like a steady string of JSRs. In this case many subroutines are executed only once, but these programmers feel that it *modularizes* their program. Each module can be checked out independently of the others, and, of course, the program will surely work when all the modules are put together. It sounds good in theory.

Special operations

Jump to Subroutine (JSR):

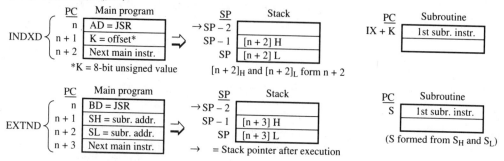

*K = 8-bit unsigned value

$[n + 2]_H$ and $[n + 2]_L$ form $n + 2$

(S formed from S_H and S_L)

→ = Stack pointer after execution

Branch to Subroutine (BSR):

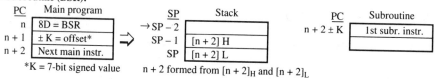

*K = 7-bit signed value

$n + 2$ formed from $[n + 2]_H$ and $[n + 2]_L$

Figure 6-5   JUMPs and BRANCHes to subroutines in the 68HC11. (*Redrawn Courtesy of Motorola, Inc.*)

E X A M P L E   6 - 9

At C100 there is a BSR 20 (8D 20). The SP is at C200. After the instruction is executed

a. Where does the subroutine start?

b. What is in the stack?

Solution

a. The BSR is a 2-byte instruction. The next instruction therefore starts at C102. Like other branch instructions, the BSR adds its displacement (20) to the address of the next instruction and branches to C122, which must be the start of the subroutine.

b. Any BSR or JSR stacks the address of the next instruction. This requires 2 bytes. The results are shown in Figure 6-6. The instruction decrements the SP twice.

Stack pointer before BSR → C200	02	
C1FF	C1	
Stack pointer after BSR → C1FE		

Figure 6-6   Effect of a BSR on the stack.

All subroutines end in a RETURN FROM SUBROUTINE (RTS) instruction. This instruction takes the 2 bytes it finds in the stack and puts them into the program counter. In this way the program resumes at the instruction following the BSR or JSR. The RTS also unstacks the return address by incrementing the SP twice. The action of the RTS is shown in Figure 6-7.

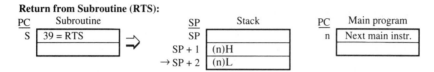

Return from Subroutine (RTS):

Figure 6-7 **Action of a return-from-subroutine (RTS) instruction.** (*Redrawn Courtesy of Motorola, Inc.*)

For the RTS to work, it is imperative that the SP be where it was at the end of the BSR or JSR. Otherwise the RTS will unstack the wrong return address and chaos will result. This means that any PUSHes or DES instructions that occur in the subroutine must be compensated for by PULLs or INS instructions so that the *stack is balanced* when the RTS is executed.

E X A M P L E   6 - 10

Using the register set of Figure 6-2, a programmer erroneously wrote

    C007   JSR   C200

                ⋮

    C200   PSHA

                (Subroutine Instructions)

                RTS

Where is the stack, what is in it, and what happened when the RTS was executed?

Solution

The stack is shown in Figure 6-8. The JSR extended is a 3-byte instruction, so the return address is C00A. Then the contents of A (44) are pushed onto the stack. The SP, however, is now at C047. The RTS will therefore place the contents of C048 and C049 (address 44C0) in the PC and get lost at this address. This occurs because the PSHA decrements the SP and there is no compensating PULL or INS.

	Location	Contents
	C04A	0A
	C049	C0
	C048	44
SP →	C047	

Figure 6-8 The stack for Example 6-10.

Because subroutines are so important, we summarize their action:
- Subroutines are routines that are generally executed more than once and can be entered *from different points* in the main program.
- Subroutines are entered by using a JSR or BSR. The instructions *stack the address of the next instruction* before going to the subroutine.
- Subroutines end with an RTS. This instruction transfers the address it finds in the stack to the PC. This allows the program to return to the address after the JSR or BSR when the subroutine finishes and resumes operation.

E X A M P L E   6 - 11

This example illustrates the use of subroutines in a program. Assume that an investor owns shares of stock in several companies. The number of shares he owns in the first company is given in C100, the number of shares in the second company is given in C101, and so on. The number FF denotes the end of the list. This is sometimes called a *flagged list*. The list is of variable size and a flag or special character is used to denote the end of the list. Here the flag is the character FF.

Also, assume that the price per share of each company is in a list starting at C200 and the dividends per share are in a list starting at C300.

Write a program to calculate the value of each company and put it in a list starting at C250. Put the total value in C2E0. The program must also calculate the dividends each company pays and list this in C350. The total value of all dividends should be found in C3E0.

Solution

After a little thought we realize that the value of each company is the number of shares times the price per share, and its dividend payout is the number of shares times the dividend per share. Either calculation involves many multiplications and additions, but they are very similar. In this case the same subroutine can work for both of them.

Let us choose C400 for the location of the subroutine and C5FF for the initial location of the stack. We will write the subroutine first:

```
    ;           CALCULATION SUBROUTINE
    ; Subroutine to calculate the value per share or the value
    ; of the dividends and add them up:

    C400    LDAA    0,X     Load the number of shares into A.
    C402    LDAB    0,Y     Load the price per share into B.
    C405    MUL             Multiply them.
    C405    STD     50,Y    Store the value.
    C408    ADDD    E0,Y    Add to the sum.
    C40B    STD     E0,Y    Store the sum.
    C40E    INY             Increment Y.
    C410    RTS             Return from subroutine.
```

We start the main program at C020.

;   MAIN PROGRAM

C020	LDS	#C5FF	Because of the subroutine, the stack will be used. Load it first.
C023	CLR	C2E0	Clear both bytes of the value sum.
C026	CLR	C2E1	
C029	CLR	C3E0	Clear both bytes of dividend sum.
C02C	CLR	C3E1	
C02F	LDX	#C100	Point X to the start of the stock list.
C032	LDY	#C200	Point Y to the value list.

;   End of Initialization

;   STOCK VALUE CALCULATION

START	C036	LDAA	0,X	
	C038	CMPA	#FF	End of list?
	C03A	BEQ	START1	If yes, go to dividend calculation.
	C03C	JSR	C400	No, go to calculation subroutine.
	C03F	INX		Get next stock and repeat.
	C040	BRA	START	

;   DIVIDEND VALUE CALCULATION

START1	C042	LDY	#C300	Point Y to dividend list.
	C046	LDX	#C100	Reset X to the start of the list.
START2	C046	LDAA	0,X	
	C048	CMPA	#FF	End of list?
	C04A	BNE	CONT	
	C04C	SWI		Yes. Stop.
CONT	C04D	JSR	C400	No. Calculate dividends.
	C050	INX		Get next stock and repeat.
		BRA	START2	

;   End of Program

This program first calculates the value of the stock and then the value of the dividends. It uses the same subroutine for both calculations. Because the calculations involved different lists, Y pointed to the value list during the first part of the program and to the dividend list during the second part of the program.

6-3.4 Nested Subroutines

It is possible for a subroutine to jump to a second subroutine, and so forth. These are called **nested subroutines** because the second subroutine can be nested inside the first subroutine. The basic procedure is as follows:

1. The main program does a JSR (or BSR) to the first subroutine. This stacks the return address for the main program.

2. During its execution the first subroutine does a JSR to the start of the second subroutine. This stacks the return address to the first subroutine below the return address to the main program.

3. When the second subroutine ends, its RTS unstacks the return address to the first subroutine.

4. When the first subroutine ends, its RTS unstacks the return address to the main program.

This procedure can be continued. The second subroutine can call a third subroutine, and so forth. Subroutines can be nested to any level provided there is enough room in the stack for all the return addresses.

The nesting of subroutines is illustrated in Figure 6-9. The main program encounters a JSR at C040. It stacks the return address, C043, and jumps to the subroutine at C1B0. When this subroutine reaches C1C3, it jumps to the second subroutine at D000 while stacking its return address, C1C6. The RTS

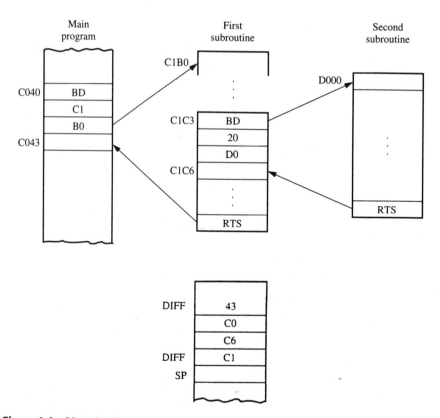

Figure 6-9   Nested subroutines.

for the second subroutine returns the program to the first subroutine at C1C6. It continues until its RTS, which causes a return to the main program at C043.

6-3.5 Registers and Parameter Passing During Subroutines

During its execution a subroutine will use the accumulators and may also use X and Y. In many cases the values of A, B, X, and Y must be preserved so that they will be unchanged when the main program resumes. In these cases the subroutine often stacks the registers it will use when it is entered and unstacks them before returning to the main program. This can be done by pushing registers at the start of the subroutine and pulling them just before the RTS. To restore the values properly, *the registers must be pulled in the reverse order from the way they were pushed*.

E X A M P L E   6 - 12
_____

If it is necessary to preserve A, B, X, and Y for the main program, what instructions should be in the subroutine?

Solution

The subroutine should look like the following:

```
START      PSHA
           PSHB
           PSHX
           PSHY
             ⋮
Body of the subroutine
             ⋮
           PULY
           PULX
           PULB
           PULA
           RTS
```

Observe how the subroutine pushed the registers at its start and then pulled them in reverse order at the end.
_____

Some programmers will use a comment at the start of their subroutines such as "Destroys A, B, and X." This means that the subroutine will use A, B, and X without preserving them. Of course, the main program can also preserve A, B, and X by pushing them before the JSR and pulling them after the JSR (see Prob. 6-16).

Sometimes programmers pass parameters (data that the subroutine needs) to the subroutine by pushing them onto the stack. This is more advanced programming and is beyond the scope of this book.

6-4   Time Delay Subroutines

It is sometimes necessary to have the computer wait or delay for a specific length of time. This often occurs when the computer is controlling or monitoring external events. One way to do this is by using a *software time delay* routine or subroutine.

A simple example might be the control of a traffic light. The program might be

Turn the traffic light green.

Wait 30 seconds.

Turn the traffic light red.

Wait 30 seconds.

Repeat.

Notice that we have had to perform identical action (wait 30 seconds) twice during the program. This strongly suggests the use of a subroutine.

The heart of a time delay subroutine is a program that decrements a register, tests to see if it is zero, and branches back if it is not. This is the time delay loop and looks like the following:

```
LOOP    DEX
        BNE     LOOP
```

The delay time depends on the number of cycles required by each instruction in the loop and the time of each cycle. Cycles were introduced in Section 3-3.1. Most **68HC11**s are driven by a 2-MHz clock, which means that each cycle takes 0.5 μs. For the time delay loop that has been discussed, both instructions take three clock cycles, so the loop takes six cycles or 3 μs to complete. If 3 μs is an inconvenient number, the user might want to add two NOPs inside the loop. Now the loop will take ten clock cycles, or precisely 5 μs to complete.

Of course, an accumulator can be decremented as well as an index register, but an accumulator can only be decremented 256 times, whereas an index register, being 16-bits long, can be decremented 65,536 times. For this reason index registers are generally preferred.

E X A M P L E   6 - 13

Write a time delay subroutine for

a. 1 millisecond (ms)   **b.** 30 seconds.

Solution

a. If the 5 μs loop is used, the program should loop 200 times for a 1-ms delay. It can look like the following:

```
T1   PSHX                Save the old value of X.
     LDX     #00C8       Load X with 200 (decimal)
L1   DEX
     NOP                 5-μs delay loop.
     NOP
     BNE     L1
     PULX                Restore the old value of X.
     RTS                 Return.
```

The subroutine is arbitrarily called T1. First, it saves the old value of X for the user by pushing it onto the stack. Then it delays for 1 ms and retrieves the old value of X for the user before returning. Because it preserved X and did not use any other registers, this subroutine destroys nothing.

b. A 30-s time delay subroutine can be written using the 1-ms subroutine developed in part (a). This subroutine will be called T2. Because 30 s means 30,000 times through a 1-ms delay loop, the subroutine becomes

```
T2   PSHX
     LDX   #             Load X with the hex equivalent of
                         decimal 30,000.
L2   JSR   T1            Jump to the 1-ms delay.
     DEX
     BNE   L2            Go around 30,000 times.
     PULX                Restore X for the user.
     RTS                 Return.
```

Note that T2 is the first subroutine and T1 is a nested subroutine called by T2.

6-4.1 Precise Time Delay Loops

The calculation of the 1-ms time delay loop in Example 6-13a was imprecise because it did not take into account those instructions that were not part of the ten-cycle, 5-μs delay loop. These instructions, which are not part of the main loop, still require time for their execution. They are called *overhead*. They are listed here along with the number of cycles each requires, as obtained from Appendix A.

```
PSHX               4
LDX (Immediate)    3
PULX               5
RTS                5
```

```
                     *Delay for 10 ms at E = 2 MHz
E78E 3C                        PSHX
E78F CE 0D 06                  LDX      #$0D06       6 ~ * 3334 = 20,004 * 0.5 MHz
E792 09             BULKDLY    DEX                   3 ~
E793 26 FD                     BNE      BULKDLY      3 ~
E795 38                        PULX

                     *Stop programming
E796 7F 10 3B                  CLR      PPROG
E799 32                        PULA
E79A 39                        RTS
```

Figure 6-10   A time delay loop in the EVB monitor. (*Courtesy of Motorola, Inc.*)

Thus the loop contains 17 overhead cycles and will take an extra 8.5 μs to complete, so it takes 1.0085 ms.

If precise timing is important, an NOP can be added to the overhead loop. This will give 20 overhead cycles, exactly the time for two loops of 5 μs each. Then two can be subtracted from the number originally loaded into X by making it LDX #00C6, to make the timing precise.

E X A M P L E   6 - 14

Show that the time delay of the subroutine is precise if the changes described in this paragraph were made.

Solution

The loop would be traversed C6 (decimal 198) times. At 5 μs per loop this would take 990 μs. The 20 overhead cycles will add another 10 μs to give precisely 1 ms.

6-4.2   The Time Delay Routine in the Evaluation Board Monitor

Figure 6-10 shows the code for a 10-ms time delay loop already written in the monitor of the EVB board. An examination of the code shows that X is loaded with 0D06 (3334 decimal) and the loop takes six cycles, giving 20,004 cycles or 10.002 ms. The loop is imprecise; it does not take overhead into account (see Prob. 6-21).

We should also note that the subroutine of Figure 6-10 cannot be used directly because *the stack is unbalanced.* The PULX compensates for the PSHX, but nothing compensates for the PULA at E799. Consequently, if the subroutine is entered by a JSR E78E, it will not return to the main program properly after the RTS. Furthermore, the routine cannot be changed because it is in ROM. Nevertheless, clever programmers can use the subroutine. Suppose, for example, that C400 is a vacant area in memory. To use the subroutine, one can write JSR C400. In C400 we can have

```
PSHA
JMP     E78E
```

Remember, the JMP does not affect the stack. Now the PSHA compensates for the PULA and the program will work.

6-5   Subroutines in the Evaluation Board Monitor

There are several subroutines that exist in the EVB monitor that may prove useful to programmers. Unlike the timing routine described in the previous paragraph, these subroutines *can* be accessed by a JSR and *will return properly.*

Table 6-8 is a table of the various subroutines available in the EVB and a brief discussion of their functions. Most of the subroutines involve input from the keyboard or output to the terminal or monitor. This input and output must be in the form of ASCII characters, which have been discussed in Section 5-2.4. For example, if an ASCII character is already in A, it can be displayed on the terminal by using the OUTA subroutine. If a hex number is in A, it can be converted to two ASCII characters by using OUTLHF, which will convert the upper nibble to ASCII and display it, and by using OUTRHF, which will do the same for the lower nibble.

Table 6-8   Utility Subroutine in the Evaluation Board Monitor
OUTLHLF and OUTRHLF do not preserve the contents of A.
OUTA does not preserve the contents of A.

UPCASE	If character in accumulator A is lowercase alpha, convert to uppercase.
WCHEK	Test character in accumulator A and return with Z bit set if character is whitespace (space, comma, tab).
DCHEK	Test character in accumulator A and return with Z bit set if character is delimiter (carriage return or whitespace).
INIT	Initialize I/O device.
INPUT	Read I/O device.
OUTPUT	Write I/O device.
OUTLHLF	Convert left nibble of accumulator A contents to ASCII and output to terminal port.
OUTRHLF	Convert right nibble of accumulator A contents to ASCII and output to terminal port.
OUTA	Output accumulator A ASCII character.
OUT1BYT	Convert binary byte at address in index register X to two ASCII characters and output. Returns address in index register X pointing to next byte.
OUT1BSP	Convert binary byte at address in index register X to two ASCII characters and output followed by a space. Returns address in index register X pointing to next byte.
OUT2BSP	Convert two consecutive binary bytes starting at address in index register X to four ASCII characters and output followed by a space. Returns address in index register X pointing to next byte.
OUTCRLF	Output ASCII carriage return followed by a line feed.
OUTSTRG	Output string of ASCII bytes pointed to by address in index register X until character is an end of transmission ($04).
OUTSTRG0	Same as OUTSTRG except leading carriage return and line feed is skipped.
INCHAR	Input ASCII character to accumulator A and echo back. This routine loops until character is actually received.
VECINIT	Used during initialization to preset indirect interrupt vector area in RAM. This routine or a similar routine should be included in a user program that is invoked by the jump to $B600 routine of BUFFALO.

Table 6-9   Addresses of the Utility Subroutines

$FFA0	JMP	UPCASE	Convert character to uppercase.
$FFA3	JMP	WCHEK	Test character for whitespace.
$FFA6	JMP	DCHEK	Check character for delimiter.
$FFA9	JMP	INIT	Initialize I/O device.
$FFAC	JMP	INPUT	Read I/O device.
$FFAF	JMP	OUTPUT	Write I/O device.
$FFB2	JMP	OUTLHLF	Convert left nibble to ASCII and output.
$FFB5	JMP	OUTRHLF	Convert right nibble to ACSII and output.
$FFB8	JMP	OUTA	Output ASCII character.
$FFBB	JMP	OUT1BYT	Convert binary byte to two ASCII characters and output.
$FFBE	JMP	OUT1BSP	Convert binary byte to two ASCII characters and output followed by space.
$FFC1	JMP	OUT2BSP	Convert two consecutive binary bytes to four ASCII characters and output followed by space.
$FFC4	JMP	OUTCRLF	Output ASCII carriage return followed by line feed.
$FFC7	JMP	OUTSTRG	Output ASCII string until end of transmission ($04).
$FFCA	JMP	OUTSTRG0	Same as OUTSTRG except leading carriage return and line feed is skipped.
$FFCD	JMP	INCHAR	Input ASCII character and echo back.
$FFD0	JMP	VECINIT	Initialize indirect vectors in RAM.

Table 6-9 shows where each subroutine starts. This table is actually a *jump table* written in the monitor's ROM. Notice that the entries are all three locations apart. There are 3 bytes in the ROM for each entry. The first byte is the JUMP Op code, and the next 2 bytes tell it where to jump to in order to find the proper routine. All the routines must end with an RTS to return the program properly.

E X A M P L E   6 - 15

Write a code segment to convert the hex character in accumulator A to two ASCII characters and display them.

Solution

The code segment is simply

```
JSR    FFB2    Jump to the OUTLHF subroutine.
JSR    FFB5    Jump to the right half routine.
```

The first instruction is a JSR to FFB2. At FFB2 it will find a JUMP to the OUTLHF routine, which will output the upper nibble and end with an RTS. The second instruction is similar.

The monitor listing (the list of instructions written into the monitor) shows that at FFB2 there is a JMP E4C3. The program then goes to E4C3 and executes the subroutine it finds there. The JUMP table could have been bypassed by doing a JSR to E4C3 directly.

Summary

This chapter concluded the explanation of the **68HC11** instruction set. It started with a discussion of signed and absolute numbers and explained which BRANCH instructions should be used with which type of numbers. It then covered the BRSET and BRCLR instructions that can be used to check the status of bits in a register and branch accordingly and instructions that affect the condition codes.

The stack was introduced in Section 6-3. Its use with PUSH and PULL instructions and with subroutines was considered. Several problems that used subroutines were presented. Finally, the use of the subroutines in the EVB's monitor was discussed.

Glossary

Absolute numbers   Numbers that are only positive.

Flagged list   A list terminated by a specific byte instead of at a specific location. The contents of the terminating byte(s) are called the flag.

Nested subroutine   A subroutine called by another subroutine.

NOP   No OPeration. An instruction that does absolutely nothing but consume time and take up one location in memory.

PULL   Writing from the stack to a register. PULLs increment the stack pointer.

PUSH   Copying a register onto the stack. PUSHes decrement the stack pointer.

Signed numbers   Positive and negative numbers that are expressed in 2s complement form.

Stack   An area of memory set aside for transient data.

Stack Pointer (SP)   A register that holds the address of the stack.

Subroutine   A routine designed to be executed more than once. It can be entered from any point in the main program by a JSR or BSR. It uses the stack to hold the correct return address.

Unsigned numbers   *See* absolute numbers.

Problems

Section 6-1

6-1 For the instruction CMPA #A0, find the condition codes if A contains
   a. E0   **b.** A0   **c.** 90   **d.** 60

6-2 For each case in Problem 6-1, which of the following instructions will result in a branch? Fill in the table using Y or N.

A contains	E0	A0	90	60
BLE				
BLT				
BGT				
BGE				
BCC				
BCS				
BHI				
BLS				
BRA				

6-3 There is a set of byte-sized numbers in a memory buffer for C080 to C3CC. Write a program to move the address of all the numbers in this buffer that are higher in absolute value than A0 into a list starting at C500.
   (*Hint:* Using both index registers may help.)

6-4 Repeat Problem 6-3 for signed numbers.

Section 6-1.3

6-5 Write instructions, starting at C110, to go to C0CC if bits 0 and 2 of D000 are clear.

6-6 Write instructions, starting at C110, to go to C0EE if bits 1, 3, and 6 of D400 are all set.

6-7 Rewrite the program of Example 6-5 to save an instruction by replacing the BRSET by a BRCLR.

6-8 A buffer is between C200 and C4FF. Write a program to
   a. Count the number of numbers divisible by 16 in the buffer.
   b. Make a list of the addresses of these numbers, starting in D100.
   c. Store these numbers in a list starting at D150.
   (*Hint:* The four LSBs of a number divisible by 16 are all 0. A BRCLR might help.)

Section 6-2

6-9 Write an instruction (or instructions) to do the following:
   a. Set the carry flag.
   b. Clear the V flag.
   c. Set the H flag.
   d. Set all the flags.

Section 6-3.2

6-10 Consider the following program segment:

```
LDS      #C100
PSHA
PSHY
TSX
LDAA     5,X
```

   At the end of the program, the contents of what location was loaded into A?

Section 6-3.3

6-11 If the stack pointer (SP) is at C5FF and X contains C200, write the contents of the stack and the starting address of the subroutine if the instruction at C050 is

   a. BSR   * +26   8D   24
   b. BSR   * −23   8D   DB
   c. JSR   C023   BD   C023
   d. JSR   23   9D   23
   e. JSR   23,X   AD   23

6-12 Consider the following program:

C020	LDS	#C132
C023	LDAA	#28
C025	LDAB	40
C027	ABA	
C029	BSR	12

a. What is the starting address of the subroutine?

b. The contents of location 40 are 79. The first few instructions of the subroutine are

PSHA	PSHA
TPA	PSHB

At this point list the exact contents of memory from C12A to C135. The contents of some of the locations in this area cannot be determined from the given data. In that case write X.

6-13 The angle A (from 0° to 180°) is located in C100, and the angle B is located in C101.

A sine subroutine has already been written and starts in C200. It expects the angle in accumulator A and puts the sine of the angle in accumulator B.

A cosine subroutine has also been written. It starts in C300. It expects the angle in accumulator A and puts its results in accumulator B.

a. Write a program to calculate sin(A + B) by adding A and B. Put the result in C400.

b. Write a program to calculate sin(A + B) using the trigonometric identity sin(A + B) = sin A cos B + cos A sin B. Put the results in C401. Ignore the eight LSBs of the result.

6-14 The following program is designed to take 10 bytes, which consist of 20 nibbles, and convert each nibble to its equivalent ASCII byte. Answer the questions following the program.

Nibble	Location	ASCII Character
0	C200	30
1	C201	31
2	C202	32
3	C203	33
4	C204	34
5	C205	35
6	C206	36
7	C207	37
8	C208	38
9	C209	39
A	C20A	41
B	C20B	42
C	C20C	43
D	C20D	44
E	C20E	45
F	C20F	46

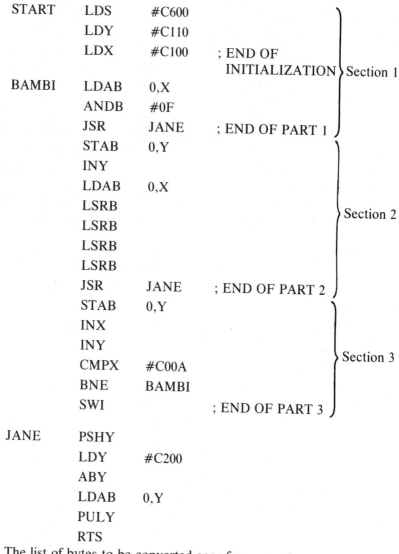

```
START     LDS     #C600                              ⎫
          LDY     #C110                              ⎪
          LDX     #C100     ; END OF                 ⎬ Section 1
                            INITIALIZATION           ⎪
BAMBI     LDAB    0,X                                ⎪
          ANDB    #0F                                ⎪
          JSR     JANE      ; END OF PART 1          ⎭
          STAB    0,Y                                ⎫
          INY                                        ⎪
          LDAB    0,X                                ⎪
          LSRB                                       ⎪
          LSRB                                       ⎬ Section 2
          LSRB                                       ⎪
          LSRB                                       ⎪
          JSR     JANE      ; END OF PART 2          ⎭
          STAB    0,Y                                ⎫
          INX                                        ⎪
          INY                                        ⎪
          CMPX    #C00A                              ⎬ Section 3
          BNE     BAMBI                              ⎪
          SWI               ; END OF PART 3          ⎭

JANE      PSHY
          LDY     #C200
          ABY
          LDAB    0,Y
          PULY
          RTS
```

a. The list of bytes to be converted goes from ____to ____.
b. The resulting ASCII list goes from ____to ____.
c. What does section 1 do?
d. What does section 2 do?
e. What does section 3 do?
f. What does subroutine JANE do?
g. At the end of the program, where is the stack and what is in it? Use specific numbers.

Section 6-3.4

6-15 A main program finds a JSR C300 at C055. The subroutine at C300 pushes A and X. It then goes until C344, where it finds a BSR to C366.

If the SP were initially at C500, what would be in the stack when the program arrives at C366?

6-16 If a subroutine's comment says "Destroys A, B, and Y," what instructions must be used before and after the JSR to preserve them for the main program?

6-17 Which registers were used (and therefore destroyed) by the subroutine in Example 6-11?

Section 6-4

6-18 Consider the following loop:

```
LOOP    DEX
        PSHX
        PULX
        BRN
        BNE
```

   a. How many cycles and how much time does it take to traverse the loop?
   b. Write a 1-s time delay subroutine using this loop.

6-19 Using the 1-s time delay subroutine of the previous problem, write a program that alternately delays for an hour and an hour and a half.

Section 6-4.1

6-20 Precisely how long does the 30-s time delay of Example 6-13b take?

Section 6-4.2

6-21 Precisely how long does the routine of Figure 6-10 take from the time it starts at E78E until it finishes the RTS?

6-22 Assume the SP contains C4FF. Using the explanation of Section 6-4.2, show where the stack is and what is in it for each instruction in Figure 6-10 if the following instruction is found at C009:
   a. JSR   E78E
   b. JSR   C400

Section 6-5

6-23 Write a program to send the message 'HELLO WORLD' to the terminal. Use the subroutines shown in Table 6-9.

6-24 Write a program to accept two decimal inputs from the terminal, add them up, and put their decimal sum out of the terminal.

7

Assemblers and Simulators

INTRODUCTION

Assemblers are computer programs that translate source code programs into machine language programs that the **68HC11** can use. **Simulators** use a computer to simulate the actions of the **68HC11** so that the user can follow the progress of the program and eliminate errors. Both of these are tools for the programmer. They simplify the writing and debugging of programs, especially long and complex programs.

INSTRUCTIONAL OBJECTIVES

After reading the chapter, the student should be able to

- Write programs using source code.
- Use directives in preparing a source program for an assembler.
- Create an object file and load it into the EVB.
- Use a simulator to debug a program.

SELF-EVALUATION QUESTIONS

Watch for the answers to the following questions as you read the chapter. They should help you to understand the material presented:

1. What is the difference between a source file, a list file, and an object file?

2. Which lines of source code need labels?

3. Why is using a label preferable to using a specific address for the target of a BRANCH or JUMP instruction?

4. What is the difference between an instruction and a directive?

5. How can space for stacks be allocated using the RMB (Reserve Memory Block) directive?

7-1 Introduction to Assemblers

As we have previously stated, computers can only understand instructions written in *machine language*. To run a program, the PC (Program Counter) must be loaded with the starting address and the program must be installed in memory at this address as a series of hexadecimal numbers that are the Op codes and addresses of the instructions in the program.

Because writing a program in machine language means constantly looking up Op codes and calculating offsets, it is very tiresome and time-consuming; thus it is rarely done. Instead, programmers use *assemblers* to simplify writing the code. An *assembler is a computer program* that operates on a *source code file* and converts it into machine language. This allows the programmer to write programs in the *source code* format, which is much simpler than machine language. Then both the assembler and the source code file are entered into a computer (usually a PC or other computer that is more powerful than the **68HC11**). The assembler operates on the source code and translates its instructions into an *object file* that contains machine language instructions. The object file can then be loaded into the **68HC11** and the program will run.

The EVB monitor contains a *one-line assembler* that greatly simplifies program writing. If the programmer must write a BRANCH instruction, for example, he or she does not need to look up the Op code for the branch and calculate the offset. Instead, the user can simply write something like BCC C020 and *the one-line assembler will calculate both the Op code and the offset*. It will then write something like 24 EE into the computer's memory. A one-line assembler does not generate an object file. Rather, it writes the code it calculates directly into memory.

So-called "full-blown" assemblers are more sophisticated than one-line assemblers and are preferred by most programmers. They take a *source code* file and generate a *list file* that lists the program and contains error messages, and an *object file* that can be entered into the computer and run. This chapter emphasizes the AS11 assembler because that assembler is provided free by Motorola and can run on any IBM-PC or compatible.

7-2 Writing Source Code Programs

When using an assembler, programmers start by writing their programs in source code, thereby creating a **source code file**. These programs must be written in a *specific format* so that the assembler can understand and operate on them. Each line of source code consists of four fields:

1. The label field

2. The operation field

3. The operand field

4. Comments

Assemblers use spaces (one or more) to separate the fields from each other. Therefore a space character cannot occur within a field; it will be interpreted as a *delimiter* or separator. A typical line of source code might be

    SAM   LDAA   #$FF   Put −1 in A.

Here SAM is the label. The mnemonic LDAA is in the operation field. The operand field specifies what is to be loaded into A. Notice that it specifies both the number and the mode of the instruction. The rest of the line is comments and does not affect the operation of the assembler.

7-2.1 Labels

Labels are used to identify lines in a source program. They have the following characteristics:

1. A label must begin at the *first character position* of a line; it must be flush left. Many lines or statements in a source program do not use labels. In that case the first character position must be a space. The assembler will then interpret the first nonspace character as the start of the operation field.

2. A label can consist of from one to six alphanumeric characters, but the first character of a label must always be an alphabetic character. Thus B1 is a proper label, but 1B is not because it starts with a number. The underline symbol (__) is sometimes used in a label where a space would normally be. Thus GO FAR is not a valid label because of the space, but GO__FAR is a valid label.

3. Some single letters are *reserved* and should not be used as labels. Typically, these are the letters A, B, X, and Y, which are used to designate registers.

4. All labels must be unique. The same label cannot refer to two or more locations in the same program.

Labels are used to designate where the code starts and must be used if a statement is the object of a BRANCH or JUMP instruction. For example, if the unlabeled statement

    BRA   SAM   Branch to SAM

occurs in a program, some other statement must start with the label SAM so that the assembler knows where to branch to. Notice that this statement does not have a label; it does not start flush left. Therefore the assembler will realize that BRA is not a label, but, rather, is an op code.

The EVB one-line assembler does not allow the use of labels; the target of every branch must be an absolute address. If a change is made anywhere in the program that changed the target address, the instruction that branched to it also has to be changed. This problem is eliminated if the target address is specified by a label.

7-2.2 The Operation Field

The operation field starts at the *first nonspace character following the first space character on a line*. It never starts flush left because that column is reserved for labels. The operation field contains either a mnemonic for a **68HC11** Op code or a directive. Directives are discussed in Section 7-3.

7-2.3 The Operand Field

The operand field follows the operation field. It holds the operand associated with the instruction. It usually contains a number (e.g., #$FF), an address, or a label. If it contains a number, the user should be careful. In our previous discussions involving the EVB all numbers were in hexadecimal; the EVB *defaults* to hex. The assembler, however, defaults to decimal. If only a number is found there, the assembler assumes it is a decimal number and converts it to hex. If a hex number is required, it must be preceded by a dollar sign ($). The two instructions of Figure 7-1 show this clearly. The instruction ADDA #25 is coded to 8B 19. The assembler converts the decimal 25 to hex 19. The next instruction, ADDA #$25, is coded as 8B 25.

```
010A 8B 19    ADDA #25
010C 8B 25    ADDA #$25 ADD SECOND NUMBER
```

Figure 7-1   The difference between decimal and hex numbers as seen by the assembler.

There are four symbols that can be used to define numbers:

1. No symbol   Decimal number
2. $ Hexadecimal number
3. @   Octal number
4. %   Binary number

Numbers that start with numerics are acceptable to the assembler. Hex numbers that start with letters, however, must be preceded by a $ symbol or else the assembler will interpret them as labels. Do not write STAA C020, for example. This works on the EVB, but the assembler will look for the label C020 and produce an error message when it cannot find the label. When writing for an assembler, use STAA $C020; this will be accepted.

7-2.4 The Comment Field

The fourth field on a line is the comment field. Comments should be used to clarify the program. Anything written to the comment field will be echoed to the output (list) file exactly as it is written. The assembler will copy it but will not try to interpret it. Entire lines can be designated as comments by starting a line with an * in the first position. Comment lines can include spaces.

7-3   Directives

A *source code program*, the program that the assembler operates on, consists of two parts: *instructions* and *directives*. The instructions are the program instructions that we have been writing in the previous chapters. A **directive** is an *instruction to the assembler* and is *not* part of the program. It directs the assembler to perform certain operations and defines the data and variables used in the program. The directives used in the Motorola assemblers are discussed in this section.

7-3.1   Control Directives

Control directives name the program, define its starting address, and define the end of the source program. There are three control directives for the **68HC11**:

1. The NAM directive

2. The ORG directive

3. The END directive

The NAM (name) directive should be the first line of a source program; it gives a name to the program. NAM must appear in the operation field of the first line (the line is unlabeled). The operand field will contain the name assigned to the program. The assigned name should describe the purpose of the program and thus help the user identify the program.

The ORG (origin) directive tells the code where to start in memory. ORG must appear in the operation field and the address must appear in the operand field. A program can contain several ORG statements if several sections of the program are to be assigned to different areas of memory. Often the stack is assigned its own area of memory by a separate ORG directive.

The END directive is the last line in the source code program and tells the assembler to stop. It consists simply of the word END in the operation field.

7-3.2   The Data Directives

The data directives define the data used in a program and allocate memory for that data. The data directives are
- **FCB**   Form Constant Byte
- **FDB**   Form Double Byte
- **FCC**   Form Constant Character
- **EQU**   EQUal
- **RMB**   Reserve Memory Block

The EQU directive require a label. The RMB does not necessarily require an label, but it does require an operand.

The FCB directive labels and reserves a single byte of memory. For example, the code

SAM   FCB $23

sets aside a memory location that the program will refer to as SAM. It also declares that its initial value will be $23. Note that the directive does not specify the memory address of SAM. This will be determined later by the assembler.

The FDB directive is similar to FCB, except that it defines 2 bytes. An example is

MAX   FDB $1234

This will set aside two consecutive memory locations, MAX and MAX + 1, and initially load the number $1234 into these locations.

The FCB and FDB directives do not necessarily take an operand. If MAX is to be used to store an address, it does not need to be initialized. The code can simply be

MAX   FDB

        STX   MAX

Note that a double byte is needed because the content of X is a 16-bit number.

The code of the previous paragraph may give the erroneous impression that the directive is written immediately preceding the instruction. This is not correct. The directives are usually written at the start or the end of the program and are not interspersed within the code.

The FCC directive requires both a label and an operand. The operand starts and ends with a single quote ('). The characters within the quotes are converted to ASCII and stored in memory. Thus the directive

JANE   FCC   'HELLO WORLD'

will store 11 ASCII characters (remember, the space between the words must be stored also) in consecutive memory locations starting at JANE.

E X A M P L E   7 - 1

How do the directives

        ORG $C000
JANE   FCC   'HELLO WORLD'

affect the memory?

```
C000                                        ORG   $C000
C000 48 45 4C 4C 4F 20     JANE  FCC        'HELLO WORLD'
     57 4F 52 4C 44
                                            END
Errors: 0
```

Figure 7-2 The memory locations for Example 7-1.

Solution

The solution is shown in Figure 7-2. The 11 ASCII characters required for the message are stored in locations $C000–C00A. The next free location is $C00B.

The directive EQU does not set aside a memory location for a label, but it equates the label to the number. An example is

CINDY   EQU $44

Whenever the assembler sees the name CINDY, the assembler will replace the name by its assigned value.

The EQU is often used when specific memory locations are set aside for specific functions. An example might be a data direction register for port C that is assigned (as it is in the **68HC11**) to memory location $1007. Then the program would typically include the directive

DDRC   EQU $1007

The user can now write DDRC rather than the specific memory location when referring to this memory location. Because it is a mnemonic value, DDRC is usually easier to remember.

The RMB directive instructs the assembler to set aside or reserve a block of memory. It may or may not take a label. The RMB directive is usually followed by a number, which is the number of bytes to be reserved for the memory block. If no number follows the RMB, the assembler will reserve 1 byte (the default value) in memory.

7-3.3 Listing Control Directives

There are several directives used for listing control or controlling the way the file appears on paper.

The PAGE directive tells the assembler to start at the top of a new page.

The **SPC** directive, written as SPC n, causes the printer to skip n lines. This is useful for breaking up the program to enhance readability. SPCs should be used after BRANCH or JUMP statements, where there is a break in the program flow. They certainly should be used before and after a subroutine to set the subroutine off from the main program and the other sub-routines.

The AS11 assembler does not really respond to an SPC. It treats an SPC like a comment by echoing SPC, but it does not cause the printer to skip any lines. The user, however, may achieve the same effect by skipping lines in the source file.

Many assemblers also have an option that allows the user to print a *symbol table*. A symbol table lists each symbol used by the program (STACK, JANE, BAMBI, etc.) and gives the value or memory location assigned to that symbol by the assembler.

7-3.4   A Source Code Example

Figure 7-3 is an example of a source code program. It does not do anything practical, but it was written to show the use of the various directives. The line numbers were not generated by the programmer and do not exist in the source file. They are included in Figure 7-3 to clarify the explanation.

```
 1:             NAM      TEST
 2:             ORG      $100
 3: ED1         EQU      25
 4: ED2         EQU      $25
 5: LIL         RMB      2
 6: START       LDA      $00
 7: LOOP        ADDA     #ED1       GET FIRST NUMBER
 8:             ADDA     #ED2       ADD SECOND NUMBER
 9:             STAA     $C020
10:             STA      3333
11:             CMPA     C1         COMPARE TO LIMIT
12:             BGE      START
13:             JMP      SAM+2
14:             SPC      3
15: * THIS IS START OF PART 2
16:             ORG      $07FE
17: C1          FCB      $A0
18: C2          FCB      $02
19: MAX         FCC      'THIS IS THE END'
20: SAM         FDB      200
21:             LDD      #SAM
22:             SUBD     #MAX
23:             STD      LIL
24:             END
```

Figure 7-3   A sample source file.

A line-by-line explanation of Figure 7-3 follows:

Line 1   The top line should be the NAM directive. The name TEST was given to this particular program.

Line 2   This ORG directive defines the origin for the first section of code.

Lines 3 and 4   They equate the labels ED1 and ED2 to decimal 25 and hex 25, respectively.

Line 5   The RMB directive reserves 2 bytes in memory for a label LIL. Because this is the first directive or instruction requiring memory, we can expect the assembler to assign the bytes to the first available memory locations. As the origin directive causes the first memory section to start at 100, these bytes will probably be assigned to locations 100 and 101.

Lines 6 through 13   This is a section of code. It simply adds ED1 and ED2 and stores the result in two locations. It then compares the results with the contents of C1. Notice that C1 is a symbol for a location and is not a number (a number would be specified by #$C1 and a location would be specified by $C1). Of course, the symbol must be further defined. In this program C1 is defined by the FCB on line 17. The program continues to loop and add until the contents of accumulator A become greater than A0. It then jumps to the second part of the program, to the label SAM + 2. SAM is the label of an FDB, which will reserve 2 bytes for data. The code resumes after the FDB. The assembler will add 2 to the label SAM and will jump the program there.

Line 14   There is a break in the program here caused by the JUMP. This line will add three line spaces to help indicate the break in the flow.

Line 15   The * in the first column indicates that this is entirely a comment line.

Line 16   The second part of this program has been assigned to a different area of memory, $07FE, as indicated by the ORG directive.

Lines 17 and 18   These are two 1-byte memory locations. In this demonstration program C2 was not used, but C1 was referred to on line 12.

Line 19   MAX is the location of the ASCII message and is defined by the FCC directive.

Line 20   SAM is the address of a double byte, which is set equal to 200 (decimal) by the FDB directive.

Lines 21 and 22   These lines contain code. They accomplish something. By loading in the address of the byte after the end of the message (SAM) and then subtracting the address of the start of the message (MAX), the length of the message will be placed in the double accumulator.

Line 23   The contents of the double accumulator are written into LIL, which was defined in line 5.

Line 24   The end of the program.

Although this is not a useful program it can be used as a model source program because it is in the proper format and shows how to use the directives in the **68HC11** assembler.

7-4 The Assembler Listing

A source program, such as that discussed in the previous paragraph, must be put through an *assembler*. Several different assemblers are available with various capabilities. We used the AS11 assembler supplied by Motorola. It has some limited capabilities, but it runs on an IBM-PC or compatible.* Any assembler will operate on the source code and generate two files: a *list file* and an *object file*.

7-4.1 A Sample List File

List files are discussed in this section. Figure 7-4 shows the list file for the previous program as generated by the AS11 assemblers.† We have added a column of line numbers for clarity. The actual list file contains seven columns. They are

Column 1 The addresses for the instructions and for the ORG, FCB, FDB, and FCC directives. It contains data for the EQU directive. Notice the difference in the data on lines 3 and 4, which is the difference between decimal and hex numbers.

Column 2 This column contains the Op codes for all instructions, but it also starts the data for the FCB, FDB, and FCC directives.

Column 3 This column contains the operands (1 or 2 bytes) associated with each instruction. It also contains data for the FDB and FCC directives.

Columns 4 through 7 These columns are the same as the source file.

The list file shows the results of the assembler program. Some comments on the various lines follow:

Line 5 The assembler reserved two memory locations, at 100 and 101, for LIL. It did not put any data in these memory locations.

Line 10 The source instruction for this line is STA 3333. The assembler converts the 3333 to its hex equivalent. Note the difference between line 10 and line 9.

Line 13 The assembler will accept operands like SAM+2. It will perform the arithmetic operations and branch to two locations beyond SAM in this case.

Line 14 The AS11 assembler does not accept SPC 3. It does not provide three blank lines as a full-scale assembler would. It treats the SPC directive as a comment. Comments, however, leave blank spaces in the first three columns.

Line 19 The string 'THIS IS THE END' has been converted to its ASCII equivalent by the FCC directive.

---

* Possibly the easiest way to invoke the AS11 assembler is to put the source file on the same diskette as the AS11 assembler file. We generally give the source file a suffix of .S. Then the command AS11 <filename.S> will invoke the assembler and send the list file to the screen. The list file can be sent to a file instead of the screen by giving the command

AS11 filename.S> filename.LST.

† An example command that works on the source file TEST.S is AS11 TEST.S > TEST.LST.

```
 1:                             NAM    TEST
 2: 0100                        ORG    $100
 3: 0019            ED1         EQU    25
 4: 0025            ED2         EQU    $25
 5: 0100            LIL         RMB    2
 6: 0102 96 00      START       LDA    $00
 7: 0104 8B 19      LOOP        ADDA   #ED1     GET FIRST NUMBER
 8: 0106 8B 25                  ADDA   #ED      2 ADD SECOND NUMBER
 9: 0108 B7 C0 20               STAA   $C020
10: 010B B7 0D 05               STA    3333
11: 010E B1 07 FE               CMPA   C1       COMPARE TO LIMIT
12: 0111 2C EF                  BGE    START
13: 0113 7E 08 11               JMP    SAM+2
14:                             SPC    3
15:    * THIS IS START OF PART 2
16: 07FE                        ORG    $07FE
17: 07FE A0         C1          FCB    $A0
18: 07FF 02         C2          FCB    $02
19: 0800 54 48 49 53 20 49  MAX FCC    'THIS IS THE END'
20:      53 20 54 48 45 20
21:      45 4E 44
22: 080F 00 C8      SAM         FDB    200
23: 0811 CC 08 0F               LDD    #SAM
24: 0814 83 08 00               SUBD   #MAX
25: 0817 FD 01 00               STD    LIL
26:                             END

Errors: 0
```

Figure 7-4   A sample list file.

Lines 23 and 24   The operands on these lines show that this part of the program works. On line 23 the number 80F is loaded into D. On line 24 the number 800 is subtracted. This leaves 0F, or 15 decimal, the number of characters in the string.

7-4.2   Errors Detected by the List File

The list file can detect some errors in the program and warn the user. Figure 7-5 shows the list file response to some of them. Perhaps the most common error is the undefined symbol as shown in Figure 7-5a. Here the programmer probably meant FF to be the number -1, but because it was written as #FF instead of #$FF, the assembler interpreted FF as a symbol. It then searched the FCBs and other directives to find the symbol. When it failed, it produced the error message. Note that the error message pertains to the line *below* the message, not to the line above.

Figure 7-5b shows what happens if the programmer uses a nonexistent Op code. Here SSS was deliberately inserted into the program. In response,

```
5: Symbol Undefined on Pass 2
0104 86 00                        LDA    #FF
(a) Undefined symbol
```

```
 7:   LOOP    ADDA    #ED1      GET FIRST NUMBER
 8:           ADDA    #ED2      ADD SECOND NUMBER
 9:           STAA    $C020
10:           SSS     3000
11:   *       STA     3333
12:           CMPA    C1        COMPARE TO LIMIT
13:           BGE     START
14:           JMP     SAM+2
15:           SPC     3
16:   * THIS IS START OF PART 2
17:           ORG     $07FE
18:   C1      FCB     $A0
19:   C2      FCB     $02
20:   MAX     FCC     'THIS IS THE END'
21:   SAM     FDB     200
22:           LDD     #SAM
23:           SUBD    #MAX
```

```
A:\>AS11 TEST.S
10: Unrecognized Mnemonic
Errors: 1
(b) Unrecognized mnemonic
```

Figure 7-5   Errors detected by the assembler.

the list file only produced an error message and indicated the line number of the error.

A branch out of range is another common error that will be flagged by the list file.

AS11 will allow the user to omit the NAM statements. *Warning:* Other assemblers react violently to this omission. We recommend using the NAM directive, as the name can convey some useful information, like the name of the source file.

AS11 will also allow the user to omit the ORG statement, but it will then start memory at $0000. This is usually not what the user intended, so an ORG statement should always be used.

We must also warn the reader that the error detection capability of an assembler is not a cure for all possible problems. It will not detect logical errors. For example, if the user writes the following code segment

```
START    LDAA    #$F0
LOOP
           ⋮
         DECA
         BNE     START
```

when he or she meant to write BNE LOOP, the program will probably go into an infinite loop, but the assembler will not detect an error. Assemblers are a big help, but the programmer must still be careful.

7-4.3   The Use of Stacks and the Reserve Memory Block Directive

Programs that use subroutines can be handled by the assembler. A typical source program that uses one or more subroutines is shown in Figure 7-6. The code in the main program starts with an LDS #$C500. This works if the stack is to be placed at a specific location.

In some instances the user might want to place the stack immediately after the program. The first instruction can then be changed to LDS #STACK, and STACK can be defined by an RMB directive. If the directive STACK RMB 20 is placed immediately after the RTS, a disaster will occur because the data written to the stack will decrement it and overwrite the end of the subroutine. This disaster can be averted by writing the following code after the RTS that ends the last subroutine:

```
         RTS
         RMB 20
STACK    RMB
         END
```

Here 20 bytes are reserved for the stack and the SP points to the top of stack. Data will now be stacked in the reserved area and will not overwrite

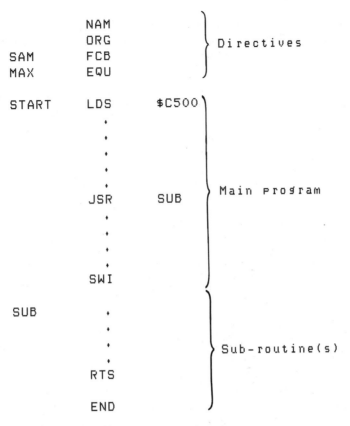

Figure 7-6   A typical source program using subroutines.

the subroutine or the main program. Of course, the size of the stack is allocated by the first RMB directive and can be changed depending on the requirements of the particular program.

7-5   The Object File and S Records

The **list file** provides a comprehensive listing of the programs, shows any errors that the assembler can detect, and gives the machine code. On some primitive computers the machine code had to be entered one byte at a time. Even then the list file would help. If, for example, the user created the program of Figure 7-3 and then assembled it to produce the list file of Figure 7-4, the code could then be read off the list file and keyed into the **68HC11** 1 byte at a time. The user could set the address at 0102 and then start keying 96, 00, 8B, 19, ... Obviously, this method is tiring, time-consuming, and prone to errors.

Fortunately, assemblers alleviate this tedium by creating an *object file* in addition to the list file. An **object file** is basically a machine code file that can be entered directly into the computer.*

7-5.1   The S-Record File

When the AS11 assembler is invoked and there are no errors that it can detect, it will write an object file to the disk. In our version that object file was called M.OUT. Figure 7-7 shows the M.OUT file for the programs of Figures 7-3 and 7-4.

```
A:\>Type M.OUT
S11701029600888198B2587C020B70D05B107FE2CEF7E081133
S11F07FEA002544849532049532054484520454E4400C8CC080F830800FD010019
S9030000FC
```

Figure 7-7   The S-record file for the program of Figs. 7-3 and 7-4.

The object or M.OUT file is written in what Motorola calls its **S-record** format. This is because each line starts with the letter S. There are three types of lines in an S-record file:

- **The S0 line**   This is a line that contains the name of the file.
- **The S1 line(s)**   These are the lines that contain the machine code for the program.
- **The S9 line**   This is a line that terminates the S-record file.

Figure 7-7 shows that the S-record file for the program of Figures 7-3 and 7-4 contains two S1 lines and an S9 line. The S0 line is omitted by the AS11 assembler but not by some other assemblers.

The S1 lines can be examined on a byte-by-byte basis. Using the first S1 line, we see that the first byte is S1; it identifies the line as an S1 line. The second byte is the number of bytes (in hex) that follow on the line. The third and fourth bytes are the starting address of the code. It can be seen that these are 0102. The code starts in the fifth byte. It is the code as shown in the first part of the program in Figure 7-4.

The code listing continues until the last byte of the S1 line. The last byte is not code; it is a *checksum*. Every S-record line ends with a checksum, which is derived by adding up all the bytes in the line (except the S1 byte) and then taking its complement.

The second S1 line contains the address and the code for the second part of the program. If the program were longer, there would be more S1 lines.

Terminating lines all look the same; S9030000FC. Simple assemblers contain the bytes 0000 in the S9 line, but more sophisticated assemblers may

---

* Using more sophisticated assemblers, object files may have to be *linked* before they can be entered into computers. Linking is beyond the scope of this book.

include the starting address of the program in these bytes. Note that the sum of the three data bytes in the S9 line is 03. The checksum is the complement of this sum, or FC, which appears as the last byte on the line.

7-5.2 Downloading the S-Record File

S-record files are of little use unless they can be loaded directly into the **68HC11**. The procedure for loading a file directly into a **68HC11** depends on the system available and the communications package.

If an EVB is being used and it is communicating with an IBM-PC or compatible using PROCOMM, the M.OUT file can be downloaded directly into memory by the following procedure:

1. The program and data should all be allocated to the EVB's RAM area, $C000–$DFFF. The ORG directives in the source file must point to locations in this area.

2. With the EVB connected to the PC, type LOAD T on the keyboard.

3. Hit the <pg up> key on the keyboard. PROCOMM will respond with a menu.

4. Type 7 for ASCII.

5. Enter the filename. For this simple system it will be M.OUT.

6. The file will then be loaded. The EVB will appear to sit there doing nothing.

7. Push the RESET button. Remember, resetting does not change the contents of memory in the C000–DFFF area.

8. The program should now be in memory. It can be examined and run by using the EVB commands discussed in Chapter 4.

Observe the power that this system gives the programmer. If a program must be changed or a section of code must be added, the simplest procedure is to modify the source code, reassemble, and download. The programmer is no longer concerned with using precise memory addresses (labels can be used), calculating the value of branch offsets, or entering the machine code. Assemblers are a significant advance over hand-coding and even over the one-line assembler in the EVB.

7-6 Simulators

A simulator is a computer program that calculates the response of a **68HC11** to a given program. A simulator is generally run in a PC or other computer. It simulates the action of a hypothetical **68HC11** and records the changes in the registers and memory as the program progresses. A simulator must be able to do the following:

- Initialize memory and the registers.
- Load the program into the simulated memory. In some simulators the program is loaded from the S-record file.
- Single step the program and monitor the **68HC11** registers and the contents of memory at each step.
- Set a breakpoint and allow the program to run until it reaches the breakpoint. Then the user must be able to examine the registers and memory.

The names of two companies that manufacture commercially available simulators are given in the references (see Appendix B).

Summary

This chapter started with a definition of a source file, which is a combination of instructions and directives. An example of a source file was shown, and it was assembled. The assembler and the procedure for assembly were explained, and error detection by the assembler was considered.

An S-record file was shown and the procedure for downloading it into a **68HC11** was discussed. The chapter ended with a description of a simulator, a powerful debugging tool for any computer.

Glossary

Assembler   A computer program that assembles or translates source code files into list files and object files.

Directive   An instruction to the assembler.

Label   A name given to a variable or a line of code to identify it.

List file   A file that contains the listing of the program, including the code.

Object file   A file containing code that can be entered into a computer and run.

Simulator   A computer program that simulates the action of the **68HC11**. It shows the results of each instruction and its effect on the registers and memory.

Source code file   A file containing source instructions and directives that can be assembled.

SPC   A directive that causes the list file to skip one or more lines.

S-record   A record of code that can be written directly into a computer. It starts with the character S.

Problems

Section 7-2.1

7-1 Which of the following labels is correct?
- **a.** ED1
- **b.** ED 1
- **c.** ONE ED
- **d.** ONE_ED
- **e.** 1ED
- **f.** STEVE
- **g.** STEPHANIE

7-2 Consider the following source code segment:

```
MAX     FCB
          ⋮
SAM     BNE     BILL
        LDAA    MAX
BILL    INCA
```

- **a.** Which of the above statements take labels? What are the labels?
- **b.** If the assembler assigns $C020 to MAX, hand code the instructions.

Section 7-3.2

7-3 The string 'I am a brilliant 68HC11 student!' is to be placed at C020. Write the ORG and FCC directive to do it. Also, write a program segment to calculate the length of the string.

Section 7-4.3

7-4 A program consists of a main program and two subroutines, labeled SUBR1 and SUBR2. SUBR1 calls a a third subroutine, SUBR3.
Draw a diagram, similar to Figure 7-6, showing the program. Include directives to place a $33-byte stack at the end of the program.

7-5 The stack in a program is to be labeled JANE and to occupy locations $C500–$C542.
- **a.** What should the first instruction in the program be?
- **b.** What directives should be written to position the stack properly?

7-6 Write the source code for Problem 5-28. If an assembler is available, assemble it.

7-7 For the following source program
- **a.** What number is assigned to the label GO_FAR?
- **b.** Write the codes for lines 14 and 16.
- **c.** Assemble it using AS11 and check the answers to parts (a) and (b).

```
 1:              NAM          JOE1
 2:   TFLG      EQU          $1023
 3:             ORG          $C200
 4:   GO_FAR    FDB
 5:   START                  LDAA      #02
 6:             LDX          $C300
 7:             LDS #CINDY
 8:*            STAA         0,X
 9:             DECA
10:             STAA         1,X
11:             BSR          SUB1
12:             SPC          1
13:             INX
14:             LDD GO_FAR
15:             SUBD $1014
16:             STD GO_FAR+2
17:             SWI
18:
19:   SUBI      LDAA #7
20:             STAA         TFLG
21:   LOOP1                  LDAA TFLG
22:             BITA         0,X
23:             BEQ          LOOP1
24:             RTS
25:             RMB          10
26:   CINDY     RMB
27:
28:             END
```
*

d. Examine and explain the code generated on line 11.

e. Where is the stack? How long is it?

7-8 For the following S-record file

```
A:\>Type M.OUT
S123C20000008602CEC300BEC233A7004AA7018D0B08FCC200B31014FDC2023F8607B7102C
S10CC22023B61023A50027F93907
S10EC00048454C4C4F20574F524C4415
S9030000FC
```

a. Where does the code on each line begin?

b. List the first few bytes of code for each line.

The Hardware Configuration and Interrupts

8

INTRODUCTION

This chapter describes the internal components of the **68HC11** and its interrupt system. The memory and I/O components are introduced and the two major modes of operation are discussed. This is followed by an explanation of the various interrupts available on the **68HC11**.

INSTRUCTIONAL OBJECTIVES

After reading the chapter, the student should be able to

- List the component circuits within the **68HC11** and describe their functions.
- Set the mode of operation of the **68HC11**.
- Design the interface between a **68HC11** in the expanded mode and its external memory.
- Construct circuitry to provide an IRQ or XIRQ interrupt.
- Write an interrupt service routine.
- Use interrupt vectors to go to the service routine in response to an interrupt.
- Preserve the contents of the stack after an SWI.

SELF-EVALUATION QUESTIONS

Watch for the answers to the following questions as you read the chapter. They should help you to understand the material presented:

1. How does one determine the differences between the various versions of the **68HC11**?
2. What is the function of ports B and C in the single-chip mode? In the expanded mode? What is the function of STRA and STRB in the expanded mode?
3. Why is the read-write line predominantly high?
4. What are the events that can cause a **68HC11** to enter RESET? What happens when it does?

5. What are the advantages and disadvantages of polling?

6. What are the differences between a JSR and an interrupt? What registers are stacked in each case?

7. What does the X bit in the CCR do?

8. What are two uses for the SWI on the EVB?

8-1 The Hardware Configuration of the 68HC11

The **68HC11** is a *microcontroller*, not just a microprocessor (μP). It contains the μP we have discussed in the previous chapters, but it also contains several memories and a large amount of circuitry dedicated to Input/Output (I/O), which will be discussed in the following chapters. This additional I/O circuitry makes the **68HC11** a microcontroller.

Inside the **68HC11** chip there is RAM (Random Access Memory, or Read-Write Memory), ROM (Read Only Memory), which cannot be written to, and **EEPROM** (Electrically Erasable Programmable Read Only Memory). The IC contains these memories as well as the microprocessor, and its associated registers, and the registers and circuits that control the I/O.

There are several *versions* of the **68HC11**. All versions contain the basic microprocessor that has been described in the previous chapters and use the same registers and instruction set. The difference between the versions is the amount of ROM, RAM, and EEPROM they contain. Table 8-1 shows some of the versions of the **68HC11**.*

The basic version of the **68HC11** is the **MC68HC11A8**. It contains 8K bytes of user-programmed ROM, 256 bytes of RAM, and 512 bytes of EEPROM. The **68HC11A1** is the version used in the EVB. It is identical to the 'A8 except that it contains no ROM. The Buffalo monitor, the ROM that controls the board, is provided in a separate ROM IC. The other versions of the **68HC11** are for special purpose uses and are not so popular as the 'A8 or 'A1. The sophisticated designer will select the version of the **68HC11** that comes closest to satisfying the requirements of the particular problem.

8-1.1 The Pinout of the 68HC11

The **68HC11** (the 'A8 or 'A1 versions) is available in the 52-pin Plastic Leadless Chip Carrier (PLCC) or a 48-pin Dual-In-line Package (DIP) as shown in Figure 8-1. The function of each pin is given in Figure 8-1, but they can be explained more clearly by referring to Figure 8-2.

Figure 8-2 is a functional block diagram of the **68HC11**. It shows that the IC has five I/O ports that take 42 pins.

---

* Because the manufacturers keep adding new versions, the list presented here may not be current or complete. The reader may want to contact the manufacturer's literature for a current list.

Table 8-1   Several 68HC11 Family Members

Part Number	EPROM	ROM	EEPROM	RAM	CONFIG†	Comments
MC68HC11A8	—		512	256	$0F	Family built around this device
MC68HC11A1	—	—	512	256	$0D	'A8 with ROM disabled
MC68HC11A0	—	—	—	256	$0C	'A8 with ROM and EEPROM disabled
MC68HC11A2	—	—	2K*	256	$FF	No ROM part for expanded systems
MC68HC811A8	—	—	8K + 512	256	$0F	EEPROM emulator for 'A8
MC68HC11E9	—	12K	512	512	$0F	Four input capture/bigger RAM/12K ROM
MC68HC11E1	—	—	512	512	$0D	'E9 with ROM disabled
MC68HC11E0	—	—	—	512	$0C	'E9 with ROM and EEPROM disabled
MC68HC11E2	—	—	2K*	256	$FF	Like 'A2 with 'E9 timer
MC68HC11D3	—	4K	—	192	N/A	Low-cost 40-pin version
MC68HC711D3	4K	—	—	192	N/A	One-time-programmable version of 'D3
MC68HC11F1	—	—	512*	1K	$FF	High-performance, nonmultiplexed 68-pin

NOTES:
* The EEPROM is relocatable to the top of any 4K memory page. Relocation is done with the upper 4 bits of the CONFIG register.
† CONFIG register values in this table reflect the value programmed prior to shipment from Motorola.

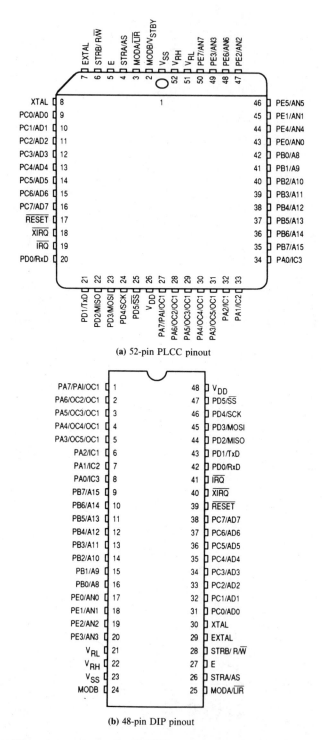

(a) 52-pin PLCC pinout

(b) 48-pin DIP pinout

Figure 8-1   Pin assignments on the 68HC11. (*Reprinted Courtesy of Motorola, Inc.*)

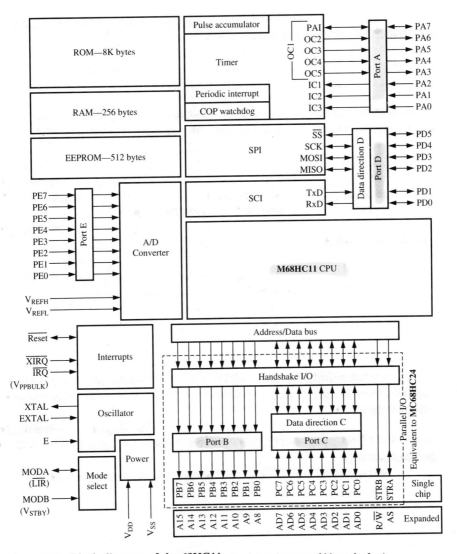

Figure 8-2   Block diagram of the 68HC11. *(Redrawn Courtesy of Motorola, Inc.)*

■ **Port A (8 pins)**   This port functions as a timer or I/O port. Port A is discussed in Chapter 10.

■ **Port B (8 pins)**   Port B functions as an address output or an output port (Sec. 9-2).

■ **Port C (8 pins)**   Port C can be an address port, a data port, an input port, or an output port. When it is used for memory interfacing it carries half the memory address and the memory data. This is discussed in Section 8-4. Its use as an I/O port is discussed in Section 9-3.

■ **Port D (6 pins)** Port D is used for serial interfacing. Two of its pins are connected to the SCI and four are connected to the SPI. Serial interfacing is discussed in Chapter 11.

■ **Port E (8 pins)** These pins can be either input or used with the Analog-to-Digital (A/D) converter. The A/D converter is also discussed in Chapter 11.

In addition to these 38 pins, there are four other pins associated with the I/O ports. These are STRB or R/W, STRA or AS, V_{RH} and V_{RL}. The first two pins are used with memory interfacing or I/O; the two latter pins are associated with the A/D converter.

Besides the 42 I/O pins, there are three pins associated with interrupts (RESET, IRQ, and XIRQ), three pins associated with the crystal clock (XTAL, EXTAL and E), two pins for mode control (MODA and MODB), and power (+5 V or V_{DD}) and ground (V_{SS}). This comprises the 52 pins on the PLCC package. The 48-pin DIP uses four fewer port E pins.

Many pins in Figure 8-1 have more than one use. Pin 40, for example, is labeled PB2/A10, which means it sometimes is used as an output pin, PB2, and sometimes as an address pin, when it is A10. At this point in the book the student is not expected to understand the function of any of the pins on the **68HC11**. They will be explained as the text progresses.

8-1.2 Input Pin Protection*

The pins that function as inputs to the **68HC11** must be protected against possibly damaging inputs, such as voltages above or below the normal range and floating inputs. The circuitry of an input pin is shown in Figure 8-3. The input, from an external source, is connected to a CMOS gate. The *thick field protection* transistor prevents damage if the input voltage goes above V_{DD} or below ground.

The input pins should also be protected against *floating inputs* (unconnected inputs). An unconnected input could cause the input gate to oscillate and cause noise and excessive power dissipation.

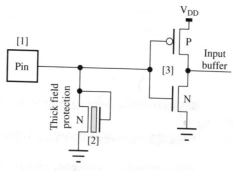

Figure 8-3 The circuitry of a digital input-pin. *(Redrawn Courtesy of Motorola, Inc.)*

* Students who have no need for the hardware details may skip this paragraph.

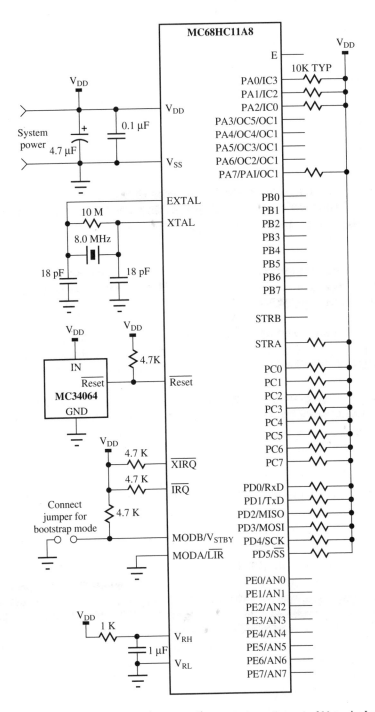

Figure 8-4 Basic single-chip-mode connections. *(Redrawn Courtesy of Motorola, Inc.)*

The simplest way to protect against floating inputs is to use *pull-up resistors* as shown in Figure 8-4. Typically, the pull-up resistors are 10K Ω, so they will only draw 0.5 mA if connected to a source that drives the input to ground. *All pins that could possibly be inputs are pulled up*. This includes four pins on port A, all pins on ports C and D, and STRA. *Port B and STRB are output only* and do not need to be pulled up. Port E is an input port, but it is designed to accommodate analog voltages, and its special circuitry makes pull-up resistors unnecessary.

8-1.3   The Memory Map of the 68HC11

The **68HC11** contains several internal memories. These are

1. The internal RAM   The 256-byte internal read/write memory usually starts at 0 and occupies locations 0000–00FF. Students using the EVB should avoid using this area because the monitor places interrupt vectors (see Sec. 8-7.2) there.

2. The input/output registers   These are the registers that control the I/O. They reside in locations $1000–$103D. Most bits of these registers are RAM, but some are EEPROM. They are definitely not available for general use.

3. The internal EEPROM   The 512-byte EEPROM on the 'A8 and 'A1 occupy memory locations $B600–$B7FF. The EEPROM will retain its information with power turned off and can be used to preserve short programs. Writing to the EEPROM is described in Section 12-2.

4. The ROM   The 'A8 includes 8K bytes of ROM, starting at $E000. These can be used for dedicated programs. The 'A1 version does not include any ROM.

8-2   Pin Functions and Modes

The functions of the pins on the **68HC11** that are not connected with I/O are described in this section.

8-2.1   The Power Pins

Power is supplied to the **68HC11** via the V_{DD} pin, connected to +5 V, and the V_{SS} pin is the ground return. Like most ICs, the power pins should be *decoupled* by placing a large capacitor and a small capacitor across the power pins as near as possible to the IC. The large capacitor absorbs current spikes and should be between 1 and 5 μF. The smaller capacitor is better at suppressing high frequency noise and should be between 0.01 and 0.001 μF. Figure 8-4 shows typical decoupling on the power pins.

8-2.2   The Interrupt Pins

The interrupt pins are RESET, IRQ, and XIRQ. RESET is discussed in Section 8-5, IRQ is discussed in Section 8-7, and XIRQ is discussed in Section 8-8.

8-2.3   The Crystal Oscillator Pins

The timing of the **68HC11** is controlled by a crystal oscillator, which determines the basic timing of the IC. There are three pins connected with the timing: XTAL, EXTAL, and E. The XTAL and EXTAL pins are inputs that must be connected to the crystal oscillator, whose frequency is four times that of the basic clock driving the **68HC11**. The E pin is an output at the basic clock frequency.

Figure 8-5 shows a typical crystal oscillator and its connections to the **68HC11**. Figure 8-6 shows a typical physical layout. The leads should be

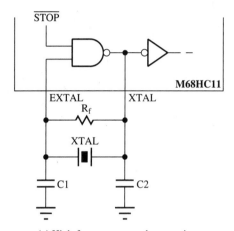

(a) High-frequency crystal connections

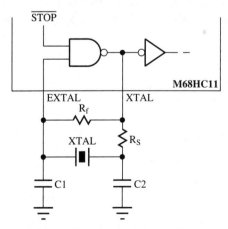

(b) Low-frequency crystal connections

Figure 8-5   Crystal oscillator connections to the 68HC11. (*Redrawn Courtesy of Motorola, Inc.*)

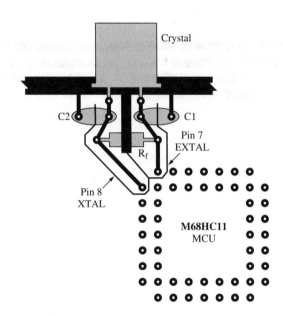

Figure 8-6 **Crystal layout example.** (*Redrawn Courtesy of Motorola, Inc.*)

kept as short as possible to minimize the effects of stray electric or magnetic fields. An oscilloscope should not be used to observe the XTAL or EXTAL waveforms because the stray capacity of the probes may distort the waveform. The slower output on the E pin can be observed.

The crystal oscillator, with typical components, is also shown in Figure 8-5. Here an 8-MHz crystal is used, and the basic clock of the μP and the E pin are at 2-MHz. These are the frequencies used on the EVB.

The **STOP instruction** has not been discussed previously because it is rarely used and should be used very carefully. The S bit in the CCR is a mask for the STOP instruction to prevent inadvertent use. The S bit is set during RESET and must be cleared; otherwise the STOP instruction is an NOP.

The STOP instruction puts the **68HC11** in a comatose state by disabling the clock. During this state power consumption is dramatically reduced but no processing can occur. The **68HC11** can be reawakened by an external interrupt or RESET.

The action of the STOP instruction is clearly shown in Figure 8-5. The STOP signal will cut off the NAND gate, thereby disabling the oscillator. There are two other precautions to be observed if the STOP instruction is to be used. The STOP instruction should be preceded by an NOP, and the oscillator may take up to 2 ms to return to full-speed operation.

8-2.4   The MOD Pins

The two MOD pins, MODA and MODB, function as inputs *only at the instant when the 68HC11 is turned on or RESET*. At that instant they determine the mode of operation of the μP. The **mode** is therefore determined at RESET and cannot be changed without restarting the **68HC11** by applying another RESET pulse.

The levels on the MODA and MODB pins and the four possible modes of operation are shown in Table 8-2. The special test and special bootstrap

Table 8-2   Hardware Mode Select Summary

Inputs:		Mode Description	Control Bits in HPRIO (Latched at Reset)			
MODB	MODA		RBOOT	SMOD	MDA	IRV
1	0	Normal single chip	0	0	0	0
1	1	Normal expanded	0	0	1	0
0	0	Special bootstrap	1	1	0	1
0	1	Special test	0	1	1	1

modes are for manufacturers' testing; they are not normally used and will not be discussed further. Figure 8-4 shows the mode inputs set up for the *normal single-chip* mode of operation (MODA = 0, MODB = 1). MODA could be changed to a 1 for the *normal expanded mode*.

In normal (not test mode) operation the **68HC11** is restricted to either the normal single chip or the normal expanded mode. In normal single-chip operation all memory references are to the memory (RAM, ROM, or EE-PROM) inside the **68HC11**. This severely restricts the size of the memory. Ports B and C, however, are available for I/O, as are the STRA and STRB signals.

In expanded mode the entire 64K-byte memory can be addressed, but this means committing port B to carry the upper half of the address and port C to carry the lower half of the address and the data. This explains the description of pin 10 of Figure 8-1, for example. Its description is PC1/AD1, which means it functions as pin 1 of port C in the single-chip mode, but it functions as both address bit 1 and data bit 1 in the expanded mode. In the expanded mode the B and C ports cannot be used for I/O and STRA becomes an address strobe (AS) and STRB becomes the read-write (R/W) line. The functions of these signals are explained in Section 8-4.

EXAMPLE 8-1

What mode is used on the EVB?

Solution

Answer A   Because the EVB can use memory between C000 and DFFF for RAM, and the Buffalo monitor occupies memory from E000 to FFFF, all memory must be available. Therefore the **68HC11** must operate in the expanded mode.

Answer B   Because ports B and C are available for I/O, the **68HC11** must be in a single-chip mode.

The answers seem to contradict each other. Actually, the **68HC11** is operated in the expanded mode and ports B and C cannot be used. But there is another IC on the EVB, the **68HC24** Port Replacement Unit (PRU). This chip performs the functions of ports B and C, including STRA and STRB. With the addition of the PRU, the user has the best of both worlds. The entire memory space can be used, and all the ports can be used for I/O.

As previously stated, the levels on MODA and MODB are only read at RESET. The level on MODA, which selects between the single-chip and expanded modes, is stored and remembered in bit 5 of the HPRIO register. This is one of the I/O registers and is at location $103C. The bit can be read by the EVB.

After RESET, MODA becomes an output and carries the Load Instruction Register (LIR) signal. This signal drives low during the first cycle of each instruction. It can be helpful if the output of the **68HC11** is being monitored by a logic analyzer, because it distinguishes those clock cycles that start an instruction.

After RESET port B can be connected to a battery or other source of power. It is designated as V_{STBY}, the standby voltage. The standby or auxiliary voltage can be used to retain the contents of RAM in the event that V_{DD} goes down for any reason.

8-3   Memory Cycles and Timing*

Figure 8-7 shows the interface between a standard computer and a memory. The interface consists of the R/W line, the address bus, and the data bus. In most microprocessor (μP) systems there are several banks of memory and other peripherals on the bus. The bus can be thought of as originating at the μP and then going to each of the memory and peripheral ICs.

---

* A more thorough description of memory ICs and memory organization is given in J. D. Greenfield, *Practical Digital Design Using ICs* (Englewood Cliffs, NJ: Prentice-Hall, latest ed.).

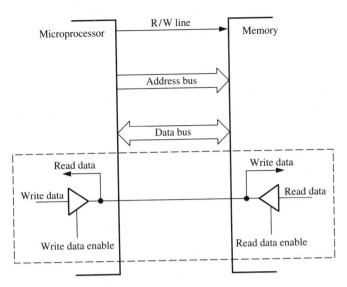

Figure 8-7   A standard memory interface.

8-3.1   The Read-Write Line

The R/W line is controlled by the μP. It tells all the devices connected to the bus whether the μP is reading or writing data. For almost all memories a 0 on the line indicates a write operation. If this line is observed on an oscilloscope, it will be high most of the time because read cycles predominate. All instruction fetches, for example, are read cycles.

8-3.2   The Address Bus

The address bus consists of 16 lines for the **68HC11**. This limits the memory to 65,536 (2^{16}) addresses. The larger 16-bit and 32-bit μPs require more address space and have larger address buses. The lower bits of the address bus determine the memory address. The higher bits determine which *bank* of memory or which peripheral IC is being accessed.

Figure 8-8 shows the system board for the IBM-PC and should help clarify the concept of memory banks. The memory consists of type **4164** ICs, which are 64K-bit RAMS. The memory ICs are shown in the upper left-hand corner of the figure. The system contains four banks of nine **4164**s, or 256K bytes of RAM. The ninth **4164** in each bank is used for parity checking. The first bank of ICs consists of ICs U37 to U45. Some PCs add more memory by installing additional memory banks on plug-in cards.

E X A M P L E   8 - 2

An IBM-PC has a 20-line address bus. If the memory is made up of 64K-bit RAM ICs, how many memory banks could there be? How many lines are required to address each memory location?

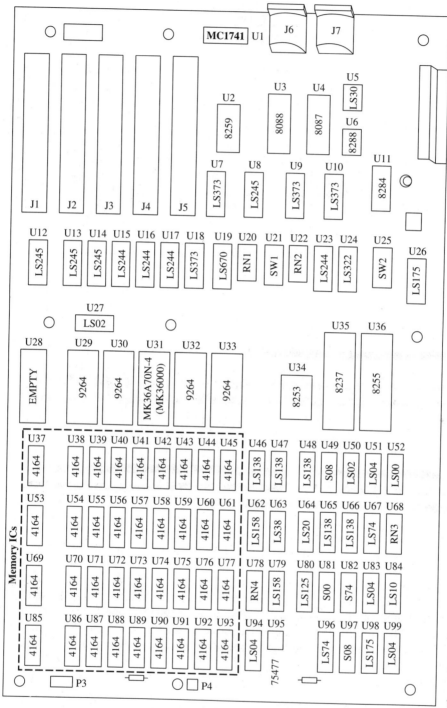

Figure 8-8   The IBM PC chip layout.   (Redrawn Courtesy of Motorola, Inc.)

Solution

With 20 lines there are 2^{20} memory locations available. Each 64K bank of memory contains 2^{16} locations and requires 16 lines to specify an address. This leaves 4 lines available to specify the bank and allows for a maximum of 16 banks of memory.

8-3.3   The Data Bus

The data bus is used for the transmission of data in a μP system. The **68HC11** contains an 8-bit data bus. Larger μPs use 16- or 32-bit data buses. The size of a μP (8, 16, or 32 bits) is determined by the size of its data bus.

The data bus is *bidirectional.* Unlike the address bus and R/W line, where the signals always originate in the μP and are received by the memories or peripherals, *data can flow in either direction.* The circuitry in the dashed box of Figure 8-7 shows the connections between the μP and the memory for a typical line on the data bus. There are two 3-state gates, each of which can drive signals onto the bus. Both these gates must never be enabled at the same time.

During a read cycle the memory decodes the R/W line and the address bus to determine that it is indeed a read cycle that pertains to that particular memory IC. It then takes control of the data bus by asserting Read Data Enable and placing the data at the specified address on the bus. The μP keeps Write Data Enable off during all read cycles so that it will not attempt to put signals on the bus. The μP can now receive and process the read data.

Read Data Enable is kept off by any write cycle. During a write cycle the μP asserts Write Data Enable and sends the data to be written onto the bus, where it can be written into memory at the address specified on the address bus.

8-3.4   Peripheral ICs

Peripheral ICs are used in conjunction with μPs to perform some specialized system functions, most commonly data transfer between the μP and the external world. These peripherals must also be connected to the address bus; they occupy specific addresses and respond only when those addresses appear on the bus. This is called *memory-mapped I/O.* Thus the peripherals are also connected to the address and data buses and the read-write line. The address must be *decoded** to select the proper IC because only one peripheral IC or memory bank may respond to any particular address. The **68HC11** does not need many peripheral ICs because of its built-in I/O capabilities.

---

* Details of decoding are also presented in J. D. Greenfield, *Practical Digital Design Using ICs* (Englewood Cliffs, NJ: Prentice-Hall, latest ed.).

8-4 The Memory Interface for the 68HC11

As previously stated, the **68HC11** accesses external memory in the expanded mode. The memory accesses are synchronized by the E clock. When the E clock is low addresses can be placed on the bus. Data is only valid when the E clock is high. In expanded mode STRA becomes a signal called **Address Strobe (AS)** and STRB becomes the R/W line. Port B furnishes the upper 8 bits of the address and port C furnishes both the lower 8 bits of the address and acts as the 8-bit bidirectional data bus. Port C can perform these two functions because it is *time-multiplexed*, which means it carries different signals at different times.

8-4.1 The Timing of the 68HC11 Memory Interface

A timing chart for both read and write memory cycles is shown in Figure 8-9. Notice that the E clock is one-fourth the frequency of EXTAL. A single

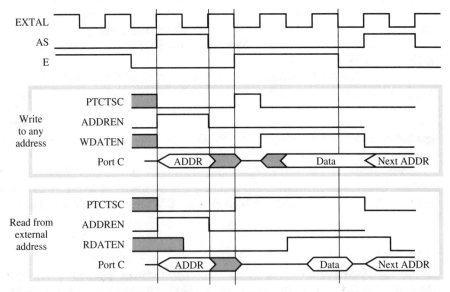

Figure 8-9 The timing for READ and WRITE memory cycles. (*Redrawn Courtesy of Motorola, Inc.*)

memory cycle requires one complete cycle of the E clock, or 0.5 μs. During the time the E clock is low AS goes high. At this time the μP is putting the memory address on ports B and C and using AS to indicate this.

During a write cycle the R/W line goes low and the data to be written is placed on port C when E goes high. The data should be written into memory at the falling edge of E.

During a read cycle the memory should place its data on the lines connected to port C as soon as E goes high. The μP will read the data on the falling edge of E.

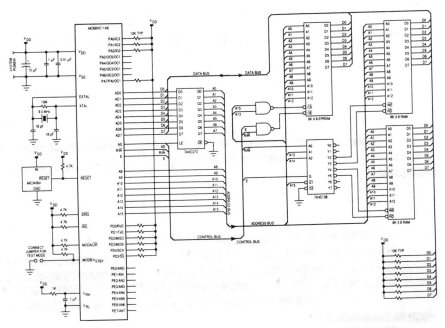

Figure 8-10   The 68HC11 in the expanded mode and its memory connections. *(Redrawn Courtesy of Motorola, Inc.)*

8-4.2   A 68HC11 Memory System

A typical **68HC11** memory system is shown in Figure 8-10. It shows the **68HC11** connected to an 8K-byte EPROM and two 8K-byte RAMs. The lower 8 bits of the address are on port C when E is low but will disappear when E is high. The memories, however, cannot function if their address changes in midcycle. Therefore the lower 8 bits of the memory address must be **latched**. The **74HC373** is an 8-bit latch that latches in data when its LE (Latch Enable) pin is high, and retains it until LE again goes high. Notice that LE is driven by AS from the µP, so the lower 8 bits are latched at the proper time. The output of the latch and port B together form the 16-bit address bus that is sent to the memories. These addresses are firm throughout the entire cycle.

The **74HC138** and the NAND gates are basically decoders. They select which memory should be accessed. Readers familiar with these ICs will notice two things:

1. For all memory accesses E must be high.

2. All the decoding is on the higher address bits. The inputs to the decoders that control the enable signals are all taken from A13, A14, or A15.

8-4.3   Memory Decoding on the Evaluation Board

The memory decoding on the EVB is essentially the same. It is discussed in more detail in Section 12-3.

8-5   RESET

When the **68HC11** is first turned on it must go through a series of steps in order to start up. This condition is known as RESET and initializes the processor.

8-5.1   Causes of RESET

There are four ways in which a **68HC11** can enter RESET.

   1. Power is applied to the 68HC11   There is an internal Power On RESET (POR) circuit that monitors the V_{DD} pin. When power is turned on and it goes from 0 to $+5$ V, this circuit causes the **68HC11** to enter RESET. During the POR the RESET pin will be held low for 4064 clock cycles before processing can begin.

   One problem with POR is that the oscillator may also have been off due to the lack of V_{DD}. When power is turned on the oscillator must also start, but it may not reach full frequency before it generates the 4064 cycles needed to take the **68HC11** out of RESET. Therefore the timing may be off during the first few milliseconds of operation following a POR. In most cases this is not a problem.

   2. The RESET pin is pulled low   This is called an *external RESET*. If the RESET pin is pulled low, the **68HC11** will enter the RESET condition. If, for example, an external voltage detector detects that the supply voltage is too low, it could pull RESET low before the low voltage causes the μP to malfunction.

   The red button on the EVB is connected to RESET. If a program inadvertently goes into an endless loop, or some other disaster occurs, the RESET button may be the only way (short of turning off power) for the user to regain control.

   3. Slow clock   If the oscillator clock slows down for any reason, the **68HC11** can detect this condition internally and cause a RESET. Any frequency higher than 200 kHz will allow the μP to continue to function. The RESET will definitely occur if the clock falls below 10 kHz. Thus the slow clock detector cannot be used to detect small variations in the clock frequency.

   4. COP RESET   The **68HC11** has a Computer Operating Properly (COP) circuit that can be used to monitor the software and cause a RESET if it detects a failure (e.g., an endless loop). The COP circuit can be disabled, and it is in the EVB. The COP is discussed in Section 12-1.

8-5.2   Events During RESET

Two major events occur during RESET:

1. The mode is established by reading in the levels on MODA and MODB.

2. The **68HC11** reads location \$FFFE and \$FFFF to find an *address*. This is the address the µP will jump to to start its program. Thus locations FFFE and FFFF must be ROM; they cannot be allowed to come up randomly.

E X A M P L E   8 - 3

A **68HC11** system is to start at \$C020 after a RESET. What must be done for this to occur?

Solution

To go to this address the µP can be presumed to be operating in the expanded mode. Therefore MODA and MODB must both be 1 at the instant of RESET. In addition, the ROM must have been written with C0 in FFFE and 20 in FFFF for the µP to start properly.

The EVB contains \$E000 in FFFE and FFFF. A RESET causes it to jump to that location, which is the start of the Buffalo monitor.

Other events that occur during RESET, but do not need to concern us immediately are

1. All interrupts are masked off.

2. The timer is reset to 0000.

3. The CONFIG register (see Sec. 12-1.2) is read.

4. Certain bits in some I/O registers may only be written to in the first 64 cycles following a RESET.

8-6   Interrupts and Polling

Any reasonable µP system must have several external devices connected to it and must be able to communicate with these devices. A typical PC, for example, receives inputs from the keyboard and the disk (in read mode), and sends outputs to a video monitor, a printer, and the disk (in write mode). Other devices that might be connected to a µP are modems, A/D converters, and special purpose controllers for industrial processes.

All of these external devices must have some sort of AVAILABLE signal telling the µP that data is available for it or that it is available for use.

Input typically operates as follows:

1. The input device receives a byte of data for the μP. On a keyboard this can be caused by the operator typing a key.

2. The input device informs the μP, via the AVAILABLE signal, that it has data.

3. The μP reads the data. In the process the AVAILABLE flag must go down, so that the external device can raise the flag again when it has another byte of data for the μP.

Output is slightly different:

1. The μP typically builds a memory buffer containing the data that must be sent out.

2. The μP examines the AVAILABLE line on the external device to see if it can send the data out. On a printer the available line is often called BUSY. If the printer is not busy, the μP sends it a block of data. The printer then raises BUSY and prints the data. The μP cannot send more data to the printer while BUSY is high. When the printer is finished printing the data block it will lower BUSY and the μP can send it the next block of data.

8-6.1   Polling

When an input device has data for the μP the μP must be made aware of the situation. One way of doing this is to have the μP *poll* the I/O device or devices to see if data is available. The AVAILABLE signal on a device, for example, can be connected to one of the many input lines on the **68HC11**. The line can be read periodically to determine if the peripheral has data. This method of polling is discussed in Section 9-4.

Polling is the act of periodically querying the AVAILABLE lines on all the input devices to see if data is available. The polling routine is illustrated in Figure 8-11; it proceeds as follows:

1. The μP stops processing its main program and enters the polling routine.

2. The first I/O device is queried for data. If there is data, the *service routine* for data from device #1 is entered. This service routine accepts the data from device #1, stores it, and performs other necessary actions.

3. Each of the other I/O devices is then queried in a similar manner.

4. When all the I/O devices have been queried and serviced, if necessary, the μP resumes its main program.

Polling can effectively assign *priorities* among the input devices by simply querying the highest priority device first.

The disadvantage of polling is that the polling routine must be entered periodically (very often if high-speed data is coming through) and there may

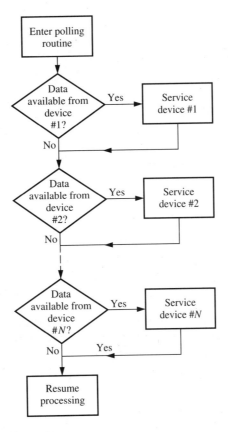

Figure 8-11 Polling in a μP.

be too little time for the μP to execute its main program. If a disk while being read supplies data every 60 μs, then the polling routine must be entered at least once every 60 μs. This leaves little time for the μP to do anything else.

8-6.2 Interrupts

There is a second method whereby a μP can communicate with an I/O device; it can allow the I/O device to **interrupt**. The interrupt procedure is shown in Figure 8-12. When the I/O device has data for the μP or requires some other service it issues an *interrupt request*. An interrupt request can come *at any time* during the execution of a program and generally comes while the μP is executing an instruction.

Figure 8-12 shows the μP executing its main program when the interrupt request occurs. In response to the interrupt request the μP

1. Finishes executing its current instruction.

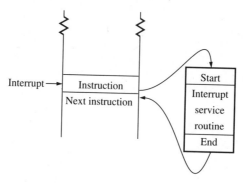

Figure 8-12 Servicing an interrupt.

2. Goes to the **interrupt service routine**, where it performs whatever actions are required by the particular interrupt.

3. Returns to the main program and resumes processing at the next instruction.

The μP does not have to waste time querying the devices in an interrupt system, but the hardware and software are more complex. They are discussed in detail in Section 8-7.

8-6.3 An Analogy

There is an analogy between polling and interrupts and the way an instructor conducts a class. The instructor can poll the class by asking each student if he or she has a question. This is rarely done because it takes too much time.

Classrooms are generally run on the interrupt system. If a student has a question and raises his or her hand, this is an interrupt request. The instructor can disable the interrupt by ignoring the student or acknowledge the interrupt by allowing the student to ask the question. The instructor's answer is analogous to an interrupt service routine. When the question has been answered, the instructor resumes the main program (the lecture).

The instructor can also prioritize interrupt requests. If several students raise their hands, the one selected has top priority.

8-7 Interrupts on the 68HC11

There are many types of interrupts available on a **68HC11**. In this section we discuss the IRQ interrupts and the response of the **68HC11** to interrupts. Most other interrupts are generated by the I/O inside the **68HC11** and will be discussed in succeeding chapters where this I/O is considered.

8-7.1   Preserving the Program Context

As shown in Figure 8-12, an interrupt can occur after *any* instruction. To respond to the interrupt, the µP must execute the interrupt service routine and then return to the main program. The **program context**, which is the contents of *all the registers* in the **68HC11**, must be the *same when the main program resumes as it was at the instant of interrupt*. This is not a trivial problem because the interrupt service routine must also use the **68HC11**'s registers and will therefore modify them.

Motorola solves this problem by *stacking* all the registers as shown in Figure 8-13. This occupies nine locations on the stack. Therefore any pro-

	Stack	
SP	PCL	— SP before interrupt
SP-1	PCH	
SP-2	IYL	
SP-3	IYH	
SP-4	IXL	
SP-5	IXH	
SP-6	ACCA	
SP-7	ACCB	
SP-8	CCR	
SP-9		— SP after interrupt

Figure 8-13   Interrupt stacking order.   *(Redrawn Courtesy of Motorola, Inc.)*

grams that use interrupts must initialize the Stack Pointer (SP) *before* any interrupts are allowed. The **RTI (Return from Interrupt)** instruction *must be the last instruction of every interrupt service routine,* just as an RTS is the last instruction in a subroutine. The RTI simply *unstacks* the six registers (CC, B, A, X, Y, PC) in that order, so the registers and the PC are exactly the same as they were when the interrupt occurred.

The action of the RTI is also shown on p. A-84 of Appendix A. Notice that the RTI affects all the flags because it unstacks the condition code register.

E X A M P L E   8 - 4
_____

The contents of the registers are given in Figure 8-14. The instruction at $C007 is LDAA #$88. If the program is interrupted while this instruction is being executed, what will be put on the stack by the interrupt and where will it be stacked?

```
>RM
P-C007 Y-7982 X-FF00 A-44 B-70 C-C0 S-0054
P-C007 C020
```

Figure 8-14   The register contents for Example 8-4. *(Courtesy of Motorola, Inc.)*

Solution

The instruction will change accumulator A to $88 and the C register to $C4 because the instruction will set the N flag. The contents of the stack are shown in Figure 8-15. LDAA in the immediate mode is a 2-byte instruction, so the following instruction will be executed at $C009. This will occur after the interrupt is serviced. Therefore the contents of the PC that is stacked is the address of the next instruction, #C009. The rest of the registers follow as shown in Figure 8-15. Note that the instruction was completed before the interrupt was allowed, so both the A and CC registers were altered before the service routine was entered.

```
            Location    Contents
Old    0054       09 ⎫
SP → 0053         C0 ⎬   PC Address of next instruction.
       0052       82 ⎫
       0051       79 ⎬   Y
       0050       00 ⎫
       004F       FF ⎬   X
       004E       88     A
       004D       70     B
New    004C       C4     CC
SP → 004B
```

Figure 8-15   The contents of the stack after an interrupt. See Example 8-4.

8-7.2   Interrupt Vectors

It may have occurred to the reader that we have not yet specified where the interrupt routine should start, or how the program gets there in response to an interrupt. The **68HC11** reserves a set of *interrupt* **vectors** in the upper end of ROM (locations FFC0–FFFF). Table 8-3 shows the locations of the vectors for each of the various interrupts on the **68HC11**. Each interrupt vector requires 16 bits (two locations).

We have already encountered the interrupt vector for a RESET, which Table 8-3 shows is in locations FFFE and FFFF. In this section we con-

Table 8-3 Interrupt Vector Assignments

Vector Address	Interrupt Source	CC Register Mask	Local Mask
FFC0, C1	Reserved	—	—
⋮	⋮		
FFD4, D5	Reserved	—	—
FFD6, D7	SCI serial system	I bit	See Table 9-3
FFD8, D9	SPI serial transfer complete	I bit	SPIE
FFDA, DB	Pulse accumulator input edge	I bit	PAII
FFDC, DD	Pulse accumulator overflow	I bit	PAOVI
FFDE, DF	Timer overflow	I bit	TOI
FFE0, E1	Timer output compare 5	I bit	OC5I
FFE2, E3	Timer output compare 4	I bit	OC4I
FFE4, E5	Timer output compare 3	I bit	OC3I
FFE6, E7	Timer output compare 2	I bit	OC2I
FFE8, E9	Timer output compare 1	I bit	OC1I
FFEA, EB	Timer input capture 3	I bit	OC3I
FFEC, ED	Timer input capture 2	I bit	OC2I
FFEE, EF	Timer input capture 1	I bit	OC1I
FFF0, F1	Real time interrupt	I bit	RTII
FFF2, F3	IRQ (external pin or parallel I/O)	I bit	See Table 9-4
FFF4, F5	XIRQ pin (pseudo non-maskable interrupt)	X bit	None
FFF6, F7	SWI	None	None
FFF8, F9	Illegal Op code trap	None	None
FFFA, FB	COP failure (reset)	None	NOCOP
FFFC, FD	COP clock monitor fail (reset)	None	CME
FFFE, FF	RESET	None	None

centrate on the IRQ interrupt, whose vectors are in FFF2 and FFF3. If, for example, they contain C030, the service routine must start at C030.

8-7.3 The Evaluation Board Vectors

The interrupt vectors are *inflexible* because they are *written in ROM and cannot be changed.* In many systems, such as the EVB, users would like to be able to write their service routines at any convenient place in memory.

The EVB and the Buffalo monitor overcome this problem by using the interrupt vector jump table shown in Table 8-4. There is an entry in this

Table 8-4 Interrupt Vector Jump Table for the Evaluation Board

Interrupt Vector	Field
Serial communications interface (SCI)	$00C4–$00C6
Serial peripheral interface (SPI)	$00C7–$00C9
Pulse accumulator input edge	$00CA–$00CC
Pulse accumulator overflow	$00CD–$00CF
Time overflow	$00D0–$00D2
Timer output compare 5	$00D3–$00D5
Timer output compare 4	$00D6–$00D8
Timer output compare 3	$00D9–$00DB
Timer output compare 2	$00DC–$00DE
Timer output compare 1	$00DF–$00E1
Timer input capture 3	$00E2–$00E4
Timer input capture 2	$00E5–$00E7
Timer input capture 1	$00E8–$00EA
Real-time interrupt	$00EB–$00ED
IRQ	$00EE–$00F0
XIRQ	$00F1–$00F3
Software interrupt (SWI)	$00F4–$00F6
Illegal Op code	$00F7–$00F9
Computer operating properly (COP)	$00FA–$00FC
Clock monitor	$00FD–$00FF

table corresponding to each interrupt source except RESET, which always causes the **68HC11** to restart the monitor. Note that the jump table is in lower memory, which is in the internal RAM area, and *can be changed*. This area is changed every time the monitor restarts, so the user program may have to reinitialize it.

Each entry in the jump table takes 3 bytes. The first byte should be the Op code for a JUMP (7E) and the next 2 bytes are the address of the start of the service routine. When an interrupt occurs, the **68HC11** first goes to the proper interrupt vector in high memory. It is then vectored to the jump table, where it finds a JUMP to the start of the service routine.

E X A M P L E   8 - 5

What address is written in locations FFF2 and FFF3 of the Buffalo monitor?

Solution

FFF2 and FFF3 are the locations of the IRQ interrupt vector. They must send the program to the IRQ entry in the jump table, which is at $00EE. Therefore 00 must be written into FFF2 and EE must be written into F3. Again, these addresses are part of the Buffalo monitor's ROM and cannot be changed.

E X A M P L E   8 - 6

The service routine for an IRQ interrupt is to start at $C234, a convenient location we have selected in user RAM. What must be done to make this happen, and what are the steps along the way?

Solution

During initialization the code 7E C2 34 (JUMP to C234) must be written into locations 00EE–00F0. When an IRQ interrupts occurs, the **68HC11** will go to FFF2 and FFF3 where it finds 00EE written. So it will load the PC with 00EE and the program will start there. But the instruction at 00EE is a JUMP to C234, and it will start executing the service routine there.

8-7.4   The Interrupt Mask

The I bit in the condition code register (CCR), which has not been previously discussed, is the so-called **interrupt mask**. If it is set, the IRQ interrupt and all interrupts above IRQ in Table 8-3 will be disabled or ignored. These interrupts will not be recognized until the I bit is clear.

The I bit can be *set* in one of three ways:

1. It is set by any RESET.

2. It is set whenever an interrupt routine is started.

3. It is set by an SEI (Set Interrupt mask) instruction.

It is cleared by a CLI (Clear Interrupt mask) instruction. A TAP instruction can also set or clear the I bit.

It is necessary to have RESET automatically set the I flag. Any pending interrupts will not affect the program on start-up until the user decides to enable them by issuing a CLI instruction. It is also necessary to have the I bit automatically set during an interrupt so that the *interrupt routine is not itself interrupted by the same interrupt signal*. During an interrupt service routine the routine must do something to remove the source of the interrupt. Otherwise when it returns it will simply be interrupted by the same interrupt.

E X A M P L E 8 - 7

An interrupt is to be recognized. What is the state of the I bit

a. Before the interrupt?

b. During the interrupt?

c. After the interrupt?

Solution

a. Before the interrupt the I bit must be 0 or the interrupt will not be recognized.

b. During the interrupt the I bit is automatically set to 1 by the **68HC11** to lock out further interrupts.

c. After the interrupt the I bit is again 0. This is because the RTI restored the registers, including the CCR, at the end of the interrupt. Since the I bit in the CCR was a 0 going in, it will be restored as a 0. Thus interrupts are enabled by the RTI as soon as the **68HC11** exits the interrupt routine.

8-7.5 Initiating an IRQ Interrupt

The **IRQ** is an interrupt generated by an external device or peripheral IC. The interrupt is initiated by pulling the IRQ pin on the **68HC11** low. The IRQ pin is **level-sensitive**; this means that the interrupt request will continue to be made as long as the pin is low. This contrasts with **edge-sensitive interrupts** that create an interrupt request by placing a rising or falling edge on an interrupt line. The difference is shown in Figure 8-16, where a negative edge is assumed to cause the request in Figure 8-16b. For edge-sensitive interrupts, it is the edge that causes the request. The level on the line after the edge is irrelevant.

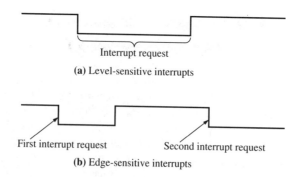

Interrupt request

(a) Level-sensitive interrupts

First interrupt request   Second interrupt request

(b) Edge-sensitive interrupts

Figure 8-16 Two types of interrupt requests.

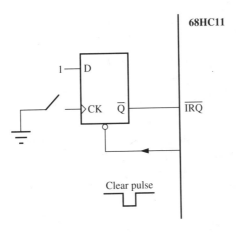

Figure 8-17 Initiating an IRQ interrupt using a switch.

Figure 8-17 shows how an IRQ interrupt can be initiated by using a switch and a flip-flop (FF).* When the switch is flipped, the $\overline{Q}$ output of the D-type FF goes low and pulls the $\overline{IRQ}$ line on the **68HC11** low, making an interrupt request. The negative-going clear pulse must be generated by the interrupt service routine to clear the FF and remove the interrupt. The **68HC11** has several methods of generating such a clear pulse. They are discussed in Section 9-2.1.

Of course, the switch input to the FF in Figure 8-17 can be replaced by any peripheral device that can provide an edge whenever it must request an interrupt. The circuit effectively converts an edge to a level-sensitive interrupt.

The timing for a typical interrupt is shown in Figure 8-18. It proceeds as follows:

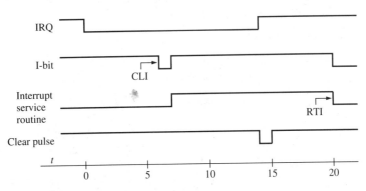

Figure 8-18 The timing of an interrupt request.

* For reliable operation the switch should be *debounced*.

1. The interrupt request is made by lowering IRQ at $t = 0$. The **68HC11** does not respond immediately because its I bit is high.

2. At $t = 6$ the main program finds a CLI instruction and clears the I bit. The interrupt is now allowed to come in.

3. At $t = 7$ the **68HC11** goes to the service routine. This causes the I bit to go high to lock out any further interrupts.

4. At $t = 14$ the clear pulse is generated by the interrupt service routine. This removes the interrupt request by making IRQ go high again. Note that the interrupt request was active during the first few instructions of the service routine (from $t = 7$ to $t = 14$), but it was ignored because the I bit was high.

5. At $t = 19$ the interrupt ends with the RTI instruction. This instruction restores the CCR that causes the I bit to go back to 0. At the end of the service routine interrupts are enabled but there is no interrupt request; the service routine has cleared it.

   The circuit of Figure 8-17 can be used to generate an IRQ interrupt on the EVB. Perhaps the best way to monitor a program with interrupts is to

1. Write a main program that loops indefinitely.

2. Write an interrupt service routine.

3. Place a breakpoint in the service routine.

   The program will then loop until the interrupt occurs. Then the breakpoint in the service routine will be reached and the user can examine the progress of the program.

8-7.6   A Program Using Interrupts

The following problem has been used as a laboratory exercise to illustrate the use of interrupts. Its requirements are

1. Write a main program to create and update a real-time clock.

2. Every time a switch is thrown, the program should be interrupted and the time of the clock should be stored in a list in a memory buffer.

   A flowchart for the main program is shown in Figure 8-19. During the initialization it clears a memory location called SECONDS, which runs from $00 to $59, and a location called MINUTES. It then goes to a 1-s time delay, after which it increments SECONDS, and, if SECONDS equals $60, it resets SECONDS and increments MINUTES.

   The interrupt service routine is not shown*. The interrupt can be initiated by a switch as shown in Figure 8-17. The service routine simply reads

---

* An entire program with its interrupt service routine is shown in Figure 9-7.

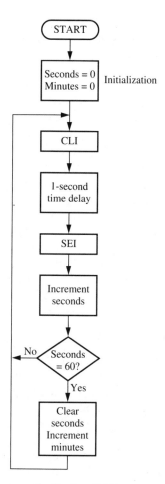

Figure 8-19   The main program for Section 8-7.6.

SECONDS and MINUTES into memory, resets the FF, and exits with an RTI.

Notice the use of the SEI and CLI instructions in the program. They prevent interrupts when SECONDS and MINUTES are being incremented but allow them during the 1-s time delay. If interrupts were allowed at any time during the program, an interrupt could occur *after* SECONDS had been incremented but *before* MINUTES could be incremented. The interrupt service routine would then store the old value of MINUTES but the new value of SECONDS. Disallowing interrupts during this time prevents the mismatch.

8-7.7   Multiple Interrupts and Nested Interrupts

Larger computer systems allow several devices to interrupt. Each device that can interrupt must be capable of pulling IRQ down when it requests an interrupt. Possibly the best way to tie all the external devices to the same IRQ pin is to use open-collector ICs between the devices and the IRQ pin.* Now, however, when the **68HC11** receives an interrupt, it doesn't know which device interrupted and there must be a different service routine for each device that can interrupt. Consequently, the first part of the interrupt routine must be to *poll* all the devices that can interrupt to find out which one caused the interrupt. It can then branch to the service routine for that device. The polling routine at the start of an interrupt service routine can be similar to the noninterrupt polling whose flowchart was shown in Figure 8-11. Each service routine must end with an RTI.

When several devices can interrupt, a priority problem may occur. Suppose that there is both a slow device (e.g., a keyboard, which can afford to wait for its interrupt to be serviced) and a fast device (e.g., a disk drive, which requires attention quickly) that can interrupt. Let us further suppose that the disk interrupt arrives just ahead of the keyboard interrupt. There is no problem. The interrupt service routine for the disk will set the I bit and lock out the keyboard. When the disk service routine finishes, the keyboard routine can take over.

It is possible, however, for the keyboard interrupt to arrive slightly before the disk interrupt. Now the disk interrupt will have to wait for the keyboard interrupt to complete. This might take too long.

This priority problem can be solved by allowing *nested interrupts*, where an interrupt service routine can itself be interrupted. A nested interrupt is similar to a nested subroutine. The first interrupt stacks the six registers pertaining to the main program and enters its service routine. The next interrupt interrupts the service routine. It then stacks the six registers pertaining to the service routine and performs its own service routine. When it finishes, the first service routine resumes and the main program is reentered when the service routine finishes. Note that the stack must be at least 18 bytes long for nested interrupts to accommodate both sets of registers.

But how can an interrupt service routine be interrupted if the I bit is set? It can't unless there is a CLI instruction in the service routine to clear the I bit. Getting back to the disk and keyboard situation, a CLI can be placed in the service routine for the low priority device. Assume the keyboard interrupts first. Then the following events will occur:

1. The **68HC11** will determine, by polling, that the keyboard has interrupted. It will then go to the interrupt routine for the keyboard.

2. The first step in the keyboard interrupt routine must be to clear the keyboard interrupt. (See Sec. 8-6.)

3. The second step in the keyboard routine must be to issue a CLI instruction.

4. The keyboard routine can now continue, but it can be interrupted by the disk. Therefore the disk will not have to wait for service.

* Open-collector ICs are discussed in J. D. Greenfield, *Practical Digital Design Using ICs* (Englewood Cliffs, NJ: Prentice-Hall, latest ed.)

E X A M P L E 8 - 8

What would happen if the keyboard routine forgot to clear the keyboard interrupt before issuing the CLI?

Solution

When the CLI is issued the keyboard interrupt request would still be active and would now interrupt the keyboard service routine. An infinite nesting of interrupts would occur, each interrupt stacking the registers until the stack gobbled up the main program and everything else in sight. Please don't let this happen.

8-8 Other 68HC11 Interrupts

Besides the IRQ, several other sources of interrupts and interruptlike instructions are available in the **68HC11**. They are discussed in this section.

8-8.1 XIRQ

The **6800** microprocessor, the predecessor of the **68HC11**, had an **NMI (Non-Maskable Interrupt)** pin on it. When this pin was pulled low an NMI occurred and the **6800** vectored to the NMI location to start the interrupt routine. Because this interrupt cannot be masked, the NMI was reserved for catastrophic events, such as an impending power failure or the loss of power in a peripheral.

The NMI had one serious drawback; if an NMI occurs before the SP can be initialized, the system may never recover. The **XIRQ** interrupt in the **68HC11** is essentially an NMI that overcomes this problem. An XIRQ interrupt is initiated by a low level applied to the XIRQ pin on the **68HC11**. It can be masked, however, by the X bit (bit 6) in the CCR. The X bit is set by RESET and will disable XIRQ interrupts. It is also set when an XIRQ interrupt is in progress. It can be cleared by a TAP instruction, or by an RTI. *Once the XIRQ bit is cleared, however, it can never again be set*. This is the difference between the IRQ and XIRQ and permits the XIRQ to function as an NMI. Once the X bit has been cleared, any low on the XIRQ pin will initiate an XIRQ interrupt, and it cannot be masked off.

E X A M P L E 8 - 9

A system is to use an XIRQ interrupt to warn of a catastrophic event. What must be done to initialize the system?

Solution

The following steps must be taken:

1. The *sensor*, the signal that senses impending doom, must be connected to the XIRQ pin.

2. The XIRQ routine that responds to the signal must be written into memory and given a starting address.

3. The XIRQ vector (see Table 8-3) at FFF4 and FFF5 must point to the starting address. If the EVB is used, a JUMP to the starting address can be placed in locations $00F1–$00F3.

4. The following sequence of instructions should be written into the initialization routine:

LDS	#STACK	Position the stack properly.
TPA		Load the flags into A.
ANDA	#BF	Zero out bit 6.
TAP		Restore the flags with the X bit cleared.

Notice that the stack was properly positioned *before* the X bit was cleared, so there could be no XIRQ interrupts until the stack was set up.

8-8.2   The STOP Instruction*

A STOP instruction can cause the **68HC11** to enter the *stopped* state, where all clocks are disabled and the μP consumes almost no power. The S (stop) bit in the CCR guards against inadvertent STOP instructions; if it is 1, the STOP instruction is ignored and treated as an NOP. The S bit can be changed by a TAP. During reset the STOP bit is set. Thus the μP cannot be stopped unless the user deliberately clears the STOP bit first.

After a STOP, the **68HC11** will resume operations in response to an unmasked IRQ, a RESET, or an XIRQ. If the X bit in the CCR is 1, an XIRQ will cause the program to resume by executing the next instruction. If X is 0, An XIRQ will stack the registers and go to the XIRQ service routine. Due to an anomaly in the **68HC11**, it is wise to precede all STOP instructions with an NOP.

8-8.3   The WAI Instruction

The WAI is a Wait-for-Interrupt instruction. When it occurs all six registers (9 bytes) are stacked as they would be in response to an interrupt. The **68HC11** then stops processing and enters a WAIT state, where it does nothing. An unmasked interrupt will cause the **68HC11** to resume. It will fetch the interrupt vector and go to the service routine.

The WAI is only useful if a very rapid response to an interrupt is required. It saves time because it stacks the registers before the interrupt occurs.

---

* This section and Sec. 8-8.3 may be omitted on first reading.

8-8.4 The SWI Instruction

The **SWI** (Software Interrupt) is an instruction that causes the **68HC11** to act as though an interrupt has occurred. The μP stacks the six registers and goes to the SWI vector FFF6 and FFF7. On the EVB these point to locations $00F4–$00F6 (see Table 8-3). A RESET loads these locations with $E53A. While a program is running it has control of the EVB. If an SWI is in the program it will cause a JUMP to $E53A and allow the monitor to take control. On the EVB the SWI is the preferred way to terminate programs and is usually the last instruction in a program.

The SWI writes the registers into the stack at the user's SP and returns control to the monitor. The user can then examine memory or the registers or take other action.

Stacking the registers at the user's SP overwrites the old contents of the stack. Sometimes, especially in a student's laboratory, the user may wish to examine the contents of the stack. If so, we suggest that the user write an LDS immediate just before the SWI to move the SP. Then the old stack will not be overwritten.

For example, suppose the user sets the SP at $C3FF and the program does a JSR, which writes the return address at $C3FE and $C3FF. If the program is terminated with the sequence

        LDS $C500

        SWI

the SWI will stack its registers at locations $C4F8–$C500 and the data in $C3FE and $C3FF will not be overwritten and can be examined.

Breakpoints in the EVB also use the SWI. *An SWI (Op code 3F) is used to replace the instruction at the breakpoint address,* so when the **68HC11** reaches that address it will execute the breakpoint, which will allow the user to take control of the μP. The SWI also replaces the SWI with the original instruction.

There is a problem that commonly occurs in the laboratory using breakpoints. Suppose, for example, that the program starts at $C050 and a breakpoint is inserted at $C080. Further suppose that due to a program bug (an endless loop, perhaps), the program never reaches $C080 and the **68HC11** must be RESET. Now the SWI has not had a chance to replace the breakpoint with the original instruction. If the program is examined, the user will find 3F in $C080 and the program will not run. The original Op code at $C080 must be rewritten for the program to operate properly.

Summary

This chapter started with a description of the various parts of the **68HC11**, including its internal memory and I/O registers. The interface between the **68HC11** and its external memory was discussed next, and then the actions of a RESET were considered.

The concept of an interrupt for a general computer was discussed in Section 8-6. Its application to the **68HC11** was considered in Section 8-7 and a specific example of an interrupt routine that could be performed in the laboratory was presented. Finally, the XIRQ and SWI and their uses were explained.

Glossary

AS—Address Strobe   This signal goes high when the addresses are valid in the expanded mode of the **68HC11**.

Edge-sensitive interrupt   An interrupt request that is initiated by an edge (a transition) on the interrupt request line.

EEPROM   Electrically Erasable Programmable Read Only Memory.

Interrupt   A request for attention initiated by an I/O device.

Interrupt mask   A bit in the CCR that allows or prevents interrupts from being recognized and acted upon.

Interrupt service routine   A software routine that the μP executes in response to an interrupt.

IRQ   An externally generated interrupt. There is an IRQ pin on the **68HC11**. Bringing this pin low causes an IRQ interrupt.

Latch   A set of flip-flops that retains information until it is needed.

Level-sensitive interrupt   An interrupt request made by bringing the IRQ pin to a low level.

Mode   A way of operating the **68HC11**.

NMI—Non-Maskable Interrupt   An interrupt request that cannot be masked, and will be recognized. The XIRQ performs this function on the **68HC11**.

Peripheral IC   An IC used in conjunction with the **68HC11** to perform specific functions.

Polling   The act of successively inquiring whether each peripheral IC has information for the μP or is available for use.

Program context   The contents of the registers at any point in the program.

RTI—Return from Interrupt   This is the last instruction of a service routine and causes a return to the main program.

STOP instruction   An instruction that causes the **68HC11** to stop processing until an interrupt is received.

SWI—Software Interrupt   An instruction that causes the **68HC11** to behave as though it has been interrupted.

Vector   Two memory locations that hold the starting address of a service routine.

XIRQ   The nonmaskable interrupt on the **68HC11**.

Problems

Section 8-1.1

8-1 What are the various functions of pin 14 of the PLCC (Plastic Leadless Chip Carrier) version of the **68HC11**?

Section 8-6

8-2 When a μP has data for a printer it sends a STROBE signal. Draw a timing chart showing the relationship between STROBE and BUSY.

Section 8-7.1

8-3 In the following table the initial conditions for the registers are given on the first line. C040 contains $57.
  a. Fill in the table with the contents of the registers after each instruction.
  b. During the STAA an interrupt is received. When the program enters the interrupt service routine, list all the locations that have been stacked and what is in each of them.

Address	Instruction		SP	X	Y	A	B	CC
C020	LDS	#C01E		C666	C222	55	33	C0
C023	LDX	#C155						
C026	SEC							
C027	LDAA	#$AB						
C029	LDAB	$C040						
C02B	JSR	$C04F						
C04F	PSHA							
C050	ROLA							
C051	ABA							
C052	DAA							
C053	INX							
C054	STAA	$C041						

8-4 The partial contents of memory are shown in Figure P8-4. If the SP is at 2A26 and an RTI is executed, what are the contents of each register in the **68HC11** after the RTI is executed?

Location	Contents
2A30	88
2A2F	77
2A2E	66
2A2D	55
2A2C	EF
2A2B	CD
2A2A	AB
2A29	34
2A28	12
2A27	CA
2A26	CE
2A25	CD
2A24	CC

Figure P8-4

Section 8-7.3

8-5 If the response to a real-time interrupt is to start at D400 on the EVB, what must the contents of FFF0, FFF1, and 00EB–00ED be?

8-6 Explain what the following code does. Why would it be used?

```
LDA    #$7E
STA    $EE
LDD    #CDEF
STD    $EF
```

Section 8-7.4

8-7 One way to keep the I bit set after an interrupt is to write the following code into the service routine:

```
PULA
ORA    #$10
PSHA
```

Explain what this code does.

Section 8-7.6

8-8 Could the number $60 ever appear in the memory list for SECONDS generated by the program and interrupt the routine of Figure 8-19? Could it appear if the SEI were omitted?

8-9 Write the code for the main program and the interrupt service routine of Figure 8-19.

Section 8-7.7

8-10 Assume an interrupt arrives when the SP is pointing to location JACK. It is then interrupted by another interrupt. What is stored at JACK-3? At JACK-14?

8-11 Device A and device B can both interrupt. Suppose device A interrupts and during the interrupt service routine, device B interrupts. Draw a timing chart similar to Figure 8-18 for this situation. Which service routine must have a CLI in it?

Section 8-8.1

8-12 Write an instruction sequence to clear S, X, and I in the CCR. Write an instruction sequence to set them.

Section 8-8.4

8-13 The routine for an SWI is to start in D444. How can we cause this to happen? If the routine is to print out the message "PROGRAM TER-MINATED BY SWI" and then return control to the monitor, write the routine.

8-14 Consider the following program:

```
C200    LDS    #C400
          ⋮
C221    JSR    SOMEPLACE
          ⋮
C240    LDS    #D000
C243    SWI
```

a. Where is the original stack and what is in it at the end of the program?

b. Where are the registers stored at the end of the program? What numbers are in the two highest locations of the registers?

9

Basic Input/Output—Ports B and C

INTRODUCTION

The five I/O (Input/Output) ports on the **68HC11** were shown in Figure 8-2 and briefly described in Section 8-8.1. To facilitate understanding them, we examine each I/O port in turn. This chapter concentrates on basic parallel I/O, which involves the transfer of data, usually in bytes, between the **68HC11** and the external world. All the ports can support some parallel I/O, but it is done predominantly through ports B and C. Their action is discussed in this chapter.

INSTRUCTIONAL OBJECTIVES

After reading the chapter, the student should be able to

- Send data out via port B.
- Send data out or read data in via port C and port CL.
- List each bit of the PIOC and explain its function.
- Use STRA and STRB to control and synchronize data transfers.
- Use interrupts in response to a signal on STRA.
- Communicate with peripheral devices using handshaking on both input and output.

SELF-EVALUATION QUESTIONS

Watch for the answers to the following questions as you read the chapter. They should help you to understand the material presented:

1. What is the function of a data direction register?
2. What is the difference between port C and port CL?
3. What are the functions of STRA and STRB?
4. When writing a byte into the PIOC, why is the data in bit 7 irrelevant?
5. What is the advantage of having STAF in the bit 7 position?
6. What is the difference between the simple strobed mode and handshaking?

9-1 Parallel Input/Output

Parallel I/O means the transmission of several bits of data at the same time between the **68HC11** and a sending or receiving device. Thus the external device must be connected to the **68HC11** by a cable containing several wires, one for each bit. This is opposed to serial I/O where data is essentially transmitted, a bit at a time, over a single wire. Serial I/O is discussed in Chapter 11. In parallel I/O data is most commonly transmitted a byte (8 bits) at a time. Port B and port C on the **68HC11** have been specifically designed to transmit this data.

9-1.1 The Pin Connections to Ports B and C

The hardware connections to ports B and C are shown in Figure 9-1. In the

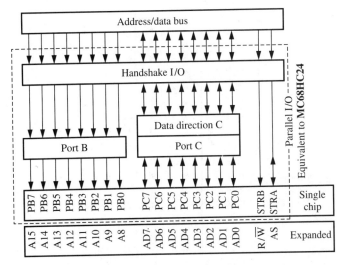

Figure 9-1 The I/O connections for ports B and C. *(Redrawn Courtesy of Motorola, Inc.)*

expanded mode these ports are used for memory interfacing, as explained in Section 8-4. To review briefly:

1. Port B contains the upper 8 bits of the address bus.

2. Port C contains the lower 8 bits of the address bus and also functions as the data bus because it is *time-multiplexed*.

3. STRB becomes the read-write (R/W) line and STRA becomes the Address Strobe (AS), which tells the external ICs when port C is carrying addresses and when it is carrying data.

Because all of these 18 lines are needed for the memory interface, none of them can be used for I/O in this mode.

In the *single-chip mode* all 18 lines are available for I/O, as shown in Figure 9-1. Port B is *output only* on lines PB0–PB7, port C can be *either input or output* on lines PC0–PC7, and STRA and STRB are available to synchronize and control the parallel I/O.

Although the **68HC11** on the EVB operates in the expanded mode, the **68HC24** PRU (Port Replacement Unit) IC *simulates* the action of ports B and C and allows them to be used for I/O as though the **68HC11** were in the single-chip mode. The I/O connector for the EVB is shown in Figure 9-2 and all the pins are clearly labeled.

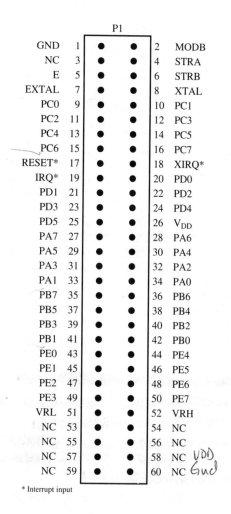

		P1		
GND	1	● ●	2	MODB
NC	3	● ●	4	STRA
E	5	● ●	6	STRB
EXTAL	7	● ●	8	XTAL
PC0	9	● ●	10	PC1
PC2	11	● ●	12	PC3
PC4	13	● ●	14	PC5
PC6	15	● ●	16	PC7
RESET*	17	● ●	18	XIRQ*
IRQ*	19	● ●	20	PD0
PD1	21	● ●	22	PD2
PD3	23	● ●	24	PD4
PD5	25	● ●	26	V_{DD}
PA7	27	● ●	28	PA6
PA5	29	● ●	30	PA4
PA3	31	● ●	32	PA2
PA1	33	● ●	34	PA0
PB7	35	● ●	36	PB6
PB5	37	● ●	38	PB4
PB3	39	● ●	40	PB2
PB1	41	● ●	42	PB0
PE0	43	● ●	44	PE4
PE1	45	● ●	46	PE5
PE2	47	● ●	48	PE6
PE3	49	● ●	50	PE7
VRL	51	● ●	52	VRH
NC	53	● ●	54	NC
NC	55	● ●	56	NC
NC	57	● ●	58	NC
NC	59	● ●	60	NC

* Interrupt input

Figure 9-2 The I/O connections on the EVB. *(Redrawn Courtesy of Motorola, Inc.)*

9-1.2   The Input/Output Control Registers

All the I/O in the **68HC11** is controlled by a set of registers that reside in locations $1000–$103D in most **68HC11**s. The entire register set is shown in Figure 9-3. The **68HC11** has many I/O capabilities, but only a few registers control each capability. Thus the registers will be discussed, a few at a time, as each function of the **68HC11** is explained so that the student will not be overwhelmed.

There are three types of registers in the set:

1. Data registers   These hold the data going in or coming out of the **68HC11**.

2. Data direction registers   These registers control the *direction* (in or out) of the data of some of the registers.

3. Control registers   These registers control the mode of operation of the data registers and otherwise regulate the operation of the **68HC11**.

There are many control registers, but most of them pertain to specific ports. To facilitate understanding, they are discussed in conjunction with the ports they control.

The registers associated with ports B and C are shown in Figure 9-4. They can be read or written to just as any other memory location. The PIOC (Port I/O Control) register at $1002 is a register that controls data transfers involving ports B and C. Every bit in it has a specific purpose. The DDRC (Data Direction Register C) at $1007 is a direction register for port C. The other three registers are data registers. The functions of these registers will be covered in the following paragraphs in this chapter.

9-2   Port B

Port B is an output only port. It drives signals onto the eight PB pins (pins 16–9) of the **68HC11**. Figure 9-5 shows the output connection to a port B pin. It consists simply of a CMOS (Complementary Metal-Oxide Silicon) inverter and a protective transistor.*

Data to be sent out via port B is *written* into the port B register at $1004. A read of this register will only read back the last data that was sent to it.

E X A M P L E   9 - 1

The number FD is to be sent out on the eight lines of port B. Write a program segment to do this.

Solution

The code is simply

        LDAA   #FD   Put data in accumulator A.
        STAA $1004   Write it to port B.

The data will now appear both in location $1004 and on the PB0–PB7 lines.

---

* A more complete description of the circuitry controlling the output pins is given in Sec. 7-3 of the *MC68HC11 Reference Manual* (Phoenix, AZ: Motorola Inc., 1990).

Address	Bit 7	Bit 6	Bit 5	Bit 4	Bit 3	Bit 2	Bit 1	Bit 0	Register	Description
$1000	Bit 7	—	—	—	—	—	—	Bit 0	PORTA	I/O Port A
$1001									Reserved	
$1002	STAF	STAI	CWOM	HNDS	OIN	PLS	EGA	INVB	PIOC	Parallel I/O Control Register
$1003	Bit 7	—	—	—	—	—	—	Bit 0	PORTC	I/O Port C
$1004	Bit 7	—	—	—	—	—	—	Bit 0	PORTB	Output Port B
$1005	Bit 7	—	—	—	—	—	—	Bit 0	PORTCL	Alternate Latched Port C
$1006									Reserved	
$1007	Bit 7	—	—	—	—	—	—	Bit 0	DDRC	Data Direction for Port C
$1008			Bit 5	—	—	—	—	Bit 0	PORTD	I/O Port D
$1009			Bit 5	—	—	—	—	Bit 0	DDRD	Data Direction for Port D
$100A	Bit 7	—	—	—	—	—	—	Bit 0	PORTE	Input Port E
$100B	FOC1	FOC2	FOC3	FOC4	FOC5				CFORC	Compare Force Register
$100C	OC1M7	OC1M6	OC1M5	OC1M4	OC1M3				OC1M	OC1 Action Mask Register
$100D	OC1D7	OC1D6	OC1D5	OC1D4	OC1D3				OC1D	OC1 Action Data Register
$100E	Bit 15	—	—	—	—	—	—	Bit 8	TCNT	Timer Counter Register
$100F	Bit 7	—	—	—	—	—	—	Bit 0		
$1010	Bit 15	—	—	—	—	—	—	Bit 8	TIC1	Input Capture 1 Register
$1011	Bit 7	—	—	—	—	—	—	Bit 0		
$1012	Bit 15	—	—	—	—	—	—	Bit 8	TIC2	Input Capture 2 Register
$1013	Bit 7	—	—	—	—	—	—	Bit 0		
$1014	Bit 15	—	—	—	—	—	—	Bit 8	TIC3	Input Capture 3 Register
$1015	Bit 7	—	—	—	—	—	—	Bit 0		
$1016	Bit 15	—	—	—	—	—	—	Bit 8	TOC1	Output Compare 1 Register
$1017	Bit 7	—	—	—	—	—	—	Bit 0		
$1018	Bit 15	—	—	—	—	—	—	Bit 8	TOC2	Output Compare 2 Register
$1019	Bit 7	—	—	—	—	—	—	Bit 0		
$101A	Bit 15	—	—	—	—	—	—	Bit 8	TOC3	Output Compare 3 Register
$101B	Bit 7	—	—	—	—	—	—	Bit 0		
$101C	Bit 15	—	—	—	—	—	—	Bit 8	TOC4	Output Compare 4 Register
$101D	Bit 7	—	—	—	—	—	—	Bit 0		
$101E	Bit 15	—	—	—	—	—	—	Bit 8	TOC5	Output Compare 5 Register
$101F	Bit 7	—	—	—	—	—	—	Bit 0		

Figure 9-3 The I/O data and control registers in the 68HC11. (*Reprinted Courtesy of Motorola, Inc.*)

	Bit 7	Bit 6	Bit 5	Bit 4	Bit 3	Bit 2	Bit 1	Bit 0		
$1020	OM2	OL2	OM3	OL3	OM4	OL4	OM5	OL5	TCTL1	Timer Control Register 1
$1021			EDG1B	EDG1A	EDG2B	EDG2A	EDG3B	EDG3A	TCTL2	Timer Control Register 2
$1022	OC1I	OC2I	OC3I	OC4I	OC5I	IC1I	IC2I	IC3I	TMSK1	Timer Interrupt Mask Register 1
$1023	OC1F	OC2F	OC3F	OC4F	OC5F	IC1F	IC2F	IC3F	TFLG1	Timer Interrupt Flag Register 1
$1024	TOI	RTII	PAOVI	PAII			PR1	PR0	TMSK2	Timer Interrupt Mask Register 2
$1025	TOF	RTIF	PAOVF	PAIF					TFLG2	Timer Interrupt Flag Register 2
$1026	DDRA7	PAEN	PAMOD	PEDGE			RTR1	RTR0	PACTL	Pulse Accumulator Control Register
$1027	Bit 7	—	—	—	—	—	—	Bit 0	PACNT	Pulse Accumulator Count Register
$1028	SPIE	SPE	DWOM	MSTR	CPOL	CPHA	SPR1	SPR0	SPCR	SPI Control Register
$1029	SPIF	WCOL		MODF					SPSR	SPI Status Register
$102A	Bit 7	—	—	—	—	—	—	Bit 0	SPDR	SPI Data Register
$102B	TCLR		SCP1	SCP0	RCKB	SCR2	SCR1	SCR0	BAUD	SCI Baud Rate Control
$102C	R8	T8		M	WAKE				SCCR1	SCI Control Register 1
$102D	TIE	TCIE	RIE	ILIE	TE	RE	RWU	SBK	SCCR2	SCI Control Register 2
$102E	TDRE	TC	RDRF	IDLE	OR	NF	FE		SCSR	SCI Status Register
$102F	Bit 7	—	—	—	—	—	—	Bit 0	SCDR	SCI Data (Read RDR, Write TDR)
$1030	CCF		SCAN	MULT	CD	CC	CB	CA	ADCTL	A/D Control Register
$1031	Bit 7	—	—	—	—	—	—	Bit 0	ADR1	A/D Result Register 1
$1032	Bit 7	—	—	—	—	—	—	Bit 0	ADR2	A/D Result Register 2
$1033	Bit 7	—	—	—	—	—	—	Bit 0	ADR3	A/D Result Register 3
$1034	Bit 7	—	—	—	—	—	—	Bit 0	ADR4	A/D Result Register 4
$1035 Thru $1038									Reserved	
$1039	ADPU	CSEL	IRQE	DLY	CME		CR1	CR0	OPTION	System Configuration Options
$103A	Bit 7	—	—	—	—	—	—	Bit 0	COPRST	Arm/Reset COP Timer Circuitry
$103B	ODD	EVEN		BYTE	ROW	ERASE	EELAT	EEPGM	PPROG	EEPROM Programming Control Register
$103C	RBOOT	SMOD	MDA	IRV	PSEL3	PSEL2	PSEL1	PSEL0	HPRIO	Highest Priority I-Bit Int and Misc
$103D	RAM3	RAM2	RAM1	RAM0	REG3	REG2	REG1	REG0	INIT	RAM and I/O Mapping Register
$103E	TILOP		OCCR	CBYP	DISR	FCM	FCOP	TCON	TEST1	Factory TEST Control Register
$103F	—	—	—	—	NOSEC	NOCOP	ROMON	EEON	CONFIG	COP, ROM, and EEPROM Enables

Figure 9-3   (*continued*)

STAF	STAI	CWOM	HNDS	OIN	PLS	EGA	INVB	PIOC $1002

Bit 7	—	—	—	—	—	—	Bit 0	PORTC $1003

Bit 7	—	—	—	—	—	—	Bit 0	PORTB $1004

Bit 7	—	—	—	—	—	—	Bit 0	PORTCL $1005

Bit 7	—	—	—	—	—	—	Bit 0	DDRC $1007

Figure 9-4   **The registers associated with ports B and C.** (*Redrawn Courtesy of Motorola, Inc.*)

9-2.1   STRB in Simple Strobe Mode

There are two modes of operation for parallel data transfer using the **68HC11**: the simple strobe mode and the handshaking mode. The simple strobe mode is covered here; handshaking is discussed in Section 9-5.

In the simple **strobe** mode a short pulse or strobe occurs on STRB whenever there is a write to port B. The mode is selected by setting the HNDS (Handshaking) bit, bit 4, of the PIOC to 0. In this mode the PLS and OIN bits are irrelevant. In the simple strobe mode *STRB becomes active for two E clocks* during each *write* to port B (to $1004). The timing is shown in Figure

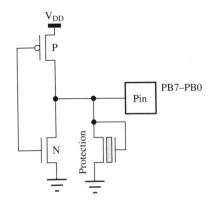

Figure 9-5   **The output connection of a port B pin.** (*Redrawn Courtesy of Motorola, Inc.*)

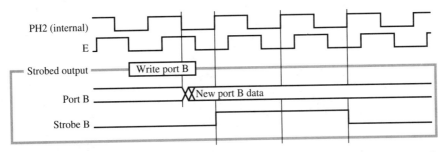

PH2 (internal)

E

Strobed output — Write port B

Port B — New port B data

Strobe B

Figure 9-6 The timing of STRB in simple strobed mode. (*Redrawn Courtesy of Motorola, Inc.*)

9-6. The pulse on STRB can be used to inform external devices that new data is available on the PB0–PB7 lines.

Figure 9-6 shows a positive pulse on STRB. This implies that the **active level** is positive and that the level on STRB will be low (inactive) at all times except during 1-μs intervals following each write to port B. The direction of the pulse can be either positive or negative and depends on the INVB bit in the PIOC. In Figure 9-6 the INVB bit was a 1 to produce a positive pulse. Clearing this bit will produce a negative pulse each time there is a write to $1004.

E X A M P L E   9 - 2*

Assume there are eight lights or Light-Emitting Diodes (LEDs) connected to port B so that they turn on when the corresponding port B bit is a 1. The number in the lights is to progress 11, 22, 33, . . . , CC. The number will start at 11 and will advance every time a switch is thrown. After the lights show CC they should go back to 11 again. Allow the switch throw to interrupt.

a. Show the hardware connections to the EVB.

b. Write the program.

Solution

a. The IRQ interrupt can be used as shown in Figure 8-17. The CLEAR pulse, which must clear the request FF, can be connected to STRB.

b. The program is shown in Figure 9-7. It was assembled using AS11 and is in three parts.

The first part is the initialization constants. Notice how the EQU statement is used to define the stack, the PIOC, and the port B register. The list of numbers to be brought out to the lights is defined by FCB (Form Constant Byte) directives.

* This example was used as a laboratory experiment at the Rochester Institute of Technology.

```
                       NAM    JOE4
               *THESE ARE THE INITIALIZATION CONSTANTS
C600           STACK   EQU    $C600
1002           PIOC    EQU    $1002
1004           PORTB   EQU    $1004
C300                   ORG    $C300
C300 01        CINDY   FCB    1
C301 00 11 22 33 44 55 66  LIST   FCB    $00,$11,$22,$33,$44,$55,$66,
C307 77 88 99 AA BB CC              FCB    $77,$88,$99,$AA,$BB,$CC

               *THIS IS THE MAIN PROGRAM

C30E 8E C6 00  START   LDS    #STACK    Initialize stack and
C311 CE C3 01          LDX    #LIST     index registers.
C314 7F 10 02          CLR    PIOC
C317 86 7E             LDAA   #$7E
C319 97 EE             STAA   $EE       Set up interrupt vector
C31B CC C2 00          LDD    #SER1     on the EVB.
```

```
C31E DD EF        SELF   STD   $EF
C320 20 FE               BRA   SELF

                  *THIS IS THE INTERRUPT SERVICE ROUTINE

C200                     ORG   $C200
C200 F6 C3 00     SER1   LDAB  CINDY
C203 3A                  ABX
C204 A6 00               LDAA  0,X      Get number from list,
C206 B7 10 04            STAA  PORTB    Put it out on Port B.
C209 5C                  INCB
C20A C1 0D               CMPB  #$0D
C20C 26 02               BNE   ENDS
C20E C6 01               LDAB  #1
C210 F7 C3 00     ENDS   STAB  CINDY
C213 3B                  RTI

                  *END OF SERVICE ROUTINE
                         END

Errors:  0
```

Figure 9-7   The program for Example 9-2.

The second part is the main program; it starts at C30E. (A better programming technique would be to start the main program farther away from the constants by using another ORG statement.)

The first instruction, at C30E, sets the SP into RAM. Then X is loaded with the address of the start of the list. This points to C300, which contains 00. This location is never displayed because the program starts sending out data by fetching the 11 in C302.

The third instruction clears the PIOC register. In this program we are only interested in clearing 2 bits in the register: the HNDS bit so that the program operates in simple strobed I/O and the INVB bit so that STRB will pulse low after each write to port B. In Section 8-7.5 we explained that the interrupt FF must be cleared during the interrupt routine. If the CLEAR input on the FF is connected to STRB, which pulses low when port B is written, this pulse can be used to clear the interrupt FF.

The instructions from C317 to C31E are used to set up the IRQ interrupt vector to point to SER1, the start of the interrupt service routine. After initialization the program goes into an endless loop, waiting for an interrupt.

The third part of the program is the interrupt service routine. It goes here after each switch throw sets the IRQ FF and causes an interrupt. Port B is loaded from location CINDY. Data is obtained from the location in the list that is the sum of the contents of B and X. It is sent out by the STAA PORTB instruction. This is the write to port B that also causes STRB to pulse low, clearing the request FF. The service routine then increments B and stores it back in CINDY so that the next interrupt will get data from the next location and returns to the infinite loop in the main program.

9-3   Port C

Port C is more versatile than port B; it can be used for input, output, or latched input. In the expanded mode port C is used to output the eight lower bits of the address when AS is high and E is low and to input or output data when AS is low and E is high. In the single-chip mode it is used for I/O.*

The pin logic for a port C pin is shown in Figure 9-8. It will not be covered in detail, but we would like to consider the connections at the pin. For output a CMOS inverter is available, just as it was for port B. Here, however, there is also a line for inputs from the pin. It goes to the inverters in the lower right portion of the figure.

The action of the CWOM (port C Wire-Ored Mode) bit (bit 5 of the PIOC) is also shown in the upper right part of Figure 9-8. When CWOM is 0, the pin functions normally. If CWOM is 1, it disables the p-type MOS transistor and provides for an *open-drain* output. If the pin is to be wire-ORed by connecting it to other open-drain transistors, then CWOM should

---

* The rest of this paragraph can be omitted by those readers who do not need to know the hardware details at this time.

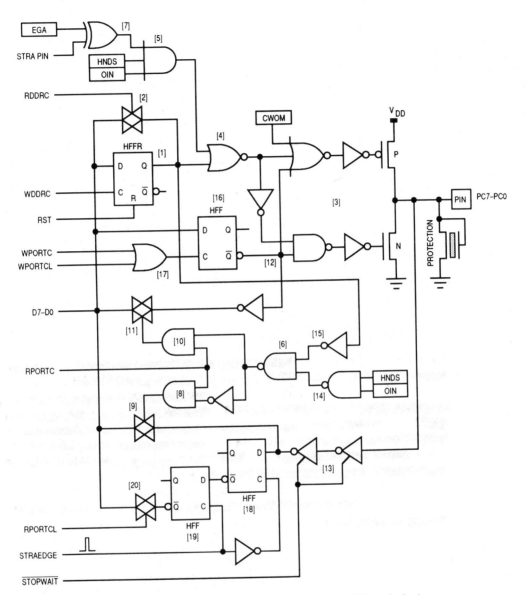

Figure 9-8 Port C single-chip-mode pin logic. (*Reprinted Courtesy of Motorola, Inc.*)

be 1 and a pull-up resistor should be used. This configuration is rarely used, and CWOM is usually 0. CWOM defaults to 0 on RESET.

The read and write timing for port C is shown in Figure 9-9. The port is read on the trailing edge of E. On a write the data is presented during the middle of the E clock and is valid shortly thereafter. There is a slight further delay if the **68HC24** PRU is used.

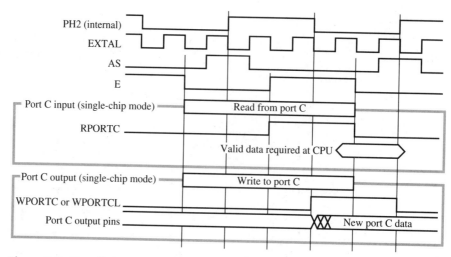

Figure 9-9   Port C timing. *(Redrawn Courtesy of Motorola, Inc.)*

9-3.1   Data Direction Register C

Data Direction Register C (DDRC) is an 8-bit register located at $1007 that determines the *direction of the data flow* for each pin of port C. Figure 9-1 shows that DDRC is behind port C. A 1 in a bit in DDRC sets the *corresponding pin* on port C for output; a 0 means the corresponding pin is input. The DDR must be initialized by writing into $1007 before any data transfer involving port C can occur. It is usually initialized to 00 (all bits input) or FF (all bits output), but DDRC allows the user the flexibility of setting the direction of the bits in any manner, as Example 9-3 shows. DDRC can be changed during a program to change the operation of port C, but this is rarely done. Most programs initialize the DDR at the start and never change it thereafter.

E X A M P L E   9 - 3

Write a code segment so that the four lower bits of port C are used for output and the four upper bits are used for inputs.

Solution

A 1 must be written into every bit of DDRC that is to be used for output and a 0 into every bit that is being used for input. The code segment is

LDAA	#$0F	Write a byte to A with 0s in the four upper bits and 1s in the four lower bits.
STAA	$1007	Store it in DDRC. This sets DDRC correctly.

9-3.2 Port C on Output

Two data registers are connected to port C: The PORTC data register at $1003 and the PORTCL data register at $1005. If port C is being used for output, a write to either one of these registers will put the data on the lines. The output will reflect the most recently written data.

E X A M P L E   9 - 4
_____

Write a code segment to send the contents of $C100 out via port C.

Solution

The code segment could be

LDAA	#$FF	
STAA	$1007	Set DDRC for output.
LDAA	$C100	Get the data from memory.
STAA	$1003	Send it to port C. This also sends it out.

9-3.3 Input via Port C

If port C is set up as input, the data on the port C lines will be read in whenever the PORTC register at $1003 is read. Thus a read of $1003 will bring the data on the lines into one of the **68HC11**'s accumulators.

E X A M P L E   9 - 5*
_____

Eight switches and eight lights are to be connected to the **68HC11** using ports B and C, as shown in Figure 9-10. Both the switches and the lights are to be read as a binary number. Write a program to cause the number in the lights to be five more than the number in the switches. Use ports B and C.

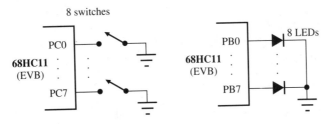

Figure 9-10 Connections for Example 9-5.

Solution

The switches must be connected as inputs and the lights must be connected as outputs. Because port B is output only, it must be connected to the lights,

_____
* This example was used as part of a laboratory experiment at the Rochester Institute of Technology.

leaving the switches to be connected to port C. The program is

	CLR	$1007	Clear DDRC—set port C for input.
LOOP	LDAA	$1003	Read the switches into A.
	ADDA	#$05	Add 5.
	STAA	$1004	Send the data out on port B.
	BRA	LOOP	

The program is in an endless loop, constantly reading the switches and writing the lights.

9-3.4   Port CL and STRA

The data coming into PORTCL (Port CL), at $1005, must be **latched**. To latch it, the STRA pin must be connected to an external signal. When an active edge occurs on STRA, data will be read from the port C lines into PORTCL (port C—latched), and it will remain there until the next edge occurs on STRA. Thus STRA *synchronizes* the data read into PORTCL. The data read into PORTC is unsynchronized; it will change whenever the data on its input changes. Example 9-5 will not work with PORTCL unless a switch that is connected to STRA is added. This switch can be flipped whenever the switches are to be read into PORTCL.

The inputs on the port C lines are latched whenever a rising or a falling edge is detected on STRA. The EGA (EdGe A) bit (bit 1) in the PIOC determines *which edge is recognized*. A 1 corresponds to a rising edge and a 0 to a falling edge. The situation is shown in Figure 9-11. The user can decide which bit edge should be active and set the EGA bit accordingly.

E X A M P L E   9 - 6

An external device is reading data into port C and provides a negative-going AVAILABLE pulse. Data should be sampled on the trailing edge of this pulse. What should EGA be?

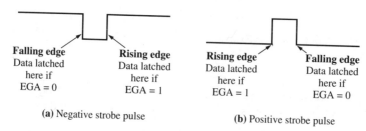

Falling edge	**Rising edge**	**Rising edge**	**Falling edge**
Data latched here if EGA = 0	Data latched here if EGA = 1	Data latched here if EGA = 1	Data latched here if EGA = 0

(a) Negative strobe pulse   (b) Positive strobe pulse

Figure 9-11   Strobes connected to STRA.

Solution

The situation is shown in Figure 9-11a. The trailing edge of the pulse is a rising edge. Therefore EGA should be set to 1 by the initialization routine.

E X A M P L E   9 - 7

An external device is sending the byte $11 to the port C pins and provides a strobe. It then changes the data to $22. What data will be placed in accumulator A by the following instructions?

a. READ PORTC   (LDAA $1003)

b. READ PORTCL (LDAA $1005)

Solution

If the PORTC register is read, the data will be $22, the current data on the port C lines. If PORTCL is read, the data will be $11, the data on the lines the last time that the strobe occurred.

9-4   The PIOC and STRA

As previously stated, the computer is strobed by an active edge on STRA. *But there must be a way for the computer to determine that a strobe has occurred.* The **68HC11** solves this problem by setting the STAF bit (Strobe A **Flag** bit—bit 7) in the PIOC *every time a strobe occurs.*

The obvious next question is: "If the STAF bit is set, how can it be reset?" The obvious answer, "Write a 0 into STAF," *will not work.* Because of their special functions, bits in the control registers do not always act the same way as bits in a memory location. In this case *the STAF bit cannot be written into* by the program, but it can be read.

The STAF bit can be reset by a two-step process:

1. Read the PIOC with the STAF bit set.

2. Read PORTCL.

This has been done deliberately to simplify data transfers. When the **68HC11** reads PIOC to determine that there is data available, as indicated by STAF being set, and then reads PORTCL, the μP assumes that the data has been accepted. It clears STAF to await the next byte.

On the EVB, if STAF is set, an MD 1000 will show that it is indeed set. The MD 1000, however, reads PIOC followed by PORTCL, which will clear STAF. Therefore, although the MD 1000 shows STAF as set, the action of the MD command has cleared it. To show this, give two MD 1000 commands in succession. STAF will always be clear on the second command. This is a peculiar case, where the act of reading a register clears a bit in the register.

9-4.1   Input Data Transfers

A typical procedure for entering data into the **68HC11** from an input device such as a keyboard is

1. The input device puts its data on the port C line to the **68HC11** and puts a strobe on STRA. This latches the data into the PORTCL register and also sets STAF.

2. The **68HC11** reads the PIOC to determine if STAF is set. If it is, the **68HC11** reads PORTCL. This resets STAF, so the μP will not read another byte of data until the next strobe has again set STAF.

The STAF bit has been deliberately placed at bit 7 of the PIOC for ease of programming. If the bit is set, any read of the PIOC will be interpreted as a negative number; if the bit is clear, any read will be interpreted as a positive number because its MSB will be zero. Therefore, after a read of the PIOC the N bit in the CCR indicates the status of STAF. This is often followed by a BPL instruction to branch back if the N bit is not set.

E X A M P L E   9 - 8
_____

A keyboard is connected to the port C lines and STRA. It sends a byte to port C and a KEYSTROBE to STRA whenever a key is depressed. Write a program segment to store $(80)_{10}$ bytes in the buffer from C100 to C14F.

Solution

The program segment is as follows:

```
BUFFER    EQU       $C100
PIOC      EQU       $1002
PORTCL    EQU       $1005
```

– – – – – – – Initialization – – – – – – – – – – – – – – – – – – –

```
          LDX       #BUFFER
          LDAA      PIOC  ⎫   These two instructions
          LDAA      PORTCL⎭   clear STAF.
```

– – – – – – – Main program – – – – – – – – – – – – – – – –

```
LOOP      LDAA      PIOC        Loop—wait for a character.
          BPL       LOOP
          LDAA      PORTCL      STAF = 1. Read the
                                character. This instruction
                                also resets STAF.

          STAA      0, X        Store the character.
          INX
          CPX       #C150       Buffer fully loaded?
          BNE       LOOP        No. Get next character.
          CONTINUE              Yes.
```

The last two instructions in the initialization section are used to clear STAF when the program starts. As long as no characters are received, STAF = 0 and the program continually reads PIOC, finds a positive number, and branches back to read it again. When the strobe occurs, it sets STAF, causing the PIOC to read as a negative number; the program falls through and stores the data.

9-4.2   Output Data Transfers

Output data transfers can also occur in the simple strobed mode. For output the receiving device can signal the **68HC11** that it is ready to receive data on STRA. The µP can then send the data and await the next signal.

The standard printer interface, shown in Figure 9-12, is an example of an output data transfer. One possible hardware configuration is shown in

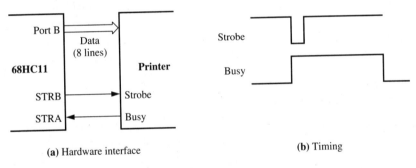

(a) Hardware interface   (b) Timing

Figure 9-12   A standard printer interface.

Figure 9-12a. The data is placed on port B and the printer STROBE is connected to STRB. The BUSY signal from the printer is connected to STRA.

The timing is shown in Figure 9-12b. When the **68HC11** has a character to send, it sends the data out on port B. But the write to port B also pulses STRB low, and STRB is connected to the strobe input on the printer. The printer sees its STROBE input go low, accepts the character for printing, and raises BUSY. The µP should not send the printer another character until busy clears. As shown in Figure 9-12, the trailing edge of BUSY becomes the active edge on STRA and sets STAF to inform the µP that the printer is again available.

E X A M P L E   9 - 9

Write a program to send the data in the buffer from locations C100–C159 to the printer using the interface of Figure 9-12.

Solution

The program resembles that of Example 9-8:

```
PIOC      EQU       $1002
PORTB     EQU       $1004
PORTCL    EQU       $1005
BUFFER    EQU       $C100
START     LDX       #BUFFER
          CLR       PIOC      Set EGA and INVB to 0.
          BRA       START2
                    End of Initialization
LOOP      LDAA      PIOC
          BPL       LOOP      Printer busy?
START2    LDAA      PORTCL    No, clear STAF.
          LDAA      0, X      Get data from BUFFER.
          STAA      PORTB     Send it out. This instruction also
                              produces the STROBE on STRB.
          INX
          CPX       #$C15A
          BNE       LOOP
          CONTINUE
```

The program starts by assuming the printer is not busy. It sends out the first byte and waits until the BUSY signal goes down, which will set STAF. The instruction LDAA PORTCL actually reads nothing; its function is to clear STAF.

9-4.3   Interrupts and STAF

The code segment in the two previous problems

```
LOOP      LDAA      PIOC
          BPL       LOOP
```

is essentially a *polling sequence* where the program does nothing but loop until the STROBE on STRA sets STAF. The **68HC11** can also *interrupt* when STAF is set. The STAF interrupt uses the same vector locations as the IRQ interrupt (FFF2 and FFF3). To allow an interrupt to occur, the user must do the following:

1. Set up the vector to the interrupt service routine. If the service routine is to start at $C100, for example, the code JUMP $C100 will be written into the EVB starting at $00EE.

2. Clear the I bit in the CCR.

3. Set the STAI bit (STA interrupt—bit 6) in the PIOC.

As we will see in Chapter 10, most flag bits have a companion **interrupt bit**, an I bit that corresponds to the F bit. The F bit (flag) will not interrupt unless the I (interrupt) bit is set. Here the pair is STAF and STAI.

E X A M P L E   9 - 10

A **68HC11** can be interrupted by both IRQ and STAF. The interrupt service routine must start at $C100. Using the EVB

a. Write the required initialization instructions.

b. The first few instructions of the service routine should be used to distinguish between the IRQ interrupt and the STAF interrupt. Write these instructions.

Solution

a. To allow interrupts from both sources, write the following instructions before an interrupt can occur:

LDAA	#$7E	
STAA	$EE	Set up the interrupt vector.
LDD	#$C100	
STD	$EF	
CLI		Clear the I flag in the CCR. This allows IRQ interrupts to enter immediately.
LDAA	#$40	
STAA	PIOC	Set STAI in the PIOC. This now allows STAF interrupts.

b. The first few instructions of the interrupt service routine might be

C100	LDAA	PIOC
C103	BMI	Branch to the STAF routine.
C105	CONTINUE	This is the IRQ routine.

The service routine starts by loading the PIOC to determine if STAF has been set. If so, the service routine assumes it is an STAF interrupt and branches to the STAF routine. Otherwise it continues, executing the IRQ routine. This is an example of polling within a service routine to determine how the interrupt was caused and how the μP should respond.

9-5   Handshaking

Handshaking is used when total control of communications between a transmitting device and a receiving device is required. It operates as follows:

1. The receiving device indicates it is ready for a data byte.

2. The transmitting device sends the receiving device a byte and a strobe indicating data is available.

3. The receiving device changes its signal to indicate it is not ready to receive additional data.

4. After the receiving device has had time to process the data, it changes the signal to indicate it can receive another byte.

    On the **68HC11** handshaking is invoked by setting the HNDS bit (bit 4) of the PIOC to 1. The signals that control handshaking appear on the STRA and STRB lines. The 2 bits in the PIOC that have not been discussed up to now, OIN (bit 3) and PLS (bit 2), are active if HNDS is 1. OIN is for *output or input* handshaking. When it is 1, output handshaking is selected (like a DDR). PLS determines whether STRB operates in *pulsed* or *interlocked* mode. A 1 indicates pulsed mode. This distinction will be covered in the following paragraphs.

9-5.1   Handshaking on Input

For input handshaking the **68HC11** must first set STRB to its active level, which indicates it can accept inputs. The timing for input handshaking is shown in Figure 9-13, which shows the levels on STRA, STRB, and STAF.

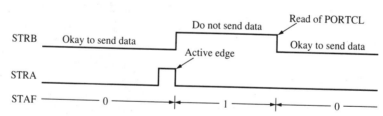

Figure 9-13   Timing for input handshaking.

Here a low level on STRB is active. This has been determined by clearing the INVB in the PIOC. The keyboard, or other device that has data for the **68HC11**, examines the level on STRB. If it is active, the keyboard presents the data on the port C pins and places a strobe on STRA. This causes STRB to go to its inactive level (high in this case), which should prevent the device from sending more data, and also sets STAF. The status of STAF can be detected by polling, or it can cause an interrupt if STAI is set. When the data is read into the PORTCL register, STAF resets and STRB goes back to its active level.

In Figure 9-13 STRB is shown operating in the interlocked mode (PLS is 0). In this mode a read of PORTCL causes STRB to go low and it will stay low until an active edge is detected on STRA. If it were operating in pulse mode, a read of PORTCL would cause STRB to pulse low for two E clocks and then to return high. Some input devices can accept this low pulse as an *acknowledgment* that data has been received, which allows them to send another data byte.

E X A M P L E   9 - 11
_____

For the timing chart of Figure 9-13, what are the contents of DDRC and the PIOC?

Solution

The chart is for input. If we assume all bits are coming in, DDRC must be 00.

For the PIOC we can assume that STAF, STAI, and CWOM are all 0. HNDS must be 1 for handshaking mode. OIN must be 0 because it is input, PLS must be 0 because we are in the interlocked mode, EGA must be 0 because the system responds to the falling edge of STRA, and INVB must be 0 because the active level on STRB is low.

Thus the program must initialize DDRC to 00 and the PIOC to 10 before the communications can start.

_____

The actions of STRA and STRB in the input handshaking mode can be observed in the laboratory by using a pulse generator and an oscilloscope as shown in Figure 9-14. The pulse generator periodically provides the proper edge on STRA, and the reaction of STRB can be observed on the scope.

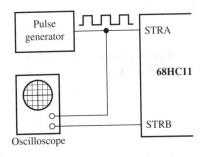

Figure 9-14 Observing STRA and STRB on an oscilloscope.

A simple program that should produce waveforms like those of Figure 9-13 is

```
LOOP    LDAA    PIOC
        BPL     LOOP
        JSR     DELAY
        LDAA    PORTCL
        BRA     LOOP
```

This is an endlessly looping program that is ideal for observation on an oscilloscope. When the program finds STAF set, it jumps to a DELAY subroutine before reading PORTCL so that STRB is inactive for the length of the delay and the action can be observed more clearly. We suggest a delay of about 1 ms. Using this setup, the student can change the PLS, EGA, and INVB bits in the PIOC and observe their effects.

9-5.2   Handshaking on Output

The **68HC11** also has a handshaking mode for output, where the **68HC11** is now the transmitting device and the external device receives the data. The hardware connections and some of the timing are shown in Figure 9-15. For

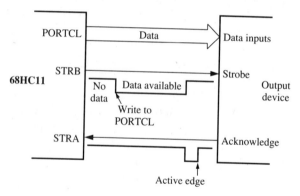

Figure 9-15   Timing for output handshaking.

handshaking the output data must be written into PORTCL. STRB serves as a DATA AVAILABLE signal; when it is active (as determined by the setting of INVB), it indicates data is available for the receiving device. STRA serves as an **acknowledge** from the receiving device. If the receiver puts an active edge on STRA, it acknowledges receipt of the byte and enables the **68HC11** to place another byte of data on the lines.

Figure 9-15 shows the output process. At first STRB is assumed to be high (inactive), which says no data is available. A write to PORTCL causes

STRB to go low, and also clears STAF. The data is placed on the port C pins, and the low level on STRB indicates data is available. The AC-KNOWLEDGE pulse sent to the **68HC11** on STRA by the receiving device returns STRB to its inactive (no data) state, and also sets STAF. The **68HC11** can monitor STAF to determine when it can send more data. If interrupts are enabled, an interrupt will occur immediately after the receiving device acknowledges. The assumption is that the service routine will then put out the next data byte by writing it to PORTCL. This will also make STRB active.

A more detailed timing chart for output handshaking is shown in Figure 9-16. The figure shows that the output starts with a write to PORTCL. If Data Direction Register C (DDRC) bits are 1s (output), the data goes onto the port C lines immediately. If the DDRC bits are 0 (input), which seems like a contradiction because we are trying to output, the data written into PORTCL will still be driven out when the strobe is low. This is indicated by the DRIVEN signal on the figure.

Figure 9-16 also shows the difference between the pulsed and interlocked modes for STRB. The active level is assumed to be high. In the interlocked (PLS = 0) mode STRB is high until the ACKNOWLEDGE pulse is received. In the pulsed mode STRB is high for exactly two E clock cycles.

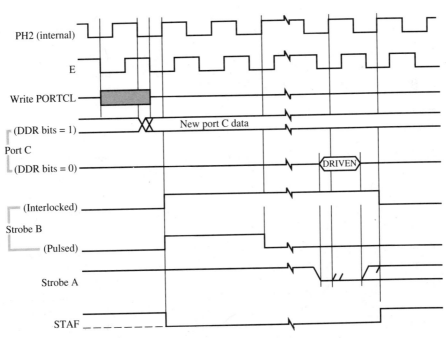

Figure 9-16   Idealized timing for full-output handshaking. (*Redrawn Courtesy of Motorola, Inc.*)

EXAMPLE   9 - 12

What should be written into the PIOC to produce the timing chart of Figure 9-16? Assume interrupts are to be enabled and the pulse mode is to be used on STRB.

Solution

Each bit of the PIOC will be considered in turn.

BIT7—STAF   This bit is set and reset by the actions on the I/O ports, not by a write to the PIOC. We will arbitrarily use a 0 in this bit position.

BIT6—STAI   Because interrupts are to be enabled, this bit must be 1.

BIT5—CWOM   We will assume a normal rather than an open-drain operation. This bit will be 0.

BIT4—HNDS   This is handshaking. Bit 4 must be 1.

BIT3—OIN   It is output handshaking. Therefore bit 3 is 1.

BIT2—PLS   Because pulse mode has been specified, this bit must be 1.

BIT1—EGA   The timing chart shows that STAF responds to a positive edge. Therefore EGA must be 1.

BIT0—INVB   The timing chart shows that the pulse on STRB is positive. Therefore INVB is also 1.

The desired results will be achieved if a $5F is written into the PIOC.

EXAMPLE   9 - 13

Using the EVB, connect a light to STRB to indicate its level. Connect a switch to STRA. Write some instructions and execute them in single step to demonstrate output handshaking.

Solution

The instructions might be

1. Write $18 into the PIOC. This will set up for output handshaking in the interlocked mode with a negative active edge on STRA and a negative active level on STRB.

2. Set DDRC to $FF (output). At this time STRB should be high and the light should be on.

3. Write to PORTCL (STORE into port CL). This should cause STRB to go low and the light should go off.

4. Using the switch, place a negative going edge on STRA. This should cause STRB to go high, and should also set STAF. STAF can be read by an MD (Memory Display) command, which reads memory at $1000. This command will actually read all the registers.

5. Now write to PORTCL again. This will clear STAF and cause STRB to go low.

By single-stepping the program, one can observe the changes in the levels on STRA and STRB after each instruction. These instructions can also be placed in a loop and the action of the circuit monitored on an oscilloscope. The user will need a pulse generator connected to STRA to simulate the acknowledge pulses.

9-6   Parallel Input/Output on Other 68HC11 Ports

The connections to other ports are often used for special purposes. If the particular system does not need these transactions, however, they can be used for I/O. The reader should refer to Figure 8-2.

9-6.1   Port A

Pins PA0–PA2 are reserved for input capture (see Sec. 10-2). If input capture is not used, these pins can function as input pins.

Pins PA3–PA6 are used for output compare (see Sec. 10-3). If this function is not used, these pins can function as output pins.

Pin PA7 has several uses and can function as either input or output. Since this pin can send data in either direction, it requires a data direction register. Bit 7 of the PACTL, a register at $1026, is the direction bit for this pin.

9-6.2   Port D

Port D consists of six pins that can be used for serial I/O. If serial I/O is not used in a particular system, however, these pins are available for input or output. The port D Data Direction Register (DDRD) at $1009 can determine which of the port D pins are assigned for input and which are assigned for output.

9-6.3   Port E

Port E consists of eight pins (four on the DIP) that can be used for A/D conversion. If A/D is not used in the system, these pins can be used as inputs.

Summary

This chapter started by discussing data transfers in the simple strobe mode using ports B and C. The action of STRA and STRB was described in this mode. Several examples of input and output data transfers were presented.

Interrupts and the handshaking mode of data transfers were considered next. The use of the PIOC to control these modes was explained. Finally, there was a brief discussion of the use of other **68HC11** ports for I/O.

Glossary

Acknowledge   A signal from a receiving device acknowledging that it has received a byte from the sending device.

Active level   In the simple strobe mode this is the level on STRB that indicates it is active, that it has just been written to. In the handshaking mode the active level indicates the μP can receive data or has data to send.

Control Register   A register that controls the operation of the I/O ports in a computer.

Direction register   A register that determines the direction of the data flow (in or out) in a corresponding data register.

Flag bit   A bit in a register that indicates an event has occurred. STAF is an example and indicates an edge on STRA has occurred.

Handshaking   Complete control over parallel communications. The receiving device must signal that it can receive data and then the sending device must signal that it is sending data.

Interrupt bit   A bit in a register that, when set, causes an interrupt every time the corresponding flag sets. STAI is an example.

Latched input   Input that is sampled or latched by a strobe edge. It will remain unchanged until the next strobe edge.

Parallel I/O   The transmission of several bits of data simultaneously over an I/O channel.

Strobe   A pulse sent out by a transmitting device to indicate valid data is available.

Problems

Section 9-2

9-1 The number in the buffer from locations $C100–$C123 are to be sent out on port B in 10-ms intervals. Write a program to do this.

Section 9-2.1

9-2 In Figure 9-7, the FCB CINDY directive must not precede the ORG directive. Why not?

9-3 In Figure 9-7, it seems that accumulator B can be initialized to $01 and then we can omit the LDAB CINDY and STAB CINDY instructions. This won't work. What is wrong with it?

9-4 A 7-segment display is shown in Figure P9-4. It consists of seven LEDs and is used to display numbers so that they can be read by humans.

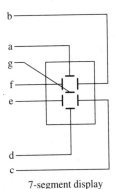

Decimal Number	Hex Number
0	3F
1	06
2	5D
3	4F
4	66
5	6B
6	5F
7	07
8	7F
9	6F

7-segment display

Figure P9-4

A table of data and the corresponding output of the 7-segment display is given above.

If the number shown in the 7-segment display is to be incremented after each switch throw, show the hardware connections and write the program. Use interrupts. Your results should be similar to Example 9-2.

Section 9-3.1

9-5 Port C is to be set up so that its two highest bits and bit 0 are outputs. The rest are inputs. Write the code segment to do this.

Section 9-3.2

9-6 Write a program to search the buffer between $C100 and $C200. It is to write the first byte it finds that is greater than CC out via port C. If no byte is found, it must write out 00.

Section 9-3.3

9-7 The **68HC11** is to read in data on the six lower lines of port C and output on the two higher lines in accordance with the following table.

Data Input	Bits Output
00–25	00
26–30	01
31–36	10
37–3F	11

Write a program to do this.

9-8 Given a 256-word external memory (RAM), the contents of locations C200–C2FF must be copied into this memory. To do so, the memory must be provided with eight address lines, eight data lines, and a

WRITE strobe, which goes low when data is to be written into the memory.

a. Show the connections between the **68HC11** and the external memory. (*Hint:* Use ports B and C and STRB.)

b. Write a program to do this.

9-9 Repeat Problem 9-8 but read from the external memory into locations C200–C2FF.

Section 9-4.1

9-10 A 10-s DELAY routine has already been written, and a light is connected to port C, pin PC1. Connect a switch to STRA. Every time the switch is thrown the light must turn on for 10 s and then must stay off for at least 10 s. Further, switch throws during the 20-s interval are to be ignored. Show the hardware and write the program.

Section 9-5.1

9-11 If the PIOC contains $17, draw a timing chart similar to Figure 9-13.

9-12 A **68HC11** is to read in from a keyboard in the handshake mode as long as a DONE flag is 0. If DONE is 1, it is to JUMP to location FINISH.

The keyboard input must interrupt. The interrupt should start at $C500. It must

a. Delay 3 ms to allow for settling.

b. Read the character into a buffer starting at $C100.

c. Set DONE if the character is an ASCII return ($0D), or if 80 characters have been sent to the buffer.

Write the initialization and the interrupt service routine.

Section 9-5.2

9-13 Write instructions to demonstrate input handshaking on the EVB in single step. The results should be similar to the results of Example 9-13.

9-14 Using the printer interface of Figure 9-12, set up to run the printer in the output handshaking mode.

a. Do not use interrupts.

b. Use interrupts.

In each case write a program to send out the contents of a set of memory locations called BUFFER to the printer.

9-15 Using the setup of Figure 9-14, observe the results of output handshaking on the oscilloscope. Write the program so that this can be done.

The Timing System and Port A

10

INTRODUCTION

This chapter discusses the timing system available on the **68HC11**. It allows the user to generate timing pulses and delays, to measure pulse widths, and to create waveforms. The timing system in conjunction with the I/O and the superior instruction set make the **68HC11** the most powerful 8-bit microcontroller currently available.

INSTRUCTIONAL OBJECTIVES

After reading the chapter, the student should be able to

- Implement time delays using TCNT, RTI, or an output compare register.
- Use a time interval to cause an interrupt.
- Use input capture registers to determine the precise time of an event.
- Use input capture registers to measure periods and pulse widths.
- Change levels on a port A pin at a given time using the output compare registers.
- Generate square waves using the output compare registers.
- Count external events using the pulse accumulator.
- Interrupt when any number of external events has occurred.

SELF-EVALUATION QUESTIONS

Watch for the answers to the following questions as you read the chapter. They should help you to understand the material presented:

1. The TCNT register is read only. What changes its contents?
2. What causes an input capture to be made? How does the **68HC11** respond to an input capture?
3. What causes an output compare flag to set?
4. How is output compare register 1 different from the other output compare registers?
5. What is the difference between external event counting and gated counting?

10-1   The Timing System

Timing is very important for the **68HC11**; it controls the rate of instruction execution and is also used to control much of the I/O. As explained in Section 8-2.3, the timing is determined by a crystal oscillator connected to the XTAL and EXTAL pins on the **68HC11**. The E clock, which determines the time of each instruction cycle, is one-quarter of the crystal frequency. For the EVB and many other **68HC11** systems the basic clock is 8 MHz and the E clock is 2 MHz. Therefore the time of each instruction cycle is 0.5 μs.

10-1.1   The Free-Running Counter

The output of the E clock is connected to a 16-bit counter, called the TCNT register, that is at locations $100E and $100F, as shown in Figure 10-1. **TCNT** is cleared at RESET. After this it is *read only;* it cannot be written but it will be incremented by every tick of the E clock.

7	6	5	4	3	2	1	0	
Bit 15	—	—	—	—	—	—	Bit 8	TCNT $100E
Bit 7	—	—	—	—	—	—	Bit 0	$100F

Figure 10-1   The TCNT register. *(Courtesy of Motorola, Inc.)*

The **68HC11** registers on the EVB can be examined by the command MD 1000. To demonstrate that the counter is operational, the user can issue two successive MD 1000 commands. The contents of all the registers will be the same except for the TCNT register, which increments every 0.5 μs. When a program reads the TCNT register, we advise that it should be read as a 16-bit number, by using an LDD, LDX, or LDY $100E instruction. If each 8-bit portion of the register is read separately, the contents of the register might change between reads and give an erroneous result. A 16-bit read momentarily disables the lower counter so that the results are always correct.

E X A M P L E   10 - 1

The following program was an attempt to read TCNT into A and B:

        LDAA $100E
        LDAB $100F

If the LDAA instruction happens to start when TCNT is 01FB, what will be in accumulators A and B when it finishes?

Solution

The LDAA $100E is a four-cycle instruction. It will read $100E during the fourth cycle. But TCNT increments with every cycle, so TCNT will be 01FE when it is read and A will receive $01. The LDAB will read $100F, four cycles later, when TCNT is at 0202, and B will get 02. Thus the results of the read will be 0102, which is not close to the correct result. Using an LDD TCNT instead of the program will eliminate this problem.

E X A M P L E   10 - 2

Write a program segment to demonstrate that the TCNT cannot be written into.

Solution

One program segment might be

LDD	#$1000
STD	TCNT
LDD	TCNT
SWI	

The contents of the D register can be examined after the SWI (Software Interrupt). If the attempt to write TCNT (STD $100E) was successful, the results in D will always be slightly more than $1000 because the LDD TCNT instruction will have occurred right after the STD TCNT instruction. If, however, the STD TCNT does not write into TCNT, the results will be random and different each time the program is run. This will indicate that TCNT cannot be written into.

The TCNT register can be used to keep track of time, as well as to implement time delays. Unfortunately, implementing time delays using TCNT may not be simple, as Example 10-3 indicates.

E X A M P L E   10 - 3

A student tried to write a 5-ms delay subroutine using TCNT. The E clock was assumed to run at the standard 2-MHz rate so that 5 ms will require 10,000 cycles. The subroutine was

SUBR	LDD	TCNT
	ADDD	10000 $(2710)_{16}$
LOOP	CPD	TCNT
	BGT	LOOP
	RTS	

The program starts by reading the current value of TCNT into D. It then adds 5-ms worth of counts to D and reads TCNT continuously until it is larger than D. It then exits. What is wrong with the program?

Solution

Sometimes finding a problem in a program like this is difficult because it works—mostly. The problem occurs when the double add causes the number to go from positive to negative. Then the program never loops and the pulse width is incorrect. Problem 10-3 should clarify the difficulty.

10-1.2   The Prescalar*

Resolution is the smallest time interval that can be measured in a system. In the **68HC11** the TCNT register is incremented once every 0.5 μs; this is the resolution of the computer.

Because the TCNT register is 16 bits, it takes 65,536 counts to cycle it around once. At the standard clock frequency of 2 MHz this requires 32.77 ms. The **68HC11** contains a **prescalar** that allows the user to *lengthen the time of each count* by dividing the E clock before it is applied to the input of the TCNT register.

Table 10-1 shows the four possible prescale values. They correspond to a division of the E clock by factors of 1, 4, 8, and 16. If a prescale factor of 16 is chosen, for example, it will take 16 times as long for the counter to cycle around, or 524.3 ms as Table 10-1 shows. In some cases this longer time may be desirable, but it also increases the resolution (the minimum time between counts) to 8 μs, so the timing is not so accurate.

Table 10-1   Crystal Frequency Versus PR1, PR0 Values

			Crystal Frequency		
			2^{23} **Hz**	**8 MHz**	**4 MHz**
PR1	**PR0**	**Prescale Factor**	One Count (Resolution)/Overflow (Range)		
0	0	1	477 ns/31.25 ms	500 ns/32.77 ms	1 μs/65.54 ms
0	1	4	191 μs/125 ms	2 μs/131.1 ms	4 μs/262.1 ms
1	0	8	3.81 μs/250 ms	4 μs/262.1 ms	8 μs/524.3 ms
1	1	16	7.63 μs/0.5 s	8 μs/524.3 ms	16 μs/1.049 s
			2.1 MHz	2 MHz	1 MHz
			Bus Frequency (E Clock)		

---

* This section can be omitted on the first reading.

Table 10-1 shows that the prescale factor is determined by 2 bits labeled PR1 and PR0. These bits reside in the TMSK2 (Timer MaSK 2) register at $1024, as shown in Figure 10-2. In normal modes, however, these bits are **pseudo-read only**. This means that they can only be written once, and only during the first 64 cycles following a RESET. Thus if these bits are to be changed, they must be changed by one of the first few instructions following a RESET. On the EVB, the monitor controls the board during this time and these bits cannot be changed. They default to 00, and the TCNT resolution is 0.5 μs.

7	6	5	4	3	2	1	0	
TOI	RTII	PAOVI	PAII	0	0	**PR1**	**PR0**	TMSK2 $1024

Reset: 0   0   0   0   0   0   0   0

PR1, PR0 — Timer Prescaler Select
  These two bits select the prescale rate for the main 16-bit free-running timer system. A prescale factor of 1 corresponds to an E divided by one rate for the main timer, whereas a prescale factor of 16 corresponds to a timer count rate of E divided by 16. In normal modes this prescale rate can only be changed once within the first 64 bus cycles after reset, and the resulting count rate stays in effect until the next reset.

Figure 10-2 **The location of the PR1 and PR0 bits.** (*Courtesy of Motorola, Inc.*)

10-1.3 Timer Overflow

Figure 10-3 shows the TMSK2 register at 1024 and the TFLG2 (Timer Flag 2) register at 1025. Every time the TCNT register rolls over from FFFF to 0000 the TOF (Timer OverFlow) bit, bit 7, is set in TFLG2. The user with an EVB has no control over TCNT, so TOF will be set every 32.77 ms.

7	6	5	4	3	2	1	0	
TOI	**RTII**	PAOVI	PAII	0	0	PR1	PR0	TMSK2 $1024

Reset: 0   0   0   0   0   0   0   0

7	6	5	4	3	2	1	0	
TOF	**RTIF**	PAOVF	PAIF	0	0	0	0	TFLG2 $1025

Reset: 0   0   0   0   0   0   0   0

Figure 10-3 **The TFLG2 and TMSK2 registers.** (*Courtesy of Motorola, Inc.*)

Many of the flag bits in the **68HC11**'s registers, including TOF, are *cleared by writing a 1 into the bit position of the flag*. Writing a 0 into the bit position has no effect on the flag.

E X A M P L E   10 - 4

Write a subroutine to clear TOF.

Solution

Perhaps the simplest subroutine is

```
SUB1    LDAA    #$80    Set bit 7 to 1.
        STAA    $1025   Write it to TFLG2.
        RTS             Return.
```

This subroutine writes a 1 into the TOF position (bit 7) of TFLG2 and clears TOF without affecting any other flags in the register.

A flag can also be cleared by using a BSET instruction, but Motorola does not recommend this because the user may inadvertently clear several other flags in the process, especially if a flag is set in the middle of the BSET. A flag can also be cleared by a BCLR instruction, with a 0 in the bit position to be cleared. This is because the **68HC11** uses the inverse mask to clear the bit. To clear TOF, for example, the following instructions would work:

```
LDX     #$1000
BCLR    $25,X       $7F
```

The mask for the BCLR has a 0 only in bit position 7.

The TCNT register can be used to create time delays. A program to cause a time delay for any amount of time is somewhat complex. Figure 10-4 is such a program, and it will be explained in detail.

For the program of Figure 10-4, N1 is the number of 32.77-ms intervals that the time delay requires and DIFF is the difference between the time of the intervals and the time of the delay. Example 10-5 should make this difference clear.

The program starts by loading the number of intervals into Y, loading TCNT, and adding the difference. If this sum is too large for the D register, Y must be incremented (see Prob. 10-8). Fortunately, if the result of the ADDD instruction exceeds the D register's capacity, the carry flag will be set.

The instructions at C035 and C039 allow for the possibility that N1 is 0. This occurs whenever the delay is shorter than 32.77 ms.

LOOP3 waits for TOF. When TOF is set, LOOP3 clears TOF by using

```
                                      NAM     DELGEN

* This is a general subroutine for any time delay.

   C020                               ORG     $C020
   C020  00  00            N1         FDB
   C022  00  00            DIFF       FDB
   100E                    TCNT       EQU     $100E
   1025                    TFLG2      EQU     $1025

   C024  18  FE  C0  20    SUB1       LDY     N1
   C028  BD  C0  50                   JSR     SUB2
   C02B  FC  10  0E                   LDD     TCNT
   C02E  F3  C0  22                   ADDD    DIFF
   C031  24  02                       BCC     LOOP1
   C033  18  08                       INY

   C035  18  8C  00  00    LOOP1      CPY     #0
   C039  27  0E                       BEQ     LOOP2
   C03B  7D  10  25        LOOP3      TST     TFLG2
   C03E  2A  FB                       BPL     LOOP3
   C040  36                           PSHA
   C041  BD  C0  50                   JSR     SUB2
   C044  32                           PULA
   C045  18  09                       DEY
   C047  26  EC                       BNE     LOOP1

   C049  1A  B3  10  0E    LOOP2      CPD     TCNT
   C04D  22  FA                       BHI     LOOP2
   C04F  39                           RTS

   C050  86  80            SUB2       LDAA    #$80
   C052  B7  10  25                   STAA    TFLG2
   C055  39                           RTS

                                      END

Errors: 0
```

Figure 10-4   A general time delay program.

the SUB2 subroutine and decrements Y. If Y is now 0, it goes to LOOP2 and finishes the count.*

Also, observe the BHI at C049. This cannot be a BNE since an exact equality between TCNT and the D register may not occur. This is because the two instruction loops at locations 31 and 32 take ten cycles, so TCNT will advance by ten cycles between each compare and exact equality may never occur.

EXAMPLE   10 - 5

Use the program of Figure 10-4 to generate a 100-ms time delay.

Solution

To create a 100-ms delay, we first divide 100 ms by 32.77 ms, the time for TCNT to count around. This gives three cycles (3 × 32.77 ms = 98.31 ms) plus 1.69 ms, or 3380 counts of the E clock. If N1 is set to 3 and DIFF is set to $0D34, the hex equivalent of 3380, the program will generate a 100-ms time delay.

EXAMPLE   10 - 6

Write a program to set PB0 high for 100 ms and then low for 100 ms. Use the concepts developed in the previous examples. Note that this is a way to test the delay subroutines

Solution

The solution is shown in Figure 10-5. It consists of two programs, LOOP5 and TOGGLE. TOGGLE is a subroutine to toggle PB0 by executing an EOR each time through. LOOP5 simply calls DELGEN, the delay program, toggles PB0 after the delay, and repeats.

```
LOOP5     JSR    $C024    Jump  to DELGEN subroutine,
          JSR    $C600    Jump to toggle subroutine,
          BRA    LOOP5    Branch back and repeat,

TOGGLE    LDAA   $1004
          EORA   #$01     Toggle PB0,
          STAA   $1004
          RTS
```

Figure 10-5   The program used to test timing routines.

* A subtle point—LOOP2 will only be entered when TCNT is a small value unless N1 = 0 and the sum of DIFF and TCNT does not cause overflow. These facts allow the program to work.

The program was tested by connecting an oscilloscope to PB0 to ascertain that it was indeed changing every 100 ms.

The program was also tested in the laboratory for short delays. We found that the LOOP5 and TOGGLE added about 35 μs to the delay. This is a 3 percent error if the delay is 1 ms. The program is not valid for short delays with high accuracy. In such cases the output compare registers should be used (see Sec. 10-3).

Bit 7 of TMSK2, the register at location $1024, is TOI (Timer Overflow Interrupt). This is the interrupt bit that is the companion to TOF. IF TOI is set, the **68HC11** will interrupt whenever TOF sets. Table 8-3 shows that the timer overflow interrupt vector is at FFDE and FFDF. On the EVB the 3 bytes between D0 and D2 are used for the jump-to-the-service-routine instruction in response to a TOI.

10-1.4 Real-Time Interrupts

The **68HC11** periodically generates real-time interrupts (RTI). Each real-time interrupt sets the RTIF flag in TFLG2 (see Fig. 10-3). The RTI rate of interrupts is shown in Table 10-2. It depends on the setting of the RTR1 and

Table 10-2   RTI Rates Versus RTR1, RTR0 for Various Crystal Frequencies

			Crystal Frequency		
			2^{23} Hz	8 MHz	4 MHz
RTR1	RTR0	$E \div 2^{13}$ Divided By	Nominal RTI Rate		
0	0	1	3.91 ms	4.10 ms	8.19 ms
0	1	2	7.81 ms	8.19 ms	16.38 ms
1	0	4	15.62 ms	16.38 ms	32.77 ms
1	1	8	31.25 ms	32.77 ms	65.54 ms
			2.1 MHz	2 MHz	1 MHz
			Bus Frequency (E Clock)		

RTR0 bits in the PACTL register at $1026 (see Fig. 10-20). If both RTR bits are 1, the RTI rate is 32.77 ms, the same rate as TOF. This may be redundant. The RTI is more useful when one of the faster rates is selected.

A real-time interrupt does not actually interrupt; it just sets the RTIF flag in TFLG2. If, however, the companion interrupt enable bit, RTII (bit 6 in TMSK2), is set, the RTII will interrupt. The interrupt vector is at FFF0 and FFF1 and the pseudo-vector is at 00EB–00ED on the EVB. When interrupts are being used, the RTIF flag must be cleared by the interrupt service routine before the RTI (ReTurn from Interrupt) instruction is executed.

E X A M P L E   10 - 7

Using the EVB, a MD 1000 command is given. What should be the contents of TFLG2 (see Fig. 10-3)?

Solution

The lower 4 bits of TFLG2 are not used and should always be 0. The two upper bits should be 1s because enough time will usually have elapsed so that both TOF and RTIF are set. Bits 4 and 5 have not yet been discussed. If they are assumed to be 0, the contents of TFLG2 ($1025) will be $C0.

E X A M P L E   10 - 8

Write a program to turn a light on and off for approximately 10 s (10 s is approximately 300 loops around at a 32.77-ms rate). Use the RTI interrupts and connect the light to pin 0 of port B.

Solution

The program follows the format we have been using previously:

1. Set up the definitions.

2. Write the initialization for the main program.

3. Write the rest of the main program.

4. Write the interrupt subroutine.

The program is shown in Figure 10-6. It is divided into subsections as specified above. Notice how the RTII pseudo-vector for the EVB is set into memory using the ORG, FCB, and FDB directives.* The main program is simply an endless loop, but it could do useful work. The interrupt routine increments Y and clears the RTIF flag. If Y reaches 300, it toggles port B, which should be connected to the light, and resets the counter by clearing BAMBI. The DELGEN program (see Fig. 10-4) could also have been used to produce this delay.

---

* This method works, but the user should be wary. The vector will be changed by the monitor if the EVB is RESET. In that case the vector can be reloaded by MM (Memory Modify) commands.

```
        NAM    DEL10S
* This is a program to cause a delay of approximately 10 seconds
* using the RTI interrupt.

*---------START OF DEFINITIONS -------------------------------------------
PORTB   EQU    $1004
TMSK2   EQU    $1024
TFLG2   EQU    $1025
PACTL   EQU    $1026
        ORG    $00EB
        FCB    $7E
        FDB    $C100           Put RTII vector into memory.


*---------START OF MAIN PROGRAM--INITIALIZATION -------------------------
        ORG    $C018
BAMBI   FDB    0
BEGIN   LDS    #$C400          Set stack pointer.
        LDAA   #03             Set bit 0 and 1 of PACTL to 1. This
        STAA   PACTL           selects the 32.77 ms rate.
        LDAA   #$40            Clear real time interrupt.
        STAA   TFLG2           Flag.
        STAA   TMSK2           Set the real time interrupt mask
                               to allow interrupts.
*
        CLI
*---------END OF INITIALIZATION --------------------------------------------
```

Figure 10-6 A program to turn a light on and off for 10 seconds using RTIs. (*continues*)

```
* MAIN PROGRAM--INFINITE LOOP

LOOP    NOP                     Place the main program in an
        NOP                     infinite loop, waiting for
        BRA     LOOP            interrupts.

*--------INTERRUPT SERVICE ROUTINE --------------------------------

        ORG     $C100
        LDY     BAMBI
        INY
        STY     BAMBI
        CPY     #300            Has the program looped 300 times?
        BNE     CLRFLG          No, Branch to clear the flag,
        CLR     BAMBI           Yes, Clear the counter,
        CLR     BAMBI+1         Both halves,
        LDAA    #01       ⎫
        EORA    PORTB     ⎬     Toggle PB0,
        STAA    PORTB     ⎭
CLRFLG  LDAA    #$40
        STAA    TFLG2           Clear TFLG2,
        RTI

        END
```

Figure 10-6 (continued)

E X A M P L E   10 - 9

In Figure 10-6 the Y register was loaded from a location called BAMBI, incremented, and then rewritten. What is wrong with simply incrementing Y?

Solution

Assume Y was 0 at the start of the interrupt. This value would be stacked before the interrupt routine was entered. During the interrupt routine the INY instruction would make Y a 1, but the RTI would unstack the previous value of Y and set it back to 0. The contents of the registers inside an interrupt routine would all be overwritten by the RTI and lost unless they are preserved in memory.

10-2   Input Capture Registers

This section starts a discussion of port A of the **68HC11**. The function of each of the eight pins on port A will be explained.

Pins PA0, PA1, and PA2 are *input only* and can serve as general purpose input pins or be connected to one of the three input capture registers in the **68HC11**. The input capture registers TIC1, TIC2, and TIC3 are all 16-bit registers, as shown in Figure 10-7. They are connected to pins PA2, PA1, and PA0, respectively. They make a *capture* when an edge occurs on one of these input pins. When a capture is made on one of the registers, the *time*

7	6	5	4	3	2	1	0	
Bit 15	—	—	—	—	—	—	Bit 8	TIC1 $1010
Bit 7	—	—	—	—	—	—	Bit 0	$1011
Bit 15	—	—	—	—	—	—	Bit 8	TIC2 $1012
Bit 7	—	—	—	—	—	—	Bit 0	$1013
Bit 15	—	—	—	—	—	—	Bit 8	TIC3 $1014
Bit 7	—	—	—	—	—	—	Bit 0	$1015

Figure 10-7   The Input Capture Registers on the 68HC11. (*Courtesy of Motorola, Inc.*)

of the capture (**input capture**) is copied from the TCNT register into the input capture register.

Input captures are controlled by the *edge bits* in TCTL2 and the table shown in Figure 10-8. TCTL2, the register at $1021, contains three pairs of edge bits, one for each register. The function of the edge bits is given in the table. If both edge bits are 0, the capture is disabled; there is never a capture and the pin is free to act as an input pin.

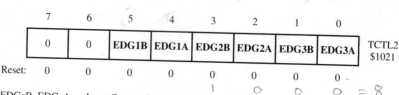

7	6	5	4	3	2	1	0	
0	0	EDG1B	EDG1A	EDG2B	EDG2A	EDG3B	EDG3A	TCTL2 $1021

Reset: 0   0   0   0   0   0   0   0

EDG*x*B, EDG*x*A — Input Capture Edge Control (x = 1, 2, or 3)

These pairs of bits determine which edge(s) the input-capture functions will be sensitive to. These bit pairs are encoded as shown in the following table:

EDGxB	EDGxA	Configuration
0	0	Capture disabled
0	1	Capture on rising edges only
1	0	Capture on falling edges only
1	1	Capture on any edge (rising or falling)

Figure 10-8   TCTL2 and control for the input capture registers. (*Courtesy of Motorola, Inc.*)

E X A M P L E   10 - 10

The time of a falling edge on pin PA1 must be detected. How must TCTL2 be set up to do this?

Solution

PA1 is connected to IC2 (Input Capture register 2). The table shows that the edge bits must be 1 and 0, respectively, to capture a falling edge. Thus the code

```
LDAA    #08
STAA    $1021
```

will set up TCTL2 to capture the time of a falling edge on PA1.

10-2.1   Input Flags and Interrupts

Bits 0, 1, and 2 of the TFLG1 register, at $1023, are used to signal an input capture on registers 3, 2, and 1, respectively, as shown in Figure 10-9. If a

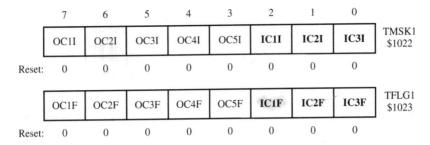

ICxI, ICxF — Input Capture Interrupt Enables and Input Capture Flags ($x = 1, 2,$ or 3)

The ICxF status bit is automatically set to 1 each time a selected edge is detected at the corresponding input-capture pin. This status bit is cleared by writing to the TFLG1 register with a 1 in the corresponding data-bit position. The ICxI control bit allows the user to configure each input-capture function for polled or interrupt-driven operation but does not affect the setting or clearing of the corresponding ICxF bit. When ICxI is zero, the corresponding input-capture interrupt is inhibited, and the input capture is operating in a polled mode. In this mode, the ICxF bit must be polled (read) by user software to determine when an edge has been detected. When the ICxI control bit is 1, a hardware interrupt request is generated whenever the corresponding ICxF bit is set to 1. Before leaving the interrupt service routine, software must clear the ICxF bit by writing to the TFLG1 register.

Figure 10-9 **Input capture and output compare flags and masks.** (*Courtesy of Motorola, Inc.*)

capture is made, the flag will set. As usual, the flag can be cleared by writing a 1 into its bit position.

The corresponding interrupt bit is in TMSK1. If this bit is set, a capture will cause an interrupt in accordance with Table 10-3.

Table 10-3 **Input Capture Register Vectors**

Register	Pin Number	68HC11 Vector	EVB Pseudo-Vector
IC1	PA2	FFEE, FFEF	E8–EA
IC2	PA1	FFEC, FFED	E5–E7
IC3	PA0	FFEA, FFEB	E2–E4

10-2.2 Determining the Period of a Square Wave

Input capture registers can be used to measure the period of a square wave. If a pulse generator is connected to PA0, as shown in Figure 10-10, the time between edges of the pulse can be determined by using the input capture feature of the **68HC11**.

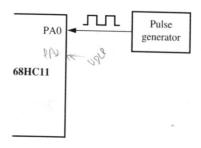

Figure 10-10   The 68HC11 monitoring a pulse generator.

E X A M P L E   10 - 11

Write a program to determine the period of a square wave connected as shown in Figure 10-10 by finding the time between the positive edges.

Solution

The solution is shown in Figure 10-11. The program uses the technique of setting the X register to $1000 and then offsetting each of the registers by indexing them. This also allows the use of the BRCLR instruction.

First, TCTL at $1021 is set up to detect rising edges on PA0. Then the program jumps to SUB2, which clears the input capture flag and waits until it is set again. The time of its setting is stored in TIC3, at $1014 and $1015. This time is saved in memory and SUB2 is executed again to get the time of the next positive edge. The difference between the times is a measure of pulse width.

The program of Example 10-11 works for any period less than 32.77 ms. If the period is longer, the TCNT register will completely roll over at least once. In this case the number of times TCNT rolls around must be counted by counting the number of times TOF sets. Thus TOF must be cleared at the beginning of the program, and it must be monitored as well as the input capture flag.

The time between pulses can be measured as follows:

1. If the time of the second pulse is greater than the time of the first pulse, subtract them to find the difference. Then add 32.77 ms times the number of TOFs.

2. If the time of the second pulse is less than the time of the first pulse, subtract the time of the first pulse from the time of the second pulse. Then subtract 1 from the number of TOFs. Then multiply this result by 32.77 ms and add it to the previous result.

```
        NAM     PERIOD

* This is a program to determine the period of a square wave
* connected to PAO. PAO is connected to IC3.

TCTL    EQU     $21
TFLG1   EQU     $23
TIC1    EQU     $14

        ORG     $C100
FIRST   FDB
RESULT  FDB
BEGIN   LDX     #$1000
        LDS     #$C400
        LDAA    #$01                Set the edge bits for IC3 to
        STAA    TCTL,X              capture on rising edges.
        JSR     SUB2
        LDD     TIC1,X              Save time of first edge.
        STD     FIRST
        JSR     SUB2
        LDD     TIC1,X              Get time of next edge.
        SUBD    FIRST               Subtract to get the time
                                    difference.
        STD     RESULT              Store the result.
        SWI                         End of main program.
*
SUB2    LDAA    #$01                Clear the input capture flag.
        STAA    TFLG1,X
LOOP    BRCLR   TFLG1,X $01 LOOP    Wait until it sets.
        RTS
        END
```

Figure 10-11   A program to determine the period of a square wave.

E X A M P L E   10 - 12

The period of a wave is exactly 32.898 ms. This is exactly the time of one complete revolution of the count plus 128 μs. Note that 128 μs takes $(256)_{10}$ or $100 counts. Find the count at the second edge and the period of the wave if the first pulse edge arrives when TCNT equals

a. $5000   **b.** $FFF0.

Solution

In each case the time will be exactly the time to wrap around once plus $100 counts.

a. TOF will have been set once and the count will be 5100. Subtracting, we have 5100 − 5000 or 100 counts The second time is greater than the first time, so the period of the wave is 32.77 ms plus 100 hex or 256 decimal counts. But 256 counts takes 128 ms, so the time is correctly calculated as 32.898 ms.

b. TOF will set after 16 counts ($10-16). It will now go through a complete revolution and TOF will set again. Then it will go to 00F0 before the second edge arrives. Now the second time is less than the first time, but TOF has been set twice. Subtracting as before, we have 00F0 − FFF0 = 100 counts. If 1 is deducted from the number of TOFs, leaving 1, the results are the same as before.

10-2.3 Measuring Pulse Widths

Input capture registers can be used to measure the width of a single pulse as well as the period of a repetitive waveform. One way to do it, using two input capture registers, is shown in Figure 10-12. The pulse input is connected to both PA0 and PA1, or input capture registers 2 and 3.

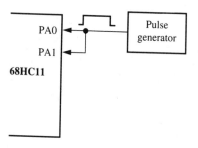

Figure 10-12   Measuring pulse widths with two input capture registers.

A program to measure pulse widths is shown in Figure 10-13. The pulse is assumed to be positive. It uses capture register 2 to determine the time

```
        NAM    PULSEWIDTH
* This program measures the width of a pulse. The input pulse is
* applied to both IC2 and IC3.
TCTL2   EQU    $21
TFLG1   EQU    $23

        ORG    $C100
FIRST   FDB
RESULT  FDB
BEGIN   LDS    #$C400
        LDX    #$1000
        LDAA   #$06              Set TCTL2 to recognize
        STAA   TCTL2,X           a positive edge on PA1
                                 and a negative edge on PA0.
*                                Clear the flags.
        JSR    SUB3

LOOP1   BRCLR  TFLG1,X 02 LOOP1  Wait for the positive edge.
        LDDA   #$02 STAATFLG1,X  Get time in TIC2
        STD    FIRST             and put it away.
        NOP
        LDAA   #$03 STAATFLG1,X  Clear the flags.

LOOP2   BRCLR  TFLG1,X 01 LOOP2  Wait for falling edge.
        LDDA   #$01 STAATFLG1,X  Get falling edge time.
        SUBD   FIRST             Subtract rising edge.
        STD    RESULT            Store result.
        SWI

SUB3    LDAA   #$03
        STAA   TFLG1,X
        RTS

        END
```

Figure 10-13   A program to measure pulse widths.

of the positive, or leading, edge of the pulse and capture register 3 to find the negative or trailing edge. The difference between the times is the pulse width. This program works for pulses less than 32.77 ms. For longer pulses TOF must be used also as in the previous section.

E X A M P L E   10 - 13

The input of Figure 10-12 is connected to a square wave generator operating at 10 KHz. What is the time difference that the program of Figure 10-13 should calculate?

Solution

A 10-KHz wave produces a pulse that is high for 50 μs. This is 100 counts or $64 counts. Thus the difference in the counts should be $64. This circuit can be set up and verified in the laboratory.

10-3   Output Compare Registers

There are five 16-bit **output compare** registers in the **68HC11**. Output compare registers can be used to signal the **68HC11** that an event has occurred,

7	6	5	4	3	2	1	0	
Bit 15	—	—	—	—	—	—	Bit 8	TOC1 $1016
Bit 7	—	—	—	—	—	—	Bit 0	$1017
Bit 15	—	—	—	—	—	—	Bit 8	TOC2 $1018
Bit 7	—	—	—	—	—	—	Bit 0	$1019
Bit 15	—	—	—	—	—	—	Bit 8	TOC3 $101A
Bit 7	—	—	—	—	—	—	Bit 0	$101B
Bit 15	—	—	—	—	—	—	Bit 8	TOC4 $101C
Bit 7	—	—	—	—	—	—	Bit 0	$101D
Bit 15	—	—	—	—	—	—	Bit 8	TOC5 $101E
Bit 7	—	—	—	—	—	—	Bit 0	$101F

Figure 10-14   The output compare registers in the 68HC11. (*Courtesy of Motorola, Inc.*)

and can also be used to control the level on output pins PA3–PA7. The output compare registers are labeled TOC1 (Timer Output Compare 1) to TOC5 and occupy locations $1016 through $101F as shown in Figure 10-14. *Whenever the time in TCNT equals the number in one of those registers a compare has occurred and a flag will be set.* Each output compare register will have its flag set once every 32.77 ms when TCNT equals the number in the register.

The TFLG1 and TMSK1 registers have already been shown in Figure 10-9. The input capture registers occupied the lower 3 bits of these registers and the output compare registers occupy the upper 5 bits. When the number in the TOC register equals the number in TCNT, the flag will set in TFLG1. If the corresponding bit is set in TMSK1, a compare will cause an interrupt. Each output compare register is also associated with an output pin on the port A register as Table 10-4 shows.

Table 10-4   Output Compare Register Locations

Register	Pin Location	68HC11 Vector	EVB Vector
OC1	PA7	FFE8, FFE9	DF–E1
OC2	PA6	FFE6, FFE7	DC–DE
OC3	PA5	FFE4, FFE5	D9–DB
OC4	PA4	FFE2, FFE3	D6–D8
OC5	PA3	FFE0, FFE1	D3–D5

E X A M P L E   10 - 14

An EVB has just been turned on and no signals are connected to any port A pins. What should be in TFLG1 when it is examined by an MD command?

Solution

Because no pins are connected, no edges will be detected by the input capture registers and bits 0, 1, and 2 will be 0. The output compare registers, however, set their bits once every 32.77 ms, so these bits should all be set. Thus the contents of TFLG1 should be F8, indicating that the top 5 bits, which pertain to output compares, are all set.

E X A M P L E   10 - 15

When a program reaches a certain point, it must cause an interrupt 15 ms later and go to a service routine at $C100. Write the code to do this.

Solution

OC5 will be used. Prior to entering the interrupt service routine, a JMP to $C100 must be written into location D3–D5—the EVB pseudo-vector for OC5—and a CLI must be issued. The program segment might be

LDD	TCNT	Place current time in D.
ADDD	#30000	Add 15-ms worth of counts.
STD	$101E	Store it in OC5.
LDAA	#$08	
STAA	TFLG1	Clear the OC5 flag.
STAA	TMSK1	Set the mask bit to allow interrupts.

The program simply finds the current value in TCNT and adds 30,000 (decimal) counts to it. It then clears the flag and sets the mask to allow interrupts. The flag must be cleared before the interrupt bit is set, or it will interrupt immediately.

This program can be checked by storing the value of TCNT in memory after the LDD instruction and then storing the value of TCNT as soon as the interrupt routine is reached. The difference in time should be approximately 30,000 counts.

10-3.1   Control of the Output Pins

Output compare registers 2 to 5 all operate in the same way. Output compare register 1 acts differently and will be discussed in the next section.

Output compare registers 2 to 5 control pins PA6 to PA3 respectively, in accordance with the bit settings in the TCTL1 register (Timer ConTroL register 1) at $1020 and the table of Figure 10-15. TCTL1 contains four pairs of bits that control these output pins. If the setting for the pins is 00, the pin is disconnected from the output compare register and can be used for general purpose output. The three other bit combinations allow the user to toggle, set, or clear the corresponding pins on each comparison.

E X A M P L E   10 - 16

Using the MM command on the EVB, a student writes $04 into $1020. What should happen?

Solution

The number $04 sets OM4 and OL4 to 01, respectively. This should cause PA4 to toggle on each output compare. Because these compares are occurring every 32.77 ms, a square wave of twice this period should appear on PA4, and can be viewed on an oscilloscope.

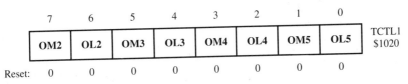

7	6	5	4	3	2	1	0	
OM2	OL2	OM3	OL3	OM4	OL4	OM5	OL5	TCTL1 $1020

Reset: 0   0   0   0   0   0   0   0

OM*x*, OL*x* — Output Compare Pin Control (*x* = 1, 2, 3, 4, or 5)
    This pair of bits determines the automatic actions that occur on the port A timer output pin when there is a successful output compare. Each OC5–OC2 function corresponds to a specific pin of port A. Each pair of bits controls the action for the corresponding output-compare function. These bit pairs are encoded as shown in the following table:

OMx	OLx	Configuration
0	0	OCx does not affect pin (OC1 still may)
0	1	Toggle OCx pin on successful compare
1	0	Clear OCx pin on successful compare
1	1	Set OCx pin on successful compare

Figure 10-15   TCTL1   Control of pins PA3-PA6. *(Courtesy of Motorola, Inc.)*

E X A M P L E   10 - 17

Generate a 50-Hz square wave on pin PA6.

Solution

A 50-Hz square wave takes 20 ms, which means the wave must be up for 10 ms and down for 10 ms. PA6 is connected to OC2 and the output compare registers can handle this problem very nicely.

    The program is shown in Figure 10-16. It clears the OC2F flag and waits for it to set. OC2F will set when the contents of TCNT equal the number in TOC2. The equivalent of 10 ms, 20,000 counts, is then added to the value in TOC2 and the program repeats. Each time the flag sets the output toggles, producing the required square wave.

10-3.2   Output Compare Register 1

A compare at TOC1, the compare register for output compare 1 (OC1), can affect the five output pins PA3 to PA7. The OC1M (Output Compare 1 Mask) register at $100C (Fig. 10-17) determines *which pins will be affected* when a compare between TCNT and TOC1 occurs. The OC1D (Output Compare 1 Define) register determines *how those pins will be affected*.

    All the pins that have a 1 in the corresponding position in OC1M will be affected by an OC1. If the corresponding bit in OC1D is a 1, the bit will set; otherwise it will clear.

    PA7 is different from the other four pins. They are output only, but PA7 can be input or output. It uses bit 7 of the Data Direction Register A

```
* Program to generate a 50 Hz Square wave.
          NAM    SQWAVE

          ORG    $C020

BEGIN     LDX    #$1000
          LDAA   #$40        Set OM2 and OL2 in TCTL1 to
          STAA   $20,X       01 so PA6 will toggle on each
                             compare.

LOOP1     LDAA   #$40        Clear OC2F in TFLG2
          STAA   $23,X       at $1023.

LOOP2     BRCLR  $23,X #$40 LOOP2    Wait until OC2F sets. This
                             will toggle the output.
          LDD    $18,X       Get count in TOC2.
          ADDD   #20000      Add 10 ms to it.
          STD    $18,X       Store in TOC2.
          BRA    LOOP1

          END
```

Figure 10-16 A program to generate a 50 Hz square wave.

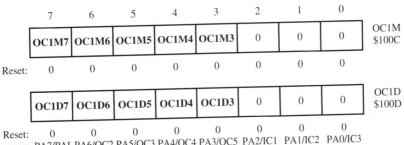

7	6	5	4	3	2	1	0	
OC1M7	OC1M6	OC1M5	OC1M4	OC1M3	0	0	0	OC1M $100C

Reset: 0   0   0   0   0   0   0   0

OC1D7	OC1D6	OC1D5	OC1D4	OC1D3	0	0	0	OC1D $100D

Reset: 0   0   0   0   0   0   0   0
References: PA7/PAI PA6/OC2 PA5/OC3 PA4/OC4 PA3/OC5 PA2/IC1 PA1/IC2 PA0/IC3

Figure 10-17 **The registers for OCl.** (*Courtesy of Motorola, Inc.*)

(DDRA7—bit 7 of the PACTL register at $1026) to determine its direction. It must be set for output (DDRA7 must be 1) if it is to be controlled by OC1.

E X A M P L E   10 - 18
_____

When OC1 occurs pins PA3, PA5, and PA7 are to be set, PA4 is to be cleared, and PA6 is not affected. Write the program segment to do this.

Solution

The program segment is

LDAA	#$80	Set DDRA7 for output on pin 7.
STAA	$1026	
LDAA	#$B8	Set OC1M to affect pins 3, 4, 5, and 7.
STAA	$100C	
LDAA	#%10101000	Set pins 3, 5, and 7. Clear pin 4.
STAA	$100D	The level of bit 6 is irrelevant.

    Pins PA3 to PA6 can be affected by both OC1 and their normal output compare. This allows the user to produce very short pulses as Example 10-19 shows.

E X A M P L E   10 - 19
_____

Write a program to generate a 0.5-μs negative-going pulse on PA4 whenever TCNT rolls over.

Solution

The program is shown in Figure 10-18. Whenever TCNT is 0 an OC1 will occur and clear PA4. On the next count an OC4 will occur and set PA4. The program produces a negative 0.5-μs pulse every 32.77 ms that can be

```
LDAA      #$10        Set OC1M so that OC1
STAA      $100C       affects OC4.
CLR       $100D       Set OC1D so that it clears.
LDAA      #0C         Set TCTL1 so that an OC4
STAA      $1020       sets PA4.
CLR       $1016
CLR       $1017       Set TOC1 to 0000.
CLR       $101C
CLR       $101D
INC       $101D       Set TOC4 to 0001.
SWI
```

Figure 10-18   The program for Example 10-19.

seen with an oscilloscope. Using this program, PA4 can serve as a marker than shows whenever TCNT resets to 0. Note that the program sets the registers and then halts but the pulses continue. The pulses depend only on the proper setting of the registers and the program does not have to continue to run. The same effect can be achieved on the EVB by setting the registers to the values in the program by using the MM command.

10-3.3   A Note on the Evaluation Board

The TRACE instruction on the EVB uses RTI and OC5. If a program is to be traced, we recommend

1. Avoid the use of OC5. Its definition in $C020 may be changed.

2. Do not set bit 3 in $1022 (TMSK1). For example, the following program can be run, but if an attempt is made to trace it, the program will hang the EVB and force the user to reset.

```
C010    LDAA    #$FF
C012    STAA    $1022
C015    NOP
C016    SWI
```

3. The trace instruction will clear the RTII bit in $1024.

10-3.4   Forced Output Compares

There may be times in a program when the user requires that an output compare occur immediately instead of waiting for a match between TCNT and the proper TOC register. An output compare can be **forced** by writing a 1 into the CFORC register at $1008 (Fig. 10-19) for each channel to be forced to compare. The 1 will cause the compare action to occur and then disappear; the CFORC register will always be read as all 0s.

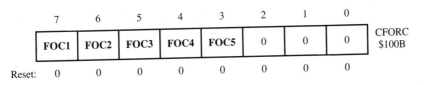

7	6	5	4	3	2	1	0	
FOC1	FOC2	FOC3	FOC4	FOC5	0	0	0	CFORC $100B

Reset: 0   0   0   0   0   0   0   0

FOCx — Force Output Compare (x = 1, 2, 3, 4, or 5)

These bits may be used to force an output compare rather than wait for a match between the output-compare register and the free-running counter. The automatic pin actions programmed for the output compare happen as if a match had occurred, but no interrupt is generated (OCxF is not set). To force one or more output-compare channels, write to the CFORC register with 1s in the bit positions corresponding to the channels to be forced. The logic-high state of these bits is transitory, and the CFORC register will never be read as anything other than zero. The force mechanism is synchronized to the timer counter clock. As many as 16 E-clock cycles could occur between the write to CFORC and the compare force if the largest prescale factor is set for the timer system (PR1, PR0 = 1:1 to ÷ 16).

Figure 10-19 Forced output compares. (*Courtesy of Motorola, Inc.*)

E X A M P L E   10 - 20

Using the EVB, write a program segment to demonstrate the CFORC action. Use OC4.

Solution

The following program was run on the EVB in the laboratory. Before it was run TCTL1, at $1020, was set to 04. This allowed PA4 to toggle and a square wave whose half period was 32.77 ms was observed on the pin. Next TOC4, at $102C and $102D, was set to $8000, so the toggle occurred at the middle of the range of numbers in TCNT.

The program was

```
          NAM    CFORC
          ORG    $C030
LOOP1     LDD    TCNT      Loop as long as TCNT is negative.
          BMI    LOOP1     Stop when TCNT becomes positive.
          LDAA   #$10      Force a compare on OC4.
          STAA   $100B
LOOP2     LDD    TCNT      Loop as long as TCNT is positive.
          BPL    LOOP2
          BRA    LOOP1     Go back and repeat.
```

The program exits LOOP1 as soon as TCNT goes positive or when TCNT is approximately 0000. It then forces a compare, and repeats. With the program running there are compares when TCNT is 0 (the forced compare) and when TCNT is 8000 (the contents of TOC4). This can be seen on an oscilloscope as a square wave whose half period is 16.38 ms.

10-4   The Pulse Accumulator

Pin PA7 is controlled by the PACTL (Port A ConTroL) register at $1026, as shown in Figure 10-20. Bit 7 is DDRA7, a data direction register for this 1 bit. As is usual, a 1 indicates the direction is outward.

We have already seen two uses for PA7 as an output pin (DDRA7 = 1): It can function with output compare register 1 (OC1), as determined by bit 7 of OC1D, or it can be a general output pin if OC1M7 is 0, disabling the output compare on this pin.

The **68HC11** has an 8-bit *pulse accumulator* in it. The **pulse accumulator** is the PACNT (Port A CouNT) register at $1027, also shown in Figure 10-20. The PAEN bit of PACTL determines whether the pulse accumulator is being used.

When DDRA7 is 0, PA7 is being used for input. There are also two uses for PA7 on input: If PAEN is 0, the pulse accumulator is disabled and PA7 can be used as a general purpose input pin. When PAEN is 1, the pulse accumulator is enabled. There are two modes for its use, as determined by the PAMOD bit in the PACTL. The PEDGE bit determines which edge the pulse accumulator will respond to when it is in use. The RTR1 and RTR0 bits have already been discussed in conjunction with real-time interrupts discussed in Section 10-1.4.

10-4.1   External Event Counting

If the pulse accumulator is enabled (PAEN is 1) and PAMOD is 0, the **68HC11** is in an *external* **event counting mode**. Every positive or negative edge (depending on whether PEDGE is 1 or 0) will increment PACNT. In this mode PA7 is sometimes referred to as the PAI (Pulse Accumulator Input).

E X A M P L E   10 - 21

Using an EVB, what is the simplest way to determine if the pulse accumulator is working properly?

Solution

Probably the easiest way is to set up PACTL properly (placing a $50 in it will cause the pulse accumulator to respond to positive edges) and to connect a debounced switch to PA7. The PACNT register can be examined by the MD (Memory Display) command and should increment each time the switch is thrown.

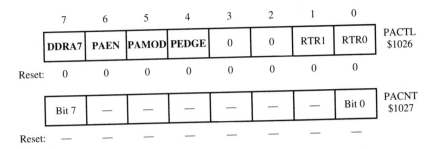

DDRA7 — Data Direction Control for Port A Bit 7

   0 = Port A bit 7 is configured for input only (output buffer is disabled).

   1 = Port A bit 7 is configured for output.

   Normally when the pulse accumulator is being used, the PAI pin will be configured as
   an input. In unusual cases, the PA7/PAI/OC1 pin can be configured as an output to allow
   OC1 or a software output to drive the pulse accumulator system. Since the input buffer is
   always connected to the pin (even when the pin is configured as an output), any output
   function that is controlling the PA7 pin will also be driving the pulse accumulator.

PAEN — Pulse Accumulator Enable

   0 = Pulse accumulator disabled.

   1 = Pulse accumulator enabled.

   When the pulse accumulator is disabled, the 8-bit counter stops counting, and pulse
   accumulator interrupts are inhibited. Though the flags cannot become set, they will
   remain set if they were 1s at the time the pulse accumulator was disabled.

PAMOD — Pulse Accumulator Mode Select

   0 = External event counting mode (pin acts as clock).

   1 = Gated time accumulation mode (pin acts as clock enable for E divided by 64 clock).

PEDGE — Pulse Accumulator Edge Select

   0 = Pulse accumulator responds to falling edges (inhibit gate level is zero).

   1 = Pulse accumulator responds to rising edges (inhibit gate level is one).

   In gated time accumulation mode (PAMOD = 1), the PEDGE bit has added meaning. In
   addition to specifying the edge polarity that causes the PAIF bit to be set, PEDGE also
   controls the inhibit gate level, which disables the internal, free-running E divided by 64
   clock to the pulse accumulator counter. The PAIF interrupts occur at the trailing edge of
   a gate enable signal; thus, selecting falling edges causes the free-running E divided by 64
   clock to be disabled while the PAI pin is low.

Figure 10-20   PACTL. (*Courtesy of Motorola, Inc.*)

   The pulse accumulator is associated with two flags in the TFLG2 register
as shown in Figure 10-21. The PAIF (Pulse Accumulator Input Flag) is set
whenever the pulse accumulator is active and a selected edge on PA7 is
detected. The PAOVF (Pulse Accumulator OVerflow Flag) is set whenever
the count in PACNT rolls over from $FF to $00. As usual, both flags are
cleared by writing 1s into their bit positions in TFLG2.

   The PAII and PAOVI bits in TMSK2 are the corresponding interrupt
enable bits. If set, the **68HC11** will interrupt whenever the corresponding

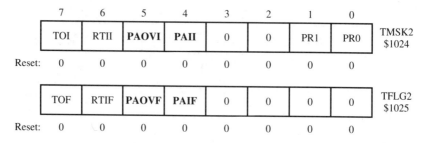

Figure 10-21   The TMSK2 and TFLG2 registers with emphasis on the pulse accumulator. (*Courtesy of Motorola, Inc.*)

flag is set. The PAII interrupt vector is at $FFDA and $FFDB, and at locations $00CA–$00CC in the EVB. The PAOVI vector is at $FFDC and $FFDD and at $00CD–$00CF in the EVB.

E X A M P L E   10 - 22

A debounced switch is connected to PA7. It must interrupt the **68HC11** after five switch throws. Write a program to do this.

Solution

First, the normal initialization must occur.

1. The SP (Stack Pointer) must be positioned properly.

2. The starting address of the interrupt routine must be selected and a JMP to it entered into $CD–$CF.

3. PACTL must be set up to enable the pulse accumulator in the event counting mode.

4. The PAOVF must be cleared and the PAOVI flag set to enable interrupts.

5. The number −5 (FB) can now be written to the PACNT register.

After five switch throws the pulse accumulator should overflow and interrupt. This can be detected by placing a breakpoint in the interrupt routine.

10-4.2   The Gated Timing Mode

The pulse accumulator can operate in a second mode, called the **gated timing mode**, which is entered if PMODA is set. The two pulse accumulator operating modes are shown in Figure 10-22. In the gated timing mode the pulses that increment PACNT are tied to the internal clock and occur at the E clock divided by 64. For a 2-MHz E clock this occurs once every 32 μs. In this

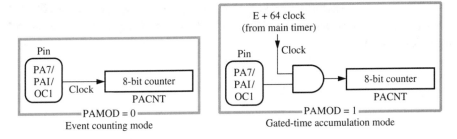

Figure 10-22 **Pulse accumulator operating modes.** *(Redrawn courtesy of Motorola, Inc.)*

mode PA7 serves as a *gate*; the pulses only increment PACNT while the level on PA7 is high. If PA7 is constantly high, PACNT will roll over once every 8.19 ms.

E X A M P L E   10 - 23

A square wave that is up for 512 µs and down for 512 µs is applied to PA7 in the gated timing mode. How many cycles of the square wave will be required for PACNT to roll over?

Solution

Because the pulses that increment PACNT occur every 32 µs, there will be 16 increments in each positive edge. It requires 256 increments to roll PACNT over, so it will take 16 cycles of the square wave. Note that the frequency of this square wave is approximately 1 KHz so that it will take about 16 ms before PACNT rolls over. This is about twice as long as the 8.19 ms that it would take if PA7 were constantly high and is consistent with the fact that PA7 is high for half the time.

Summary

This chapter discussed the powerful and versatile timing available on the **68HC11**. The basic timing is controlled by TCNT, a 16-bit register that increments every E clock and is used to synchronize the other timing routines. Time delays can be implemented on the **68HC11** using TCNT, real-time interrupts, or output compares.

The input capture registers were discussed next. These store the time of an external event (edge) and can be used to measure the period of a waveform or the length of a pulse.

Output compare registers are also available. These registers can cause events to occur when TCNT reaches a specified count. They can also be used to generate pulses or waveforms of a specific frequency.

Finally, the pulse accumulator was explained. The primary use of the pulse accumulator is to count external events, but it can also be used for gated timing.

Glossary

Event counting mode   A mode of input pin PA7 that causes a counter within the **68HC11** to increment each time it detects an edge on PA7.

Forced compare   An immediate compare caused by writing a 1 into the CFORC register.

Gated timing mode   A mode that causes a counter within the **68HC11** to increment at a specific rate whenever the gate—the input level on PA7—is high.

Input capture   The recording of the time of an event (edge) on one of the input capture pins (pins PA0 to PA2)

Output compare   The setting of a flag within the **68HC11** whenever TCNT reaches a specific count (the count within the output compare register).

Prescalar   A divider that can be applied to the clock to slow it down.

Pseudo-read only   Bits in a register that can only be written during the first 64 cycles following a RESET.

Pulse accumulator   An 8-bit register that can increment on each pulse applied to PA7.

Resolution   The smallest time interval that can be measured in a system.

TCNT   A 16-bit register that increments at every E clock.

Problems

Section 10-1.1

10-1 Write a program to copy TCNT into X and then copy TCNT into Y. What should the difference between the two values be?

10-2 In the following program the LDX occurs when TCNT is FFFA. What numbers are in A, B, X, and Y at the end of the program?

```
LDX      TCNT
LDAA     TCNT
LDAB     TCNT + 1
LDY      TCNT
```

10-3 In Example 10-3, how long is the delay if TCNT, when the LDD is executed, is
a. $2000 **b.** $4000 **c.** $7000.
(*Hint:* Be careful with part (c).)

Section 10-1.3

10-4 In Figure 10-4, why not use an LDAA TCNT instead of a TST TCNT?

10-5 In Figure 10-4, how long is the delay if
a. N1 = 2, DIFF = $2000.
b. N1 = 0, DIFF = $300.

10-6 Using the program of Figure 10-4, what numbers should be used for N1 and DIFF if the delay is to be
a. 1 ms **b.** 15 ms **c.** 500 ms.

10-7 What is the maximum delay for the program of Figure 10-4?

10-8 When the program of Figure 10-4 reaches LOOP1 at C035, what are the values in D and Y if N1 is 3 and DIFF is 200, and TCNT, when read at C02B, is
a. $2000 **b.** $FF00.
Show that the delay is the same in each case.

10-9 Write a 2-s delay subroutine using TOF.

10-10 Rework Example 10-6 using interrupts. Write your new main program, which must initialize the interrupt vector on the EVB, and the interrupt service routine.

Section 10-1.4

10-11 Write a 10-min time delay routine using RTIs.

Section 10-2

10-12 Write a code segment to set up TCTL2 to detect a rising edge on PA2 and any edge on PA0. PA1 must be available for input.

Section 10-2.1

10-13 An interrupt service routine must start at $C400 in response to a rising edge on PA1. Write the code to set the registers to do this.

Section 10-2.2

10-14 The period of a waveform is 90 ms. Find the time of the second edge and the number of TOFs if the first edge arrives at
a. $1000 **b.** $F000.

10-15 Refer to Problem 10-14. Write a program segment to find the number of TOFs and the difference in the counts. Decrement the number of TOFs if the second number is less than the first number.

Section 10-2.3

10-16 Redo the program of Figure 10-13 using interrupts.

10-17 A waveform should be high for 5 ms and low for 3 ms. Write a program to monitor the waveform constantly and jump to ERROR if the time is off by more than 10 percent.

10-18 A pair of pulses is generated on a single line as shown in Figure P10-18.

Figure P10-18

a. Use IC1 to find the time of the rising edge of the first pulse.
b. Use IC2 to find the time of the rising edge of the second pulse.
c. Use IC3 to find the time of the falling edge of both pulses.
Show the connections to the **68HC11** and write a program to find the time of both pulses.

Section 10-3

10-19 Write a program to interrupt 23 ms after a negative edge has been detected on PA1. Use both input capture and output compare registers.

10-20 Repeat Problem 10-15 but interrupt 150 ms after any edge has been detected on PA1.

Section 10-3.1

10-21 Write a program to generate a 1000-Hz square wave on PA5.

10-22 Write a program to put a repetitive pulse on PA5 that is high for 4 ms and low for 7 ms.

10-23 Redo Example 10-17 using interrupts. The interrupt routine must add 10 ms to the value in TOC2.

10-24 In Example 10-17 we could have added 10 ms to the value in TCNT and stored it in TOC2. Why is it more precise to add the 10 ms to the value in TOC2?

10-25 In response to a positive edge on PA1, PA4 should
a. Wait 4 ms and then go low.
b. Remain low for 6 ms, then go high again and stay there until the next positive edge on PA1.
Write a program to do this.

10-26 Write a program to provide a 50-μs negative pulse once every 32.77 ms on PA6.

10-27 Write a program to put out an 18-ms pulse on PA4 every time a negative edge is detected on PA1.

Section 10-3.2

10-28 Write a code segment to initialize the **68HC11** so that PA3 goes to 0 and PA4 to PA7 go to 1 every time that there is a compare on OC1.

10-29 Write a program that causes PA3 to go high when TCNT is 100 and then low when TCNT is 103. The program should then go into an infinite loop but the pulses should continue. What should an oscilloscope trace look like?

Section 10-3.4

10-30 Modify Example 10-20 so that the wave is high for 100 μs and low for the remainder of the time.

10-31 PA5 must start out high. It must go low
   a. After 20 ms, or
   b. When a positive transition is detected on PA1.
      Once it goes low it must stay low for 5 ms. Write a program to do this. Include a way to show that the program is working properly.

Section 10-4.1

10-32 Write a program to interrupt and go to $C600 whenever a negative edge is detected on PA7. Set the stack at $C700.

10-33 Write a program to interrupt and go to $C600 after 25 (decimal) pulses are detected at PA7. Set the stack at $C700.

Section 10-4.2

10-34 In the gated count mode show that PACNT will roll over every 8.19 ms if PA7 is constantly high.

10-35 A square wave that is high for 1.024 ms and low for 10 ms is applied to PA7 in the gated count mode.
   a. How long will it take for PACNT to roll over?
   b. Write a program on the EVB to demonstrate that PACNT does indeed advance by 32 counts for each positive pulse. You may need an input capture register to help control the program.

11 | *Ports D and E*

INTRODUCTION

This chapter considers the D and E I/O ports on the **68HC11**. These ports control the serial communications interface, the parallel communications interface, and the A/D converter. The A/D converter, probably the most important and useful of these features, is discussed first.

INSTRUCTIONAL OBJECTIVES

After reading the chapter, the student should be able to

- Calculate the resolution of an A/D converter.
- Perform A/D conversions using the **68HC11**.
- Decode the ASCII characters in an asynchronous waveform.
- Transmit asynchronous data using the SCI.
- Receive asynchronous data using the SCI.
- State the function of each pin of the SPI.

SELF-EVALUATION QUESTIONS

Watch for the answers to the following questions as you read the chapter. They should help you to understand the material presented:

1. What determines whether ports D and E are general purpose I/O or whether they are used for special functions?
2. What determines which channel(s) of the A/D converter get converted? Where are the results found?
3. What is the function of a START bit? Why must it be a 0?
4. What are the RS-232 levels? Where are they used?
5. How can bit 9 be used to "wake up" a receiver?
6. What is the difference between the SPI and the SCI? Which one synchronizes its data with a clock?

11-1 The Analog-to-Digital Converter

Port E is connected to the analog-to-digital (A/D) converter on the **68HC11**. The A/D converter is activated when the ADPU (A/D Power Up) bit, which is bit 7 of the options register at $1039, is set. Otherwise the pins on port E function as general purpose input pins.

If the ADPU bit is 0, the logic levels on the port E pins can be found in the port E register at $100A. If ADPU is 1, the voltage levels on the port E pins are converted to their digital equivalents, and the port E register is irrelevant. Because the port E pins are input only, the port does not require a DDR.

11-1.1 Basics of Analog-to-Digital Conversion

An A/D converter accepts an analog input within a certain *range* of voltages and converts the voltage into a digital number that represents the analog voltage. Figure 11-1 shows the analog input being applied to pin PE0. It is then converted to a digital number by the A/D converter, and the result appears in an 8-bit digital register.

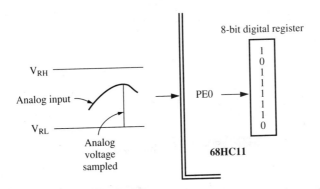

Figure 11-1 **Applying an analog input to a 68HC11.**

The **resolution**, which is the voltage represented by each digital increment, is given by Equation (11-1).

$$\text{Resolution} = \text{range}/2^n \qquad (11\text{-}1)$$

where n is the number of bits in the digital register.

In the **68HC11** the digital registers are always 8 bits and the range is determined by voltages on two input pins associated with the A/D converter. V_{RH} is the high reference voltage and V_{LH} is the low reference voltage. The difference between these two voltages is the range of analog voltages that can be converted.

There are several general restrictions on the reference voltages:

V_{RH} should not be higher than 6 V.

V_{LH} should not be lower than ground.

The difference between the two reference voltages should be at least 2.5 V.

The reference voltages allow the user, within the limitations discussed above, to select the range of analog voltages to be converted to digital. If V_{RH} and V_{LH} are close, they will only convert a small range of voltages but the resolution will be higher.

E X A M P L E   11 - 1

If V_{RH} is 4 V and V_{LH} is 1 V, what is the range and resolution of the A/D converter?

Solution

The range is simply the difference between the voltages, or 3 V.
    Because the output is an 8-bit register, the resolution is

$$\text{Resolution} = \text{range}/2^n = 3 \text{ V}/256 = 11.72 \text{ mV}$$

Thus each digital increment corresponds to 11.72 mV.

E X A M P L E   11 - 2

An analog input is applied to the A/D converter of Example 11-1 and the digital register reads $49. What is the analog voltage?

Solution

Hex 49 converts to 73 decimal. There are 73 increments of 11.72 mV, which calculates to 0.85556 V. This must be added to V_{LH} or 1 V to give the correct analog voltage of 1.85556 V.

E X A M P L E   11 - 3

The 60-Hz rectified sine wave is applied to the A/D converter on a **68HC11** as shown in Figure 11-2.

a. What should the values of V_{LH} and V_{RH} be for maximum resolution?

b. What will the output of the A/D converter be if the analog input is sampled at
    (1) $t = 0$ ms
    (2) $t = 5$ ms

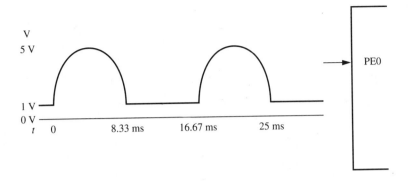

Figure 11-2 A half-wave rectified wave applied to PE0 of the 68HC11.

Solution

a. For maximum resolution V_{LH} and V_{RH} should be as close together as possible to cover the waveform. In this example they should be set at 1 and 5 V, respectively.

b. At $t = 0$ the voltage will be 1 V, which is at the bottom end of the range. The digital output would be 00.

At $t = 5$ ms, we must realize that 8.33 ms corresponds to 180°. Thus 5 ms corresponds to 5 ms/8.33 ms $\times$ 180° = 108°. The input is a sine wave whose value is therefore $\sin^{-1} 108° = 0.951$.
The digital output is $0.951 \times 255 = (243)_{10}$ or $(F3)_{16}$.

11-1.2 The OPTION Register

The A/D conversion process in the **68HC11** is controlled by the OPTION register in $1039 and the ADCTL (Analog-Digital ConTroL register) at $1030. The OPTION register is shown in Figure 11-3.

Only 2 bits in the OPTION register affect the A/D conversion. The ADPU bit, bit 7, must be set before conversions can take place. After setting ADPU, there should be a small delay to allow the internal charge pump and comparator to stabilize before starting conversions.

The CSEL (Clock SELect) bit should be set whenever there is a slow clock (less than 750 kHz) so that an internal oscillator can control the conversions. In most systems, including the EVB, the clock is high enough so that CSEL should remain clear. Fortunately, its default value is 0, so CSEL is rarely changed or set.

11-1.3 The Analog-Digital Control Register (ADCTL)

A/D conversions are controlled by the ADCTL at $1030, as shown in Figure 11-4. Conversions are initiated by writing to the ADCTL. This clears the CCF (Conversions Complete Flag) in bit 7. At an E-clock frequency of 2

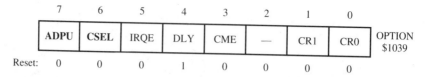

7	6	5	4	3	2	1	0	
ADPU	CSEL	IRQE	DLY	CME	—	CR1	CR0	OPTION $1039
Reset: 0	0	0	1	0	0	0	0	

The CSEL control bit also selects an alternate clock source for the on-chip EEPROM charge pump. This charge pump is separate from the A/D charge pump, but both pumps are selected with the CSEL control bit. In case of the A/D charge pump, CSEL needs to be 1 when the E clock is too slow to assure that the successive-approximation sequence will finish before any significant charge loss. In the case of the EEPROM, the efficiency of the charge pump is at issue. More details on EEPROM charge-pump efficiency are presented in **SECTION 4 ON-CHIP MEMORY**. When the E clock is at or above 2 MHz, CSEL should always be zero; when the E clock is below 750 kHz, CSEL should almost always be 1.

Figure 11-3   The OPTION Register. *(Courtesy of Motorola, Inc.)*

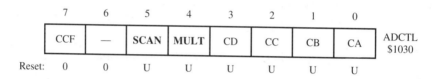

7	6	5	4	3	2	1	0	
CCF	—	SCAN	MULT	CD	CC	CB	CA	ADCTL $1030
Reset: 0	0	U	U	U	U	U	U	

Figure 11-4   The Analog-Digital Control Register (ADCTL). *(Courtesy of Motorola, Inc.)*

MHz conversions take about 16 µs. When the conversions are complete, the CCF flag sets.

It is not possible to perform a single A/D conversion. In response to a write to the ADCTL, at least four conversions are performed and the results are stored in the four A/D result registers, ADR1 to ADR4, that occupy locations $1031–$1034. These registers are written by A/D conversions but cannot be written into by the µP; they are read only. The CCF sets after the fourth register has been written into.

The SCAN bit in the ADCTL determines whether the inputs are continuously scanned. If the SCAN bit is 0, four conversions are performed, the four ADR registers are filled, CCF sets, and conversions terminate. If SCAN is 1, each channel is scanned in turn and conversions continue. The newer results overwrite the older results in the ADR registers.

When a conversion is initiated, there are two modes:

Single channel   Four successive conversions are performed on a single channel, and the four results are written into the ADR registers. Unless the analog input voltage is changing rapidly, the four registers will all have about the same number (see Example 11-6).

Multiple channel   In this mode one conversion is performed on each channel in a *four-channel group*. The four conversions fill the ADR registers.

The mode is determined by the MULT bit in the ADCTL. If MULT = 0, only the analog input on a single channel is converted. If MULT = 1, all four channels in a group are converted.

The channel(s) to be converted are determined by the four LSBs of the ADCTL as given in Table 11-1. In the single-channel mode these 4 bits

Table 11-1 Analog-to-Digital Channel Assignments

CD	CC	CB	CA	Channel Signal	Result in ADRx if MULT = 1
0	0	0	0	PE0	ADR1
0	0	0	1	PE1	ADR2
0	0	1	0	PE2	ADR3
0	0	1	1	PE3	ADR4
0	1	0	0	PE4*	ADR1
0	1	0	1	PE5*	ADR2
0	1	1	0	PE6*	ADR3
0	1	1	1	PE7*	ADR4
1	0	0	0	Reserved	ADR1
1	0	0	1	Reserved	ADR2
1	0	1	0	Reserved	ADR3
1	0	1	1	Reserved	ADR4
1	1	0	0	V_H†	ADR1
1	1	0	1	V_L†	ADR2
1	1	1	0	1/2 V_H†	ADR3
1	1	1	1	Reserved†	ADR4

* Not available in 48-pin package versions.
† These channels intended for factory testing.

determine the channel to be converted. Although there are 16 choices in the table, the actual number of channels is constrained by the number of input pins assigned to A/D conversion. In a 52-pin PLCC **68HC11**, and in the EVB, there are eight possible A/D inputs. In a 48-pin DIP version there are only four pins reserved for A/D conversion.

Perhaps the simplest way to test the A/D converter in the laboratory is shown in Figure 11-5. The input voltage can be applied to an A/D input and also measured by the voltmeter. The digital number in the ADR registers should correspond to the input analog voltage.

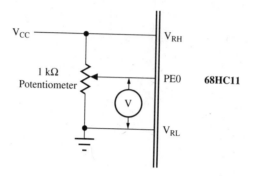

Figure 11-5   Testing the A/D converter.

E X A M P L E   11 - 4

Write a program segment to set up the ADCTL for a single conversion on channel 5, which is connected to input pin PE5.

Solution

The code segment might be

```
ADCTL   EQU       $30
        LDX       #$1000
        LDAA      #05                      Select channel 5.
                                           This sets MULT and
                                           SCAN to 0.
        STAA      ADCTL,X                  This sets up ADCTL
                                           and starts the
                                           conversions.
LOOP    BRCLR     ADCTL,X  #80  LOOP       Wait until CCF sets.
        CONTINUE
```

At this point the four conversions will be in the ADR registers. Many programs will simply take the average of these results as the correct response (see Prob. 11-5).

Multiple conversions convert a group of channels. The group is determined by the CD and CC bits in ADCTL. There are really only two four-channel groups available (one group on the 48-pin DIP). If CD and CC are both 0, channels 0 to 3 are selected; if CD and CC are 0 and 1, respectively, channels 4 to 7 are selected.

E X A M P L E   11 - 5
_____

If $34 is written into the ADCTL, what will happen? Where will the digital conversion of the input to PE5 be found?

Solution

Now both SCAN and MULT are 1. The A/D will perform conversions continuously on the four channels in group 1 because CD = 0 and CC = 1. Channel 5 is channel 1 in group 1, so the results for this channel will be found in ADR1 at $1032.

E X A M P L E   11 - 6
_____

A sine wave that varies from 1 to 5 V is given by the equation

$$V = 3 + 2 \sin 2\pi \, ft$$

If the frequency of this wave is 1 KHz and it is applied to the A/D converter whose sampling time is 16 μs, find the maximum difference in the results in the four ADR registers.

Solution

At 1 KHz one cycle or 360° corresponds to 1000 μs. Therefore each μs corresponds to 360°/1000 = 0.36°, and 16 μs corresponds to 16 × 0.36° = 5.76°. The sine wave changes most rapidly when it is near 0°, so for maximum change the four conversions would be made at −8.64°, −2.88°, +2.88°, and +8.64°. These are all 5.76° apart. The sines of these angles are approximately −0.15, −0.05, +0.05, and +0.15, respectively, so the voltages would be 2.7, 2.9, 3.1, and 3.3 V, respectively. If the range is set at 4 V (from 1 to 5 V), the resolution is 4 V/256 = 15.625 mV.

Since 2.7 V is 1.7 V above the 1 V base, the result in the first ADR will be 1.7 V/15.625 mV = $(108.8)_{10}$ or $(6D)_{16}$. This will be the result in the first ADR register. The results in the other A/D registers can be calculated similarly and will be 7A, 86, and 93, respectively.

11-2   Asynchronous Data Transmission

This section discusses the general principles of serial asynchronous data transmission. These principles apply to serial communications channels found on microprocessors and to UARTs (**Universal Asynchronous Receiver-**

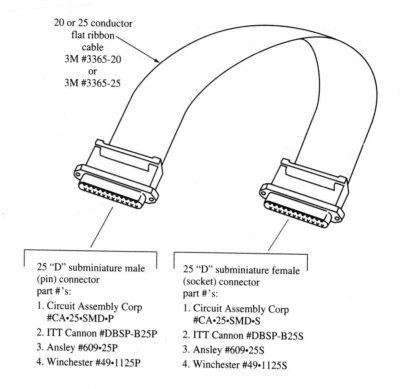

20 or 25 conductor
flat ribbon
cable
3M #3365-20
or
3M #3365-25

25 "D" subminiature male
(pin) connector
part #'s:
1. Circuit Assembly Corp
   #CA•25•SMD•P
2. ITT Cannon #DBSP-B25P
3. Ansley #609•25P
4. Winchester #49•1125P

25 "D" subminiature female
(socket) connector
part #'s:
1. Circuit Assembly Corp
   #CA•25•SMD•S
2. ITT Cannon #DBSP-B25S
3. Ansley #609•25S
4. Winchester #49•1125S

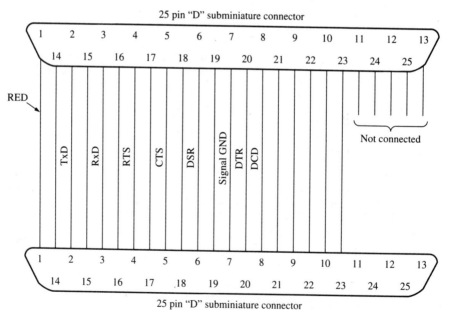

Figure 11-6 **Terminal/Host Computer Cable Assembly Diagram.** (*Redrawn courtesy of Motorola, Inc.*)

Transmitters), which are ICs that control this type of transmission. Examples of these UARTs are the **1013** or **1015** from General Devices, the Motorola **6850**, and the Intel **8251**. The UART on the **68HC11**, called the serial communications interface, will be discussed in the next section.

11-2.1 Serial Data Transmission

Serial data transmission is essentially data transmission over a single wire. Most communications channels use 25 wires, however, and terminate in a DB-25 connector. A typical communications cable is shown in Figure 11-6. This is the cable used by the EVB. Although the cable contains 25 lines, only a few are actually used. This is true of most communication systems.

Figure 11-7 shows the format of an asynchronous* character. It starts in the idle condition when no data is being transmitted. The voltage on the line will always be a logic 1 when the line is idle.

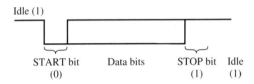

Figure 11-7 The format of an asynchronous character.

When a character is to be transmitted, it is introduced by a START bit. This is a logic 0. Every character transmitted is preceded by a START bit.

The data bits follow, with the LSB first or closest to the START bit. The data bits can be either 1s or 0s, depending on the character transmitted. The data is usually transmitted using the ASCII code (see Appendix C).

After all the data bits have been transmitted, they must be followed by one or two STOP bits. STOP bits are always 1. If there is no more data to be transmitted after the STOP bit, the line will remain in the 1 state and be idle. If another character is to be transmitted, its START bit can occur as soon as the STOP bit is finished.

The time of each bit is determined by the transmission rate of the system. Both the receiver and the transmitter must be set to operate at the same speed. Typical speeds vary from 300 bits per second (bps) to 9600 bps.

---

* The word **asynchronous** means without a clock. The data is transmitted without an accompanying clock to synchronize it. The receiver must synchronize the bits from the waveform it receives.

E X A M P L E   11 - 7

A system is operating at 4800 bps. How long is each bit?

Solution

At 4800 bps each bit takes 1/4800 s or about 208 μs.

A typical character requires 7 bits of data, typically ASCII (see Appendix C), and an eighth bit that can be used for parity.

E X A M P L E   11 - 8

The letter E is to be transmitted in an 8-bit asynchronous format with odd parity. Sketch the output.

Solution

The ASCII table shows that the letter E is coded as $45, or 1000101. This will be preceded by a START bit. There are three 1s in the code, so the parity bit will be 0 to keep the parity odd. The output is shown in Figure 11-8.

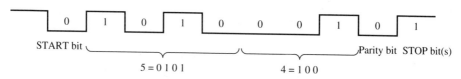

Figure 11-8   The letter E in asynchronous code.

E X A M P L E   11 - 9

The data stream of Figure 11-9 is sent to a terminal.

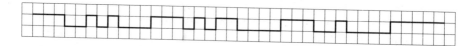

Figure 11-9   The data stream for Example 11-9.

a. What characters are they?

b. Is the parity even or odd?

c. What does the terminal do?

Solution

The bits are shown in Figure 11-10, which also shows the function of each bit. Note that only 7 bits are required to determine the ASCII character and the eighth bit is a parity bit. In this case the parity is odd because each 8-bit character has an odd number of 1s in its 8 bits.

The characters are 0A, 0D, and 42. The ASCII table shows that these are a carriage return, a line feed, and the letter B. The terminal will go to the start of the next line and print the letter B.

11-2.2 RS-232

The asynchronous voltage levels for the waveforms discussed in the previous paragraph were the standard TTL logic levels, 0 and 5V. These are the logic levels generated within the **68HC11**. Most communications interfaces, however, convert these voltage levels to the so-called RS-232 levels.* The **RS-232** voltage levels are

Logic 0—any voltage between $+3$ and $+25$ V

Logic 1—Any voltage between -3 and -25 V.

The most common RS-232 voltages are ± 9 and ± 12 V. RS-232 voltages have better noise margins and can be transmitted for greater distances than TTL voltages.

Figure 11-11 shows the connections that can be made to the **68HC11** if it were to operate in the single-chip mode. These are not the connections on the EVB, where the **68HC11** is in the expanded multiplexed mode. The asynchronous output data is connected to the RS-232C connector shown in Figure 11-6. Data to be transmitted goes from the TxD (Transmit Data) pin on the **68HC11** to the **MC1488** IC. This chip is a TTL-to-RS-232 converter. It transforms $+5$ to -12 V and 0 to $+12$ V. This is the IC that requires the $+12$ and -12 V power inputs to the EVB. Thus the voltage levels at the output, pin 3, are RS-232.

The levels coming in on pin 2 are also RS-232, and these have to be converted to TTL levels before going into the RxD (Receive Data) on the **68HC11**. The two diodes and the **74LS14** inverter make this transformation. If the input voltage is $+12$ V (logic 0), diode D1 turns on and brings the input to the **74LS14** to $+5$ V. The **74LS14** then inverts it to 0 V and places it on the RxD pin. If the input voltage is -12 V, D2 turns on and brings the input to the **74LS14** to 0 V. This is inverted and applies $+5$ V to the RxD pin. A more elegant solution to the problem of converting RS-232 levels to TTL levels is to use the **MC1489** IC. This is a companion IC to the **MC1488**

---

* A complete discussion of RS-232 is beyond the scope of this book. A more thorough explanation is given in J. D. Greenfield, *Practical Digital Design Using ICs* (Englewood Cliffs, NJ: Prentice-Hall, latest ed.). There are also complete books available such as Byron Putman, *RS-232 Simplified* (Englewood Cliffs, NJ: Prentice-Hall, 1987) or Joe Campbell, *The RS-232 Solution* (Berkeley, CA: Sybex, 1984).

Bits   Meaning

1 ⎫
1 ⎬ Idle.
1 ⎭

0   Start bit of first character.

0 ⎫
1 ⎪
0 ⎪
1 ⎬ First character
0 ⎪   0A
0 ⎪
0 ⎭

1   Parity bit--first character.
1   Stop bit--first character.
1   Idle bit (or second stop bit).
0   Start bit--second character.

1 ⎫
0 ⎪
1 ⎪
1 ⎬ Second character
0 ⎪   0D
0 ⎪
0 ⎭

Figure 11-10   The bit stream for Example 11-9.

Bits	Meaning

0   Parity bit--second character.
1   Stop bit--second character.

$\left.\begin{matrix}1\\1\end{matrix}\right\}$ Idle.

0   Start bit--third character.

$\left.\begin{matrix}0\\1\\0\\0\\0\\0\\1\end{matrix}\right\}$ Third character
                        42

1   Stop bit--third character. Parity bit.

$\left.\begin{matrix}1\\1\\1\end{matrix}\right\}$ Stop
                        Idle.

Figure 11-10   (*continued*)

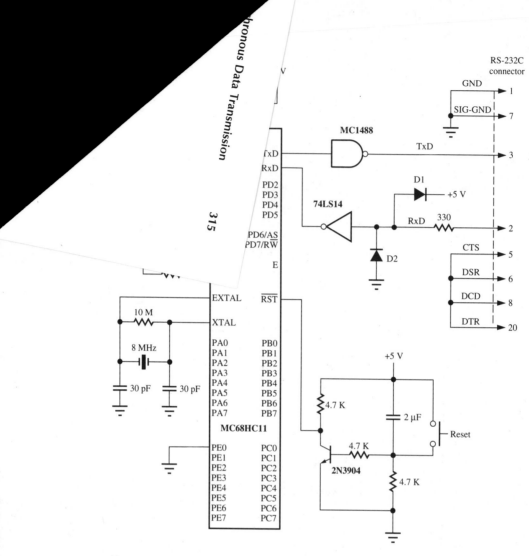

Notes:

1. Unless otherwise specified
   all resistors are in ohms, ±5%, 1/4 W,
   all capacitors are in μF,
   all voltages are DC.

2. Tie unused inputs to +5 V (V_{DD})
   or ground (GND).

Figure 11-11   Connecting RS-232 data to a 68HC11. (*Courtesy of Motorola, Inc.*)

and is specifically designed to transform RS-232 voltages to TTL voltages. It is used on the EVB for that purpose.

A RESET circuit is also shown in Figure 11-11. When the button is pushed, the transistor saturates and brings the RESET input on the **68HC11** to ground.

Figure 11-11 also shows the crystal connections and the RESET circuit, and that MODA is grounded and MODB is connected to 1 to select single-chip operation. This is the way to build a very simple **68HC11** system; the EVB is more complex and operates in the expanded multiplexed mode.

11-2.3 Transmitting Data

The method of transmitting asynchronous data is shown in Figure 11-12. The single buffer circuit is shown in Figure 11-12a. In this circuit a character of data (typically 8 bits) is sent from the μP to the transmit shift register. The circuit also causes the character to be preceded by a 0 (the START bit) and followed by a 1 (The STOP bit). It then sends the data out on the TxD line. The data is shifted out, 1 bit at a time, at the rate determined by the clock connected to the shift register.

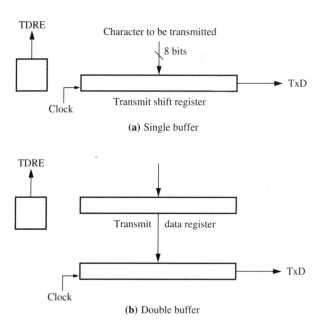

Figure 11-12   Transmitting asynchronous data.

As soon as the transmit shift register receives a character from the μP, it is full; it has no room to receive another character. This condition continues until the character is shifted out. The **TDRE** (Transmit Data Register Empty) signal is a status signal that is sent back to the μP. When the shift register is busy shifting out data, this signal is in its full state and warns the μP not to send another character. When the character has been totally shifted out, the register is again available, as indicated by the level on TDRE. The μP must wait until TDRE is empty before sending another character.

The **double buffer circuit** is shown in Figure 11-12b. Here the character is sent to a transmit data register. It is then sent to the transmit shift register, which empties the transmit data register so that it can receive another character. The transmit data register is only full if it has a character that it cannot send to the shift register because the shift register is still sending out the previous character. Again, TDRE informs the μP of the status.

The **68HC11** and most UARTs, which are ICs designed for this specific purpose, use double buffering. The **MC6850** is such an IC. It is called an ACIA (Asynchronous Communications Interface Adapter) and is used on the EVB to connect the EVB to a host terminal or PC (Personal Computer). The **MC6850** contains all the circuitry of Figure 11-11 within it. UARTs also allow the μP to receive data.

E X A M P L E   11 - 10

A μP must transmit several characters at 9600 bps. How long must it wait between characters if

a. Single buffering is used?

b. Double buffering is used?

Assume each character consists of 10 bits (8 data bits, 1 START bit, and 1 STOP bit).

Solution

Each bit requires 104 μs at 9600 bps. Thus each character will require 1.04 ms to transmit its 10 bits.

a. For single buffering the first character will be sent and the shift register will be busy for 1.04 ms. Then the second character will be sent, and so forth.

b. For double buffering the first character will be sent to the transmit data register and immediately transferred to the transmit shift register. This empties the data register, so μP can send the second character almost immediately. Now, however, both registers are full and succeeding characters will have to wait 1.04 ms before they can be sent to the transmit data register.

11-2.4   Receiving Data

The receive section of a UART receives data from an external device. The EVB is an example. When a user types a character on the PC or terminal keyboard, that character is transmitted to the EVB in asynchronous format. The character must then be received and acted upon by the EVB. The character is actually received by the **MC6850** ACIA.

The receive section works like the transmit section in reverse. It is shown in Figure 11-13. The data bits on the line are clocked into the receive shift register by the receive clock. When a character is fully received, the UART strips off the START and STOP bits and sends the character to the receive data register. This raises the **RDRF (Receive Data Register Full)** flag to inform the μP that data is available. When the μP reads the receive data register, RDRF is also reset.

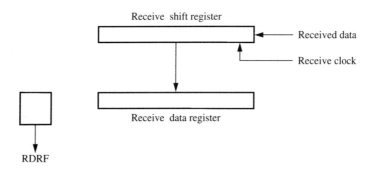

Figure 11-13   Receiving data.

Actually, the receive section of a UART is somewhat more complicated than the transmit section because it never knows when a character will arrive. The receive clock is shown in Figure 11-14. Typically, it is 16 times the **baud rate** (or bits per second*) of the data.

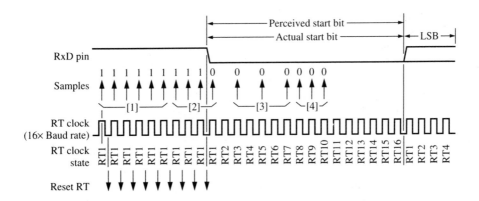

Figure 11-14   Start Bit—Ideal Case.   (*Redrawn courtesy of Motorola, Inc.*)

* The terms "baud" and "bit per second" (bps) are often used synonymously although they are not, strictly speaking, synonymous. In this context, however, they are synonymous, so both Motorola and the author prefer to avoid the distinction.

Consider the problem of the receive side of a UART. It is looking at a steady stream of 1s on the RxD line, indicating an idle condition. When the line goes down, it must assume that this is the beginning of a START bit. The receive section of the UART will expect the bits to change at the 16th, 32th, . . . , clocks because its clock is 16 times the data rate. To determine the pattern of the bits coming in, the receive side of the UART must examine them. The time at which it samples each bit is called the sample time. The receive side should sample the data at the middle of each bit for best accuracy.

E X A M P L E   11 - 11

Assume the START bit begins at clock 0. At which clock should the data for bit 0 be sampled?

Solution

The START bit lasts from clock 0 to clock 15. Bit 0 should then be on the line from clock 16 to clock 31. The line should be sampled in the middle of this time, or at clock 24.

The receive side of a UART is also capable of detecting errors. It can detect one or more of the following errors that can occur during data reception.

Parity error   If the transmission protocol uses parity, and if the parity of the received character is wrong, the receive side of the UART will report a parity error.

Framing error   Once a START bit is received, the UART knows where to expect the STOP bit. If the STOP bit is not present (there is a logic 0 when there should be a logic 1), this is reported as a framing error.

Overrun error   When a character comes in, the UART raises the RDRF. If the computer refuses to read the character and another character is received, the UART has no place to put the second character. It will then lose data. This is reported as an overrun error.

Noise error   If there is noise (spike or glitches on the RxD line), this is reported as a noise error.*

11-3   The Serial Communications Interface

The **Serial Communications Interface (SCI)** is basically a UART that is within the **68HC11**. It allows the **68HC11** to transmit and receive serial data in the asynchronous format. The SCI is associated with two port D pins (PD0 and

---

* Noise errors are beyond the scope of this book. They are thoroughly discussed in Chapter 9 of the *Motorola M68HC11 Reference Manual*, 1990. The latest edition of this manual is available from Motorola Literature Division, P.O. Box 20912, Phoenix, AZ 85036.

PD1) and eight registers. The eight registers are

The port D data register—at $1008

The port D direction register (DDRD)—at $1009

The SPCR register—at $1028

The BAUD register—at $102B

The SCI Control Register 1 (SCCR1)—at $102C

The SCI Control Register 2 (SCCR2)—at $102D

The SCI Status Register (SCSR2)—at 102E

The SCI Data Register (SCDR)—at 102F

Various bits of these registers control various functions. The best approach is to consider the SCI function by function instead of register by register.

11-3.1 Using Port D for General Purpose Input/Output

Bits 0 and 1 of port D are associated with the SCI but can also be used for general purpose I/O. If they are to be used for I/O, the SCI must be turned off. The TE (Transmit Enable) and RE (Receive Enable) bits control the transmit and receive sections of the SCI. If either of these bits is 0, that section of the SCI is disabled. These are bits 3 and 2, respectively, of SCCR2 at $102D (see Figure 11-17).

The three registers that control I/O on the six pins of port D are shown in Figure 11-15. At this time only bits 0 and 1 (PD0 and PD1) will be considered.

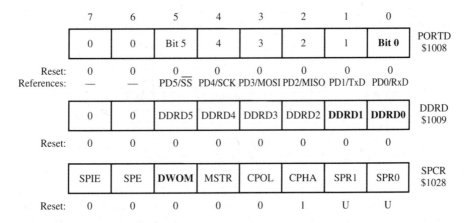

Figure 11-15 Port D I/O registers. (*Courtesy of Motorola, Inc.*)

If the SCI is disabled, the port D I/O is very much like port C I/O. First, the direction of the data must be established by writing the DDRD. Then the data can be sent out or brought in via the PORTD register. The only bit of the SPCR that affects port D is DWOM (port D Wire-Ored Mode). When this bit is a 1, the outputs from port D are open-drain and can be wire-ored. Normally, the bit is a 0.

E X A M P L E   11 - 12

Write a program segment to send out a logic 1 and logic 0 on PD1 and PD0, respectively.

Solution

The code could be

LDAA	$102D	Clear the TE and RE bits
ANDA	$#F3	in SCCR2
STAA	$102D	at 102D.
LDAA	#$03	Load 11 into the 2 LSBs of the
STAA	$1009	direction register to set it for
		output.
LDAA	#$02	Load the data (10)
STAA	$1008	into the port D data register.

11-3.2   The BAUD Register

If the SCI is going to transmit serial data, the speed of transmission or data rate, bps, must first be established. The BAUD register shown in Figure 11-16 controls the bit rate for both data transmission and reception.*

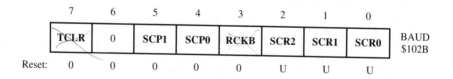

7	6	5	4	3	2	1	0	
TCLR	0	SCP1	SCP0	RCKB	SCR2	SCR1	SCR0	BAUD $102B

Reset: 0   0   0   0   0   U   U   U

Figure 11-16   **The BAUD register.** *(Courtesy of Motorola, Inc.)*

In the BAUD register the TCLR and RCKB bits are for the test mode only and will not be discussed. The two SPC bits determine the division factor for the basic crystal. Table 11-2 gives the baud rate for five different crystals. The EVB has an 8-MHz crystal, so only the second column of the

---

* Some UARTs allow the user to specify different rates for data transmission and reception. This is rarely needed and cannot be done on the SCI.

Table 11-2   Baud-Rate Prescale Selects

SCP1	SCP0	Division Factor	2^{23} Hz	8 MHz	4.9152 MHz	4 MHz	3.6864 MHz
			Crystal Frequency				
			Highest Baud Rate				
0	0	1	**131.072K baud**	125.000K baud	**76.80K baud**	62.50K baud	57.60K baud
0	1	3	43.691K baud	41.667K baud	25.60K baud	20.833K baud	19.20K baud
1	0	4	**32.768K baud**	31.250K baud	**19.20K baud**	15.625K baud	14.40K baud
1	1	13	10.082K baud	**9600 baud**	5.908K baud	4800 baud	4431K baud
			2.1 MHz	2 MHz	1.2288 MHz	1 MHz	921.6 KHz
			Bus Frequency (E clock)				

table applies. The three SCR bits are a further frequency division; each increment divides by 2. This is shown in Table 11-3. It starts with the so-called highest baud rate, which is one of the bold numbers in Table 11-2 and divides them down successively. Users of the EVB should use the last column, 9600 baud, in this table.

Table 11-3   Baud-Rate Selects

SCR2	SCR1	SCR0	Division Factor	131.072K Baud	32.768K Baud	76.80K Baud	19.20K Baud	9600 Baud
				Highest Baud Rate				
				SCI Baud Rate				
0	0	0	1	131.072K baud	32.768K baud	76.80K baud	19.20K baud	9600 baud
0	0	1	2	65.536K baud	16.384K baud	38.40K baud	9600 baud	4800 baud
0	1	0	4	32.768K baud	8192 baud	19.20K baud	4800 baud	2400 baud
0	1	1	8	16.384K baud	4096 baud	9600 baud	2400 baud	1200 baud
1	0	0	16	8192 baud	2048 baud	4800 baud	1200 baud	600 baud
1	0	1	32	4096 baud	1024 baud	2400 baud	600 baud	300 baud
1	1	0	64	2048 baud	512 baud	1200 baud	300 baud	150 baud
1	1	1	128	1024 baud	256 baud	600 baud	150 baud	75 baud

Table 11-4 is a simplified table for the coding of the BAUD register. The *Motorola M68HC11 Reference Manual* shows the rates for several crystals, but we have only included the table for the 8-MHz crystal. This table can be used directly with the EVB.

E X A M P L E   11 - 13

The SCI on the EVB is to be connected to a 300-bps terminal. What should be written into the BAUD register?

Solution

Three hundred baud is found as number 5 in Table 11-3 under the 9600-baud column. The 9600-baud rate is selected from Table 11-2 by making the SCP bits 11. Therefore a load of #$35 into the BAUD register will select a 300-baud rate. This can be checked by referring to Table 11-4.

Table 11-4 SCP and SCR Settings for Various Baud Rates Using a 2-MHz Crystal

SCP1	SCP0	SCR2	SCR1	SCR0	8 MHz
0	0	0	0	0	125.00K baud
0	0	0	0	1	62.50K baud
0	0	0	1	0	31.25K baud
0	0	0	1	1	15.625K baud
0	0	1	0	0	7812.5 baud
0	0	1	0	1	3906 baud
0	0	1	1	0	1953 baud
0	0	1	1	1	977 baud
0	1	0	0	0	41.666K baud
0	1	0	0	1	20.833K baud
0	1	0	1	0	10.417K baud
0	1	0	1	1	5208 baud
0	1	1	0	0	2604 baud
0	1	1	0	1	1302 baud
0	1	1	1	0	651 baud
0	1	1	1	1	326 baud
1	0	0	0	0	31.250K baud
1	0	0	0	1	15.625K baud
1	0	0	1	0	7812.5 baud
1	0	0	1	1	3906 baud
1	0	1	0	0	1953 baud
1	0	1	0	1	977 baud
1	0	1	1	0	488 baud
1	0	1	1	1	244 baud
1	1	0	0	0	9600 (+0.16%)
1	1	0	0	1	4800 baud
1	1	0	1	0	2400 baud
1	1	0	1	1	1200 baud
1	1	1	0	0	600 baud
1	1	1	0	1	300 baud
1	1	1	1	0	150 baud
1	1	1	1	1	75 baud
					2 MHz

E X A M P L E   11 - 14

The BAUD register on the EVB reads $21. What is the data transmission rate?

Solution

The SCP bits are 10, which means that the baud rate given in Table 11-2 is 31.250K baud, or each bit takes 32 μs. The SCR bits are 001, which indicates a division by 2. Therefore the baud rate is 15.625K baud and each bit takes 64 μs. This can also be checked by using Table 11-4.

11-3.3   Data Transmission Using the Serial Communications Interface

In Section 11-3 we stated that eight registers were involved with the SCI. The three registers shown in Figure 11-15 are only used for simple I/O, and the BAUD register simply sets the speed. Thus we only have to consider four registers and the problem is more manageable.

To transmit data using the SCI

1. The transmit side of the SCI must be enabled by making the TE bit in SCCR2 a 1.

2. The BAUD register must be set up for the proper data rate.

3. The data configuration must be set up.

4. The transmit data register must be empty.

5. Now a character can be sent to the SCDR (SCI Data Register), and it will appear as an asynchronous character on TxD, which is connected to bit 1 of port D (PD1).

The four registers involved with data transmission and reception are shown in Figure 11-17. In SCCR1 only the M bit and T8 are associated with data transmission. The M bit determines the length of the character to be transmitted. If M = 0, the character will have eight data bits or will be 10 bits long when the START and STOP bits are counted. If M = 1, the character will have nine data bits and will be 11 bits long. If M = 1, T8 will hold the ninth bit to be transmitted. Most systems use 8-bit data lengths. Some devices, however, require eight data bits plus two STOP bits. The second STOP bit can be included by specifying 9-bit data transmission and setting T8. On the EVB this register is initialized to 00 because 9-bit data transmission is seldom used.

The status register, SCSR2, contains only 2 bits that pertain to transmission. TDRE has been discussed previously. Data may not be sent to the transmit data register if TDRE = 0, which indicates that the register is full. The other bit is TC (Transmit Complete), which will be a 1 when no data is being transmitted. TC can be used to determine when power may be turned

7	6	5	4	3	2	1	0	
R8	T8	0	M	WAKE	0	0	0	SCCR1 $102C

Reset: U U 0 0 0 0 0 0

7	6	5	4	3	2	1	0	
TIE	TCIE	RIE	ILIE	TE	RE	RWU	SBK	SCCR2 $102D

Reset: 0 0 0 0 0 0 0 0

7	6	5	4	3	2	1	0	
TDRE	TC	RDRF	IDLE	OR	NF	FE	0	SCSR2 $102E

Reset: 1 1 0 0 0 0 0 0

SCDR $102F

7	6	5	4	3	2	1	0	
R7	R6	R5	R4	R3	R2	R1	R0	RDR (read)
T7	T6	T5	T4	T3	T2	T1	T0	TDR (write)

Reset: U U U U U U U U

Figure 11-17 The registers concerned with data transmission and reception using the SCI. (*Courtesy of Motorola, Inc.*)

off after a transmitting session by indicating when the last bit has been sent. When nothing is happening, SCSR2 reads C0, which indicates that there is no transmission (TC = 1), and the transmit data register can receive a character (TDRE = 1).

SCCR2 is the main control register for the SCI subsystem. The following bits in it pertain to data transmission:

SBK (bit 0)   A *break* is a character of all 0s. It can be used to signal various conditions between the transmitter and receiver. If SBK (Send BreaK) is a 1, the transmitter will send a break as soon as it finishes transmitting its current character and will continue to send break characters until SBK is reset. Every time SBK goes to 1, a stream of at least ten 0s will be sent.

TE (bit 3)   A 1 in Transmit Enable enables the transmit side of the SCI. A 0 configures PD1 as an I/O bit.

TCIE (bit 6)   A 1 in this bit will cause the **68HC11** to interrupt whenever TC in SCSR2 becomes a 1.

TIE (bit 7)   A 1 in this bit causes an interrupt every time TDRE in SCSR2 sets. It is slightly faster than waiting until TC sets if characters are sent out continuously.

There is only one interrupt that can be generated by the SCI. It vectors to C4–C6 on the EVB. There are, however, 4 bits in SCCR2 that can cause an interrupt. If one of several conditions can cause an interrupt, then SCCR2 will have to be examined to determine which bit caused the interrupt. On the EVB, SCCR2 is initialized to 0C, which indicates that both transmit and receive are enabled and all other bits are 0.

The SCDR is the data register for both transmit and receive. For transmit it is *write only*. The character to be sent is written to it and then sent out on TxD. This character cannot, however, be read by reading the SCDR.

E X A M P L E   11 - 15
_____

There are a set of ASCII characters in a buffer between $C100 and $C1FF. Write a program, using interrupts, to transmit these characters via the SCI. The baud rate should be 2400 bps.

Solution

The program is shown in Figure 11-18. It begins by defining the various registers associated with the transmit section of the SCI. The program starts by writing the interrupt vector for the SCI, defining the baud rate, and enabling interrupts on TDRE. It then goes into an endless loop, waiting for interrupts. Of course, the computer could be doing something useful while waiting for the interrupt.

The interrupt routine writes out the character and checks for the end of the data buffer.

_____

11-3.4   Data Reception Using the Serial Communications Interface

Characters that are received by the SCI must come in via pin PD0. They must have TTL levels; if they are originally RS-232, they must be converted by diode gates or a **1489**. The character format for both transmit and receive must be the same; they must both be either 8 or 9 bits long and have the same baud rate.

Characters come in a bit at a time on PD0 and are shifted into the SCDR register, where they are assembled into a byte and then can be read. Note that *the SCDR is write only for characters transmitted out and read only for incoming characters*.

The bits that affect data reception are discussed here. Some of these bits involve a "wake-up" feature. Wake-up is discussed in the next section.

On receive, SCCR1 simply provides a parking space for the ninth data bit, when 9-bit characters are being received. This is the only bit in SCCR1 that pertains to reception.

The status register, SCSR2, contains 5 bits that pertain to receive. The most important of these is RDRF (Receive Data Register Full–bit 5), which is set whenever an entire character has been received and is in the receive data register. This bit must be constantly monitored to determine if a character has been received, or it can cause an interrupt. RDRF is cleared by reading SCSR2 followed by a read of the SCDR.

```
                NAM     TRANS1
* This is a program to transmit a block of data using the SCI.
*-------START OF INITIALIZATION -------------------------
BAUD    EQU     $102B
SCCR1   EQU     $102C
SCCR2   EQU     $102D
SCSR2   EQU     $102E      Set up registers.
SCDR    EQU     $102F
STACK   EQU     $D100
POINT   FDB                Set stack.
*-------START OF PROGRAM ---------------------------------
BEGIN   LDS     #STACK
        LDX     #$C100     Set point to start of the buffer.
        STX     POINT
        LDAA    #$7E
        STAA    $C4
        LDD     #$C400     Set SCI vector on the EVB to
        STD     $C5        $C400.
        LDAA    #$32       Set baud rate to 2400 bps.
        STAA    BAUD
        CLR     SCCR1
        CLI
        LDAA    #$88       Set TE and TIE.
        STAA    SCCR2
*-------END OF INITIALIZATION ----------------------------
```

```
*-------MAIN PROGRAM  -----------------------------------------
SELF    BRA     SELF            Infinite loop--could do something
*                               useful.

*-------START OF INTERRUPT SERVICE ROUTINE  ------------------
        ORG     $C400
        LDX     POINT           Place character to be transmitted
        LDAA    0,X             in A.
        LDAB    SCSR2           Dummy read of SCSR2.
        STAA    SCDR            Put character to be transmitted in
*                               SCDR. This also resets TDRE.
        INX
        CPX     ##C200          Done?
        BNE     CONT
        SWI                     Yes.
        NOP
        NOP
CONT    STX     POINT           No. Prepare for next character.
        RTI
*-------END OF INTERRUPT SERVICE ROUTINE  -------------------
```

Figure 11-18 Program for Example 11-15.

A stream of characters is arriving at the SCI. The baud rate is 9600 bps. How often does RDRF have to be monitored to prevent losing a character and causing an overrun error? Assume each character contains eight data bits.

Solution

At 9600 bps each bit takes 104 μs. Each character is 10 bits long, so the minimum time for a character to arrive is 1.04 ms. Thus RDRF must be monitored at least every 1.04 ms to prevent overrun.

The IDLE condition occurs when the RxD line is idle for at least one character time (10 bits if M = 0, 11 bits if M = 1) of all 1s on PD0. If the receive line is active and then goes idle, the IDLE bit will be set. It can be cleared by reading SCSR2 and then reading SCDR. The IDLE bit can be useful in a system that must receive data and cannot transmit until the sending device stops. The system can monitor IDLE and start transmission when IDLE is a 1.

The other 3 bits in SCSR2 are error bits; they are OR (overrun error), NF (noise flag), and FE (framing error). These were discussed in the previous section.

SCCR2 is the main control register for both transmit and receive. The bits that pertain to reception are (in order of importance)

RE (Receive Enable)   A 0 in this bit allows PD0 to be used for normal I/O. It can be input or output depending on bit 0 of DDRD. A 1 in this bit enables the receive section of the SCI, and PD0 is strictly an input pin.

RIE (Receive Interrupt Enable)   A 1 in this bit will cause the SCI to interrupt whenever RDRF goes high. A 0 means the receive side must be in the polling mode to detect when RDRF goes high.

ILIE (Idle Line Interrupt Enable)   Setting this bit will cause the SCI to interrupt whenever the IDLE bit goes to 1.

a. Write a program to determine if a character has been received on the SCI.

b. If a character has been received, check for errors. JUMP to a subroutine called ERROR if any errors occur.

c. Place the character in a buffer. Stop if $(80)_{10}$ characters have been received or if the character is a RETURN (0D).

Solution

The program could be

```
        LDX     #$C100          Initialize.
        LDY     #$1000
```

```
LOOP1   BRCLR   2E,Y 20   LOOP1     Wait for TDRF.
        LDAA    2E, Y               Load the status register
        BITA    #$0E                and check for errors.
        BNE     ERROR
        LDAA    2F,Y                Load the character. This
        STAA    0,X                 also clears RDRF.
        CMPA    #$0D                Is it 0D?
        BEQ     DONE
        INX                         Increment character count.
        CPX     #$C051              If not finished, go back and
        BNE     LOOP1               wait for the next character.
DONE
```

11-3.5 Wake Up*

The **wake-up feature** in the SCI might better be called a go-to-sleep feature. It is designed for a large system, where one computer is transmitting a message to several other computers using the same asynchronous bus. Although all devices on the bus receive the same message, the message may be destined only for a particular device and should be ignored by all other devices. Thus some sort of addressing scheme must be used to designate the proper receiver. Usually, the first byte in the message contains the address in this type of system.

The RWU (Receiver Wake-Up) bit, bit 1 in SCCR2, determines the wake-up status of the receiver. If this bit is 0, the receiver is in its normal mode, awake. If the bit is 1, the receiver is sleeping, which means it is ignoring the current message on the line because that message did not have the computer's address and is, presumably, for another device. While the receiver is sleeping, it ignores all the signals associated with the receive side of the SCI, such as RDRF and the errors, so they will not disturb the μP, which is then free to perform other functions.

When wake-up is being used, the receiver typically is awake or wakes up when the first character of a message is received. It decodes this address. If it is not the designated recipient, it goes to sleep by setting RWU.

There are two ways to wake up a sleeping receiver. These are selected by the WAKE bit, bit 3 of SCCR1. If WAKE is a 0, the receipt of an IDLE character will reset RWU and wake up the receiver. Using this protocol, the transmitter must separate all messages with at least one IDLE character. The situation is

1. The transmitter transmits an IDLE character. This wakes up all the μPs on the line.

---

* This is an advanced topic that may be omitted at the first reading.

2. All the μPs decode the first character sent.

3. The designated addressee stays awake; it does not set RWU. All other μPs set RWU and go to sleep.

4. The message is transmitted as a steady stream of characters. Only the μP(s) that are awake monitor it.

5. At the conclusion of the message the transmitter sends at least one idle character. This wakes up all the μPs so that they can await the next message.

If WAKE is a 1, address mark protocol is being used. This protocol generally uses 9-bit characters; a 1 in the MSB of the character indicates that the character is an address. All the μPs on the bus wake up and decode the address. Those μPs that should ignore the message set RWU and go to sleep. All the data characters in the message are sent with an MSB of 0.

The transmitter starts the next message by sending out another address with an MSB of 1. This wakes up all the μPs so that they can decode the address.

E X A M P L E   11 - 18

A μP is being used in the wake-up idle line protocol. It is to interrupt on RDRF. Describe the major points of the program.

Solution

During initialization WAKE must be cleared and RIE set to enable interrupts.

When a character interrupts, if it is the first character of a message it will be decoded by comparing the character with the address for that particular μP.

If the message is not for that μP (the address did not match), the μP will set RWU and go to sleep. The μP will not respond to any characters (RDRF will not set or cause interrupts) until it is awakened by an idle line character. Then the next idle character received will cause the μP to interrupt again.

11-4   The Serial Peripheral Interface

The **Serial Peripheral Interface (SPI)** enables several **68HC11s** in the same system to communicate rapidly using serial data. Because any system that uses several **68HC11**s is a large, advanced system and this is supposed to be a small, introductory book, the SPI will not be described in great detail.*

The SPI uses pins PD2 to PD5 for I/O and is controlled by the two registers shown in Figure 11-19. The SPCR, in $1028, is the SPI control

SPCR—SPI Control Register

This register, which may be read or written at any time, is used to configure the SPI system. The DDRD register must also be properly configured before SPI transfers can occur.

7	6	5	4	3	2	1	0	
SPIE	SPE	DWOM	MSTR	CPOL	CPHA	SPR1	SPR0	$1028

Reset: 0   0   0   0   0   1   U   U

SPSR—SPI Status Register

This read-only register contains status flags indicating the completion of an SPI transfer and the occurrence of certain SPI system errors. The flags are automatically set by the occurrence of the corresponding SPI events; the flags are cleared by automatic software sequences.

7	6	5	4	3	2	1	0	
SPIF	WCOL	—	MODF	—	—	—	—	$1029

Reset: 0   0   0   0   0   0   0   0

Figure 11-19 **The registers that control the SPI.** (*Courtesy of Motorola, Inc.*)

register and the SPSR in $1029 is the status register. Perhaps the most important SPI bit for our purposes is bit 6 of the SPCR. This is SPE (SPI Enable). If this bit is 0 the SPI is turned off and pins PD2 to PD5 can be used for general purpose I/O. Their direction is then determined by DDRD.

When the SPI is active (SPE = 1), one **68HC11** acts as the master and the rest are slaves. Communication between the master and its slaves uses pins PD2 to PD5, as shown in Figure 11-20. The top **68HC11** is the master and the other two are the slaves.

The four pins involved in SPI communications are

Port D pin 2—MISO Master-In Slave-Out. If the μP is the master, it accepts messages from the slave on this line. If the μP is a slave, it can send data out on this line only if DDRD2 is a 1 and $\overline{SS}$ is a 0. Otherwise the driver for this line goes to high impedance, so another slave can drive the line.

* A more thorough discussion of the SPI is provided in the *Motorola* **68HC11** *Reference Manual*, 1990. The latest edition of this manual is available from Motorola Literature Division, P.O. Box 20912, Phoenix, AZ 85036.

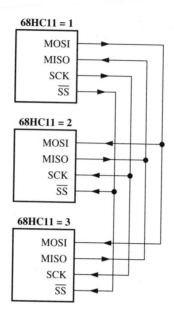

Figure 11-20   A master and two slave processors using the SPI.

Port D pin 3—MOSI   Master-Out Slave-In. The master sends its message out via this line and the slave can read it. DDRD3 must be a 1 on the master μP.

Port D pin 4—SCK   System Clock. It is controlled by the master and is an input to the slaves. DDRD4 must be a 1 on the master μP.

Port D pin 5—$\overline{SS}$ Slave Select. A low on this pin designates the device as an active slave, which means it sends data out via MISO. If the μP is a master and DDRD5 is 0, the μP is also being selected as a slave. This causes an SPI mode-fault error. If DDRD5 is 1 in the master, the $\overline{SS}$ level is ignored.

Data transmission is initiated when the master

1. Selects a slave and sets $\overline{SS}$ to 0 for that slave only.

2. Writes a byte into the SPDR (SPI Data Register) at $100A.

3. Clocks the data out by activating SCK.

While data is being clocked out on MOSI, data is also coming in from the selected slave on MISO. This data enters the SPDR, where it can be read.

Besides SPE, the SCPR (SPI Control Register) contains the following bits:

SPIE (bit 7)   The SPI Interrupt Enable bit causes an interrupt when certain conditions occur in the status register.

DWOM (bit 5) If this bit is a 1, all outputs on port D are open-drain. Many systems that use the SPI will set DWOM so that any μP may drive one of the port D lines.

MSTR (bit 4) This bit determines whether the μP is the master or the slave. A 1 makes the μP a master.

CPOL, CPHA, SPR1, SPRO (bits 3–0) These 4 bits determine the frequency and phase of SCK. See the *Motorola M68HC11 Reference Manual* for further details.

The SPI Status Register (SPSR) contains only 3 bits. They are

SPIF (bit 7) SPI transfer complete Flag. It is set when the current data transfer is complete to indicate that another transfer may begin.

WCOL (bit 6) Write COLlision. This bit is set whenever a write to the SPDR is attempted while a transfer is in progress.

MODF (bit 4) MODe Fault. If a master is being told by SS that it is a slave, a problem exists. The MODF bit indicates this.

SPIF or MODF will cause an SPI interrupt if SPIE is set in the SPCR.

Summary

This chapter started by discussing the A/D converters on the **68HC11**. Up to eight A/D converters are available. Single channels can be converted or they can be converted in two groups of four each.

The general principles of asynchronous data transmission and reception were examined next, including RS-232. Then the use of asynchronous data transmission using PD1 and reception using PD0 were considered. This is the function of the SCI on the **68HC11**. The wake-up feature of the SCI for reception was also discussed.

Finally, the SPI was introduced. It is used in large systems for serial communications between several μPs.

Glossary

Asynchronous The word means unclocked. It is used for data transmission without a synchronizing clock accompanying it.

Baud rate Used here synonymously with the transmission rate; the number of bits per second (bps).

Double buffer circuit The ability of a circuit, generally a UART, to hold two data characters during data transmission or reception.

Framing error An error that occurs when a STOP bit does not appear in proper relation to the START bit.

Overrun error An error that occurs when the computer does not respond to RDRF.

RDRF—Receive Data Register Full   A bit indicating there is a character in the receive data register that should be read by the μP.

Resolution   The smallest voltage that can be distinguished by the A/D converter.

RS-232   A set of signals and levels used in serial data transmission.

SCI—Serial Communications Interface   Used for transmitting serial data, generally between a μP and a terminal.

Serial data transmission   Data transmission a bit at a time, usually over a single wire. Data transmission in both directions, however, may occur simultaneously.

SPI—Serial Peripheral Interface   Used in a large system for transmitting serial data among several μPs.

TDRE—Transmit Data Register Empty   A bit indicating a character to be transmitted can be sent to the transmit data register.

UART (Universal Asynchronous Receiver-Transmitter)   An IC that converts parallel data, usually a character, into serial data for asynchronous transmission. It also converts serial data that it receives into parallel data for the computer.

Wake up   The ability of a receiver to wake up and decode the address of a message. The receiver will then go to sleep (ignore the message) if it is not the addressee, and wake up when the address of the next message is presented.

Problems

Section 11-1.1

For the following problems assume V_{RH} is 4.5 V and V_{LH} is 0.5 V.

11-1 What would the digital output be if the analog input is
    a. 3 V   **b.** 4 V.

11-2 Find the analog input voltage if the digital output is
    a. 44   **b.** CC.

11-3 What is the digital output voltage for the waveform of Figure P11-3 at $t = 7.5$ ms?

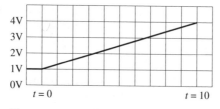

Figure P11-3

Section 11-1.3

11-4 What is one advantage of having the CCF flag in bit 7 of the ADCTL?

11-5 **a.** Set up the ADCTL for continuous conversions on channel 2.
   b. Write a program to average the four conversions as soon as possible.
   c. Write a program to average the four conversions once every 65.55 ms. (Use the RTI interrupt.)

11-6 The input to PE4 can vary from 0 to 5 V. Write a program to turn a light on whenever the input is greater than 2 V.

11-7 Using the EVB, write a program to detect how long it takes CCF to set after a write to ADCTL.

11-8 The inputs to PE1 and PE2 can both vary from 0 to 5 V. Write a program to turn on a light whenever the input on PE1 exceeds 1.5 V or the input on PE2 exceeds 4 V.

11-9 Repeat Example 11-6 if the input frequency is 10 kHz.

Section 11-2.1

11-10 Repeat Example 11-8 if the input character is
   a. C   **b.** D   **c.** 9.

11-11 What message is given by the data stream of Figure P11-11? Is the parity odd or even?

Figure P11-11

Section 11-2.2

11-12 Sketch the letter E (see Fig. 11-8) after it has been transformed to RS-232 levels.

Section 11-2.3

11-13 Sketch TDRE if a system has just started to transmit data at 300 bps, with one START bit, eight data bits, and two STOP bits. (Use double buffering.)

Section 11-2.4

11-14 The letter A is being received. Sketch the letter along with the clock. Show which clock samples each bit.

Section 11-3.1

11-15 Write a program to bring in whatever data is on PD0 and send it out on PD1.

Section 11-3.2

11-16 What should be written into the BAUD register if transmission at 2400 baud is required?

11-17 What is the data rate if the BAUD register contains
   a. 15   **b.** 24   **c.** 36.

Section 11-3.3

11-18 For continuous transmission explain why sending data whenever TDRE is set is faster than sending data whenever TC is set.

11-19 Write a program to transmit data in a buffer from $C100 until either
   a. Eighty (decimal) characters have been transmitted, or
   b. An ASCII $0D is found in the buffer. This is a carriage return character.
   Do not use interrupts, but poll TDRE. In the EVB monitor a polling subroutine, called SCI2, starts at $E461 as shown in Figure P11-19. You may use this subroutine in your program.

11-20 Explain what is happening in Figure P11-19.

11-21 Five 9-bit characters are stored in locations C000–C009. The MSB of each character is the LSB of the odd-numbered location. Write a program to transmit them.

11-22 In Figure 11-18, which instruction controls the baud rate? Which instruction controls the number of characters to be sent out?

11-23 In Figure 11-18, modify the program to
   a. Send out three characters.
   b. JUMP to the 100-ms delay subroutine of Figure 10-4.
   c. Repeat.
   Note that it is necessary to JUMP back to BEGIN after the subroutine to reload the SP.

     This problem makes an interesting laboratory experiment. The characters coming out can be seen on the oscilloscope. The program has been positioned so that there is no address conflict with the delay subroutine and NOPs have been included so that the SWI can be replaced by a JMP.

```
0951 E453 81 0D          CMPA  #$0D      If CR, send LF.
0952 E455 26 04          BNE   OUTSCI1
0953 E457 86 0A          LDAA  #$0A
0954 E459 20 06          BRA   OUTSCI2
0955 E45B 81 0A  OUTSCI1 CMPA  #$0A      If LF, send CR.
0956 E45D 26 0E          BNE   OUTSCI3
0957 E45F 86 0D          LDAA  #$0D
0958 E461 F6 10 2E OUTSCI2 LDAB  SCSR    Read status.
0959 E464 C5 80          BITB  #$80
0960 E466 27 F9          BEQ   OUTSCI2   Loop until TDRE=1.
0961 E468 84 7F          ANDA  #$7F      Mask parity.
0962 E46A B7 10 2F       STAA  SCDAT     Send character.
0963 E46D 39     OUTSCI3 RTS
```

Figure P11-19

Section 11-3.4

11-24 If the BAUD register is set at 35 and $M = 1$, how often must RDRF be monitored to prevent overrun?

11-25 A system is receiving data and has data to transmit. It cannot start transmission until 50 ms after it has completely received its data. Write a program segment to determine when transmission can start.

11-26 A system is receiving 7-bit characters with even parity and two STOP bits. Write program segments to detect
a. A parity error on reception.
b. A framing error where the first STOP bit is a 0.
c. A framing error where the second STOP bit is a 0.

11-27 Repeat Example 11-17 using interrupts when a character is received.

Microprocessor Control and Memories

12

INTRODUCTION

This chapter discusses those registers that control the operation of the **68HC11**, the memories, and the hardware configuration of the EVB.

INSTRUCTIONAL OBJECTIVES

After reading the chapter, the student should be able to

- Award a source of interrupts the highest interrupt priority.
- Place the internal registers and internal RAM at specific locations in the memory map.
- Write programs or data into the EEPROM.
- Explain the function of each IC on the EVB.

SELF-EVALUATION QUESTIONS

Watch for the answers to the following questions as you read the chapter. They should help you to understand the material presented:

1. When can bits be written into the HPRIO?
2. How can a system determine whether it has come up in the single-chip or the expanded multiplexed mode?
3. What is the difference between the CONFIG registers in the **68HC11A8** and the **68HC11A1**? What does the difference signify?
4. Why must there be ROM available in addresses $FFC0–$FFFF?

12-1 The Control Registers

There are several *control registers* that determine the operation of the **68HC11**, especially when power is first applied to the system. Many of the bits in these control registers, however, apply only to the test modes of the **68HC11** operation, a topic beyond the scope of this book. The function of these control registers is described in this section, with an emphasis on those bits that apply to the two normal modes (single-chip and expanded multiplexed) of operation.

12-1.1   The High Priority Register

The HPRIO (High PRIOrity) register at $103C is shown in Figure 12-1. The higher 4 bits of this register are determined by the mode of operation. As stated in Section 8-2.4, the mode is selected by the levels on the MODA

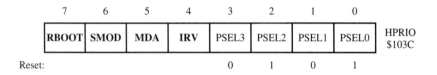

7	6	5	4	3	2	1	0	
RBOOT	SMOD	MDA	IRV	PSEL3	PSEL2	PSEL1	PSEL0	HPRIO $103C

Reset: 0   1   0   1

Figure 12-1   The HPRIO register. (*Courtesy of Motorola, Inc.*)

and MODB pins when the **68HC11** is reset. Table 12-1 shows the four modes that can be established at RESET, and the bits that are written into the HPRIO at that time. In most cases these bits are not changed during normal operation of the **68HC11**.

Table 12-1   Hardware Mode Select Summary

Inputs			Control Bits in HPRIO (Latched at RESET)			
MODB	**MODA**	**Mode Description**	**RBOOT**	**SMOD**	**MDA**	**IRV**
1	0	Normal single chip	0	0	0	0
1	1	Normal expanded	0	0	1	0
0	0	Special bootstrap	1	1	0	1
0	1	Special test	0	1	1	1

The SMOD bit is written as a 1 during RESET if a test mode (special bootstrap or special test) is being established. This occurs if MODB is 0 during RESET. During operation in a test mode this bit may be cleared and normal operation will commence. It may not, however, be changed from 0 to 1. Thus a **68HC11** operating in a normal mode may not enter a test mode without a RESET.

The RBOOT and IRV bits are writable only when SMOD is 1, which indicates the **68HC11** is in one of the two test modes. They are not used in normal operation. The MDA bit is used in normal operation and indicates whether the **68HC11** is in the single-chip or expanded mode. This bit may be read by the operating system, if it must determine in which mode the **68HC11** is operating. The MDA bit is written at RESET and can only be changed if SMOD is 1.

The four PSEL (Priority SELect) bits in the HPRIO determine which source of *maskable interrupts* has the highest priority if several devices interrupt simultaneously and there is a contention problem. The highest priority interrupt source is specified by the PSEL bits in accordance with Table 12-2. If an interrupt occurs, the priorities are determined by the following list:

The highest priority, as determined by the PSEL bits

The IRQ interrupt

The real-time interrupt

The input capture registers

Table 12-2   Highest Priority Interrupt Versus PSEL3–PSEL0

PSEL3	PSEL2	PSEL1	PSEL0	Interrupt Source Promoted
0	0	0	0	Timer overflow
0	0	0	1	Pulse accumulator overflow
0	0	1	0	Pulse accumulator input edge
0	0	1	1	SPI transfer complete
0	1	0	0	SCI serial system
0	1	0	1	Reserved (default to $\overline{\text{IRQ}}$)
0	1	1	0	$\overline{\text{IRQ}}$ (external pin or parallel I/O)
0	1	1	1	Real-time interrupt
1	0	0	0	Timer input capture 1
1	0	0	1	Timer input capture 2
1	0	1	0	Timer input capture 3
1	0	1	1	Timer output compare 1
1	1	0	0	Timer output compare 2
1	1	0	1	Timer output compare 3
1	1	1	0	Timer output compare 4
1	1	1	1	Timer output compare 5

The output compare registers

The timer overflow

Pulse accumulator overflow

Pulse accumulator input edge

SPI

SCI

The PSEL bits can only be written if the I bit in the CCR is 1. This eliminates the possibility of trying to change priorities while an interrupt is in progress or impending.

E X A M P L E   12 - 1

Write a code segment to set OC4 as the highest priority source of interrupts.

Solution

Table 12-2 shows that PSEL 3, 2, and 1 must be 1 and PSEL0 must be 0 to select OC4. The code segment could be

SEI		Disable interrupts so that HPRIO can be written.
LDAA	#$0E	Set the bits properly.
STAA	$103C	Write the bits into the PSEL bits in HPRIO.
CLI		Enable interrupts.

In this example we assumed that SMOD was 0, so the other three high-order bits could not be written. We arbitrarily used 0s in their place.

E X A M P L E   12 - 2

Explain how to test the code segment of Example 12-1 to determine if OC4 is actually getting the first interrupt.

Solution

The following procedure can be used on the EVB:

1. Clear $1022. This clears all the interrupt bits in TMSK1, so no interrupts are allowed from the input capture or output compare registers.

2. Set the vectors for interrupts OC2, OC3, OC4, and OC5 to four different locations in memory.

3. Put breakpoints in at each starting location. Only four registers are used here because the EVB only allows four breakpoints.

4. Jump to the program of Example 12-1. At its conclusion there will be no interrupts. All the interrupts are disabled because their mask bits are 0.

5. Write #$78 into TMSK1. This enables interrupts from OC2, OC3, OC4, and OC5 simultaneously.

As soon as step 5 is executed, the program should jump to the breakpoint for OC4, indicating that OC4 is the highest priority interrupt.

The HPRIO comes up on the EVB as $25, which indicates the **68HC11** is operating in the normal expanded mode and the highest priority interrupt is 5, which defaults to IRQ.

12-1.2 The Configuration Register

The CONFIG (CONFIGuration) register at $103F determines the configuration of the memory and other system attributes. It is actually an EEPROM (Electrically Erasable Programmable Read Only Memory). Only 4 bits of this register are implemented and changing them is somewhat difficult and rarely done (see Sec. 12-2.3).

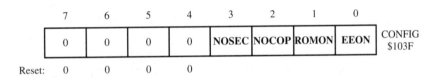

Figure 12-2   The CONFIG register. (*Courtesy of Motorola, Inc.*)

The 4 bits are

NOSEC   A 1 in this bit means no security. **Security** is an attempt to prevent a software pirate from seeing what is the ROM or EEPROM on the **68HC11**. To use security, the user must first specify a security option when the ROM is programmed. Then security can be invoked by setting this bit to 0. Most systems do not use security and this bit is usually a 1.

NOCOP   A 0 in this bit enables the COP (Computer Operating Properly) option, which is discussed in Section 12-1.4.

ROMON   A 1 in this bit enables the internal, 8K-byte, on-chip ROM.

EEON   A 1 in this bit enables the internal EEPROM (see Sec. 12-2.3) and allows it to function. A 0 disables the EEPROM.

In the **68HC11A8**, which uses the internal ROM, CONFIG is generally $0F. The **68HC11A1**, which is on the EVB, uses the Buffalo monitor ROM, which is in an external chip. Therefore the internal ROM is disabled. The CONFIG register on the EVB is $0D, which indicates no security, the COP system is disabled—an external ROM being used—and the internal EEPROM is available.

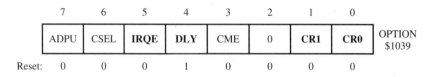

7	6	5	4	3	2	1	0	
ADPU	CSEL	IRQE	DLY	CME	0	CR1	CR0	OPTION $1039

Reset: 0   0   0   1   0   0   0   0

Figure 12-3   The OPTION register. (*Courtesy of Motorola, Inc.*)

12-1.3   The OPTION Register

The OPTION register, at $1039, also contains several bits that control the operation of the **68HC11**. It is shown in Figure 12-3. The ADPU (Analog-Digital Power Up) and CSEL (Clock SELect) bits have already been discussed in Section 11-1 on the A/D converter. The CR1 and CR0 bits are used with the COP and are discussed in the next section. This leaves 3 bits to be explained.

■ **IRQE (bit 5)**   This bit determines whether the interrupt requests presented on the IRQ pin are level-sensitive or edge-sensitive. The distinction has already been discussed in Section 8-7.5. If IRQE is clear, the IRQ pin is level-sensitive; it will attempt to interrupt whenever it is low. If interrupts from several sources are to come in via IRQ, they must be level-sensitive so that they can be wire-ORed. For level-sensitive interrupts the interrupt service routine must reset the cause of the interrupts.

If IRQE is set, an IRQ interrupt will be generated whenever there is a falling edge on the IRQ pin. Only one device can cause interrupts in this case.

■ **DLY (bit 4)**   Execution of the STOP instruction stops the clocks to conserve power. When the **68HC11** comes out of the stopped state, in response to a XIRQ or unmasked IRQ signal, the oscillator resumes but may require some time to stabilize. If DLY is 1, the **68HC11** will remain stopped for about 4000 E-clock cycles to allow this to occur. If DLY is 0 the **68HC11** will restart almost immediately.

E X A M P L E   12 - 3

Why is it advantageous to have DLY set?

Solution

When the **68HC11** enters the stopped state it is usually for an indefinite time. Allowing the delay adds 4000 cycles to this time, but that is only 2 ms. In most cases the additional 2 ms is insignificant and it is wise to allow the clock to stabilize before resuming processing. Figure 12-3 shows that DLY is normally initialized to 1.

- **CME (bit 3)** The **68HC11** contains a *clock monitor* that can cause a clock monitor interrupt if the E clock is too slow. The clock monitor interrupt vector is at $FFFC and $FFFD. The frequency will vary somewhat, but in general a frequency of 10 kHz or less will definitely cause a clock monitor interrupt and a frequency of 200 kHz or more will not.

The **CME (Clock Monitor Enable)** bit allows a clock monitor interrupt if it is set. This interrupt can be used to warn the operator that the **68HC11**'s clock is not functioning properly and the rest of the system is, therefore, also suspect.

If the clock monitor is to be used in systems where the STOP instruction may also be used, CME should be cleared before the STOP instruction is issued. CME can be set again after the interrupt that causes the system to restart.

12-1.4 The Computer Operating Properly System

The **COP** system can be invoked by clearing the NOCOP bit in the CONFIG register. This is not a simple procedure. The CONFIG register is EEPROM. A second complication is that the COP system will not be invoked until the NOCOP bit is cleared and the **68HC11** is then reset.

The COP system, when invoked, does act as a sort of traffic cop. Perhaps a parole officer would be a better analogy. The system must report periodically that it is working by resetting the COP timeout. The program that uses the COP must execute a section of code every so often; if it does not it is assumed to have failed and the cop "blows the whistle" by executing a COP interrupt. The COP failure interrupt is at FFFA and FFFB.

The COP timeout period is given in Table 12-3. It can vary between 16 ms and 1 s as determined by the CR1 and CR0 bits in the OPTION register. If the COP timeout is not reset within the specified time interval, the COP interrupt will occur.

Table 12-3 Computer Operating Properly Watchdog Rates
Versus Crystal Frequency

			Crystal Frequency		
			2^{23} Hz	8 MHz	4 MHz
CR1	CR0	E $\div$ 12^{15} Divided By	Nominal Timeout		
0	0	1	15.625 ms	16.384 ms	32.768 ms
1	0	4	62.5 ms	65.536 ms	131.07 ms
1	0	16	250 ms	262.14 ms	524.29 ms
1	1	64	1 s	1.049 s	2.1 s
			2.1 MHz	2 MHz	1 MHz
			Bus Frequency (E Clock)		

The COP timeout is reset by writing $55 into the COPRST register at $103A and then writing $AA into the same register. Most programs that invoke the COP will have a COP RESET SUBROUTINE that they must jump to periodically. Thus if the program gets hung up in an endless loop, or for some other reason, it will fail to reset the COP timeout and an interrupt will be generated. This interrupt can be used to indicate a system failure.

E X A M P L E   12 - 4

Write a subroutine to reset the COP timeout every 1 s.

Solution

During initialization CR1 and CR0 must both be set to specify a 1-s timeout, as Table 12-3 indicates.
The subroutine might be

```
LDAA    #$55
STAA    $103A       Put $55 in COPRST.
COMA                Complement A to make it $AA.
STAA    $103A       Put $AA in COPRST.
RTS
```

12-1.5   Interrupt Vectors on the Evaluation Board

On RESET the EVB vectors to $E000, the start of the monitor program that is contained in the Buffalo ROM. The clock monitor, COP, illegal Op code, and SWI vector to $FD, $FA, $F7, and $F4, respectively. During initialization, however, the monitor puts a JUMP to $E35A in each of these vectors, so they will all act as an SWI. Of course, the user can change these locations after the **68HC11** is initialized.

12-2   The Memories

As mentioned previously, the **68HC11** contains several memories within the IC chip. These consist of ROM, RAM, and EEPROM. The amount of each type of memory varies with the particular version of the **68HC11** being used. We will concentrate on the most commonly used versions, the **68HC11A8** and the **68HC11A1**, which is used in the EVB.

12-2.1   The Read Only Memory

The **68HC11A8** contains 8K bytes of internal Read Only Memory (ROM). This memory is enabled by placing a 1 in the ROMON bit in the CONFIG register. The CONFIG register for the **68HC11A8** is $0F. Of course, those who want to take advantage of the internal ROM must specify what the

ROM contains and whether the security option will be invoked at the same time. The specifications for the ROM must be given to the manufacturer and the program must be correct because it cannot be changed. This can become expensive and time-consuming. It is a valuable feature, however, when large quantities of μPs using the same program are required. The μP controlling the ECM (Electronic Control Module) in an automobile is an example.

The **68HC11A1** is basically the same as the 'A8 except that its internal ROM is disabled. Its CONFIG register is set to $0D, which indicates the ROMON bit is 0, disabling the ROM. It is used in the EVB, where an external chip, the Buffalo monitor, is the ROM instead of an internal ROM. The Buffalo monitor starts at $E000.

All versions of the **68HC11** must have ROM in locations $FFC0–$FFFF to retain the addresses of the interrupt service routines, and also to allow the **68HC11** to start properly after a RESET.

E X A M P L E   12 - 5

A user wants to specify an 8K-byte ROM that responds to addresses $A000–$BFFF. What is the problem?

Solution

The problem is that there must be ROM in locations $FFC0–$FFFF to contain the reset and interrupt vectors. Either the ROM must be made to respond to these addresses (see Problem 12-5) or additional ROM will be required.

There is also a small on-chip ROM from $B4F0 to $BFFF. This is only used in the bookstrap mode of operation.

12-2.2 The Random Access Memory

There are two read-write (RAM) memories in both the 'A8 and the 'A1—a 256-byte RAM and a 64-byte RAM that contains the control registers. Strictly speaking, however, not all bytes of the control registers are RAM; some are EEPROM and some are protected (see Sec. 12-2.4).

In the EVB the 256-byte RAM goes from $0000 to $00FF, but this RAM is used to hold the interrupt pseudo-vectors that we have been using throughout this book. They direct the interrupt routines to the user-selected starting points. They can be changed, but pseudo-vectors are written into this RAM by the monitor on RESET, so we advise against writing anything but these pseudo-vectors in this area of memory.

The locations of the RAM memories can be changed by using the INIT (INITialization) register. It is shown in Figure 12-4, and is at $103D. The higher 4 bits of INIT contain the top 4 bits of the address of the 256-byte RAM and the lower 4 bits contain the top 4 bits of the location of the registers. The remaining 12 bits are assumed to be 0 in each case.

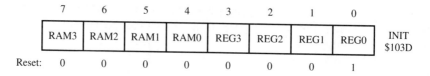

7	6	5	4	3	2	1	0	
RAM3	RAM2	RAM1	RAM0	REG3	REG2	REG1	REG0	INIT $103D

Reset: 0   0   0   0   0   0   0   1

Figure 12-4   The INIT register. (*Courtesy of Motorola, Inc.*)

The bits in the INIT register are protected (see Sec. 12-2.4). On the EVB they cannot be changed.

E X A M P L E   12 - 6

What are the contents of the INIT register in the EVB?

Solution

In the EVB the RAM starts at $0000, so the top 4 bits must be 0s. The registers start at $1000 so that the lower 4 bits must be the same as the higher 4 bits in this address or 0001. Thus the INIT register should contain $01. It can be examined by the MD command on the EVB.

The internal RAM in the 'A1 is insufficient to hold user programs and data and is committed to the pseudo-vectors. Additional RAM must be provided. The EVB, for example, contains 8K bytes of user RAM at locations $C000–$DFFF.

12-2.3   The EEPROM

Many versions of the **68HC11** contain an Electrically Erasable Programmable Read Only Memory (EEPROM). Both the 'A1 and the 'A8 have 512 bytes of EEPROM in memory locations $B600–$B7FF.

The EEPROM acts as a ROM in the sense that the data written into it will be retained when power goes down. It acts as a RAM in the sense that it can be written by the user. It must be written, however, while the machine is "offline." It cannot be written while executing a program; an instruction such as STAA $B600 will not write anything into the EEPROM location at $B600. Typical uses for the EEPROM are

Programs that will be used many times.

Subroutines that will be used over and over again.

Identification numbers that identify a particular **68HC11**.

Storing data or tables that will not change often but are used often.

The MM (Memory Modify) command on the EVB will modify EEPROM locations; the Buffalo monitor contains the code that will identify an EE-

PROM location and respond accordingly. The ASM (ASseMble) command, however, will only work on RAM locations and cannot be used to write code into the EEPROM.

E X A M P L E   12 - 7

A program or subroutine is to be written into the EEPROM on the EVB. How can this be done?

Solution

The program should first be written into RAM using ASM. Then the machine code for the program can be copied down on a piece of paper. Now the machine code can be written, a byte at a time, into the EEPROM using the MM command.

* The EEPROM operates by trapping charges on a **floating gate** within the memory. It is generally programmed in two steps:

1. The location to be programmed is erased to eliminate any charges that may be on the floating gate.

2. The new data is written to the floating gate at this location.

To erase or write the EEPROM, a programming voltage of about 20 V must be applied to the floating gate for 10 ms. This programming voltage is developed from an on-chip charge pump, and no external voltages are required.

The erasing and programming of the EEPROM is controlled by the PPROG register shown in Figure 12-5. This register is at $103B. The function of the bits is described below.

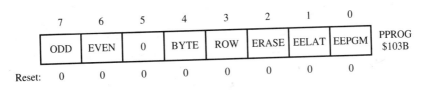

7	6	5	4	3	2	1	0	
ODD	EVEN	0	BYTE	ROW	ERASE	EELAT	EEPGM	PPROG $103B

Reset: 0   0   0   0   0   0   0   0

Figure 12-5   The PPROG register. (*Courtesy of Motorola, Inc.*)

Bit 0 is the EEPGM (EE ProGraM) bit. While it is 1 the programming voltages are applied.

* The rest of this section contains more technical details of the EEPROM operation. Those who are only interested in using the EEPROM may skip these paragraphs on the first reading.

Bit 1 is EELAT, the *latch* bit. If this bit is 0, the EEPROM operates as a ROM and cannot be written to or erased. When EELAT is 1, the EEPROM cannot be read; it should be in the programming or erase mode. EEPGM cannot be 1 unless EELAT is 1. Furthermore, they cannot both be written to 1 simultaneously (by the same command). EELAT must first be written to 1, then EEPGM can be written to 1.

Bits 2, 3, and 4 are used to control the erase of the EEPROM as described in the next paragraph. The ODD and EVEN bits are only used in the test mode and will not be described further.

If the ERASE bit in PPROG is 1, the EEPROM will be erased when EELAT is set and then EPGM is set for 10 ms. There are three possible erase modes, as shown in Table 12-4 and as specified by the BYTE and

Table 12-4   Erase Modes for the EEPROM

BYTE	ROW	Type of Erase
0	0	Bulk erase (all 512 bytes)
0	1	Row erase (16-byte row)
1	0	Byte erase
1	1	Byte erase

ROW bits in PPROG. The modes are
- **Bulk erase**   Erase the entire EEPROM.
- **Byte erase**   Erase only a single byte in the memory.
- **Row erase**   The 512-byte EEPROM is arranged in 32 rows of 16 bytes each as shown in Figure 12-6. It can be erased a row at a time.

The typical procedure for erasing and rewriting a byte in EEPROM is as follows:

1. Start erasure by storing a $16 into PPROG. This sets the ERASE and EELAT bits and sets up for byte erase.

2. Store $FF (or any data) into the location to be erased.

   This identifies the location to be erased to the EEPROM.

3. Store $17 into PPROG. This sets EEPGM to 1.

4. Delay for 10 ms.

5. Clear PPROG by loading A with 00 and storing it in PPROG at $103B. A CLEAR PPROG instruction may not work.

   This ends the erase part of the procedure. The EEPROM can now be written.

6. Write $02 into PPROG. This only sets EELAT.

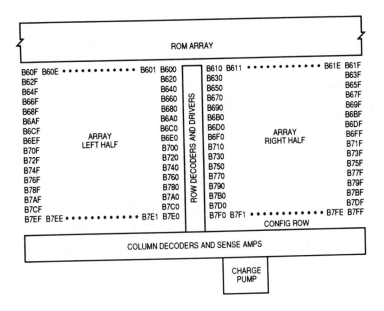

Figure 12-6 **Topological arrangements of EEPROM bytes.** (*Redrawn courtesy of Motorola, Inc.*)

7. Store the data to be written into the EEPROM location.

8. Write $03 into PPROG. This sets EEPGM.

9. Delay for 10 ms.

10. Clear PPROG as in step 5. This should complete the writing of the data. It can be verified by reading the location.

This is essentially the procedure followed in the EVB. The monitor routine called CHGBYT, at E267, identifies the address in X as an EEPROM location. If it is, the program enters CHGBYT1, which erases and writes the new data into the specified byte.

The CONFIG register (see Sec. 12-1.2) is a special register that is EEPROM. It must be erased and rewritten as an EEPROM. It can only be erased and rewritten, however, in the test modes. On the EVB it is difficult to change this register and is rarely done.

12-2.4 Protected Bits

Certain bits in certain registers in the **68HC11** are *protected*. This means that they can only be written during the first 64 bus cycles after a RESET. The protection mechanism is used to prevent inadvertent changes to the system configuration during normal operation and can be overridden in the test modes.

The protected bits are

■ **The INIT register**   The memory locations allocated for the RAM and the I/O registers must be specified at RESET and may not be changed thereafter.

■ **PR1 and PR0**   These 2 bits are the timer prescalar select in TMSK2. The timer prescalar must be selected at RESET and cannot be changed thereafter.

■ **IRQE, DLY, CR1, and CR2**   These 4 bits in the OPTION register are protected. The user must select edge- or level-sensitive interrupts, whether to delay the oscillator or to invoke it immediately, and the COP timeout rate.

In addition to the bits previously mentioned, the CONFIG register is basically protected. It is a register that reflects the hardware configuration of the system (Is the internal ROM being used? Is EEPROM available? etc.). The COP bit can only be changed in protected modes and the change is not effective until the next RESET.

In the EVB the user has no control over the protected bits because the monitor takes far longer than 64 cycles to initialize. The user can, however, read these bits to determine the system configuration.

12-3   The Hardware of the Evaluation Board*

Figure 12-7, taken from the EVB manual, shows the hardware configuration of the EVB. The **68HC11A1** is in the upper left-hand corner of the figure. Note that MODA is always pulled up to +5 V by the 10-K Ω resistor, R6, and MODB is pulled up to +5 V by the 10-K Ω resistor R1. Thus the system normally comes up with MODA and MODB both 1s, which places it in the normal expanded mode. MODB, however, is brought out on pin 2. It can be brought to 0 by grounding it. If the EVB is then RESET, it will come up in the special test mode. This pin can also be used to apply a standby voltage to preserve the contents of the memory when power goes down.

Continuing across the top of the figure, U2—the **74HC373**—is the address latch. Recall that in the expanded mode the addresses and data are multiplexed on the PC lines. The address is latched by the address strobe, U3. The **2764** is the 8K-byte by 8-bit ROM that holds the Buffalo monitor. It occupies memory locations $E000–$FFFF. U4 is a socket at location $6000. The user may add additional RAM or ROM at this location. U5, the **MCM6164**, is an 8K-byte by 8-bit RAM that occupies locations $C000–$DFFF. This is where we have been writing our programs throughout this book.

---

* The reader should have some knowledge of digital circuitry to understand this section. Most of the circuits in this section are also discussed in J. D. Greenfield and W. C. Wray, *Using Microprocessors and Microcomputers: The Motorola Family*, 2nd ed. (Englewood Cliffs, NJ: Prentice-Hall, 1988), originally published by John Wiley; or J. D. Greenfield, *Practical Digital Design Using ICs* (Englewood Cliffs, NJ: Prentice-Hall, latest ed.).

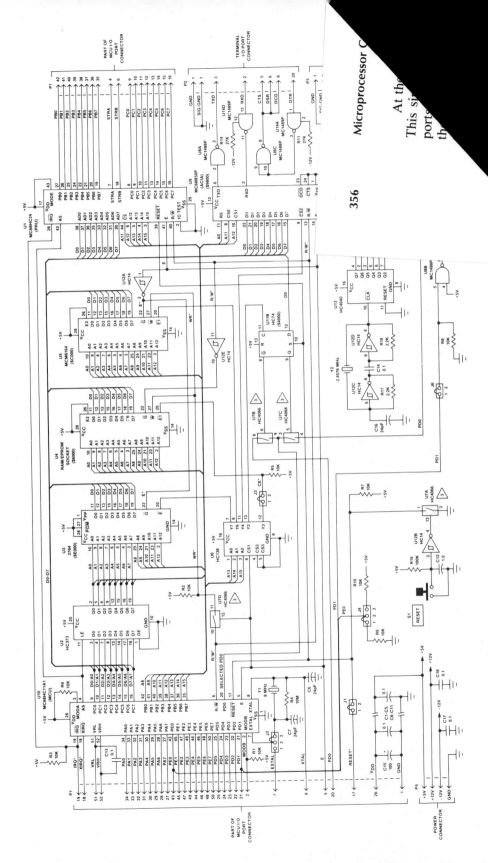

Figure 12-7 The EVB schematic diagram. (*Reprinted courtesy of Motorola, Inc.*)

far right there is U1, the **68HC24** PRU (Port Replacement Unit). mulates the operation of port B, port C, STRA, and STRB. These are now available to the user as though the **68HC11** were operating in single-chip mode, although it is actually operating in the expanded multiplexed mode.

The crystal oscillator connections (see Sec. 8-2.3) are below the **68HC11**. The **74HC138** (U6) in the center of the drawing is an address decoder, which examines address lines A13, A14, and A15 and selects the RAM or the ROM according to the programmed address. On the right side is U9, the **MC6850** Asynchronous Communications Interface Adapter (ACIA). This sends and receives information with the terminal or PC controlling the EVB. The signals are sent through **1488** and **1489** TTL-to-RS232 level converters. The crystal oscillator, Y2, the **74HC4040** counter, and the jumper J5 determine the baud rate of the ACIA. It must match the baud rate of the terminal. Programs such as PROCOMM can set their baud rate by software and can match the selected speed.

The host connector at the lower right is driven by the SCI (see Sec. 11-3). The RESET switch and circuitry is shown in the lower center of the drawing.

Summary

This chapter started with a discussion of those registers that control the operation of the **68HC11**. These registers, HPRIO, CONFIG, and OPTION, are usually preset for a particular system or configuration and are rarely changed. An explanation of the COP, a system to assure that the computer is functioning properly, was included.

The second part of the chapter was devoted to the memories on the **68HC11**. The functions of the RAM, ROM, and EEPROM were explained, and their memory locations were identified.

The chapter concluded with a brief discussion of the hardware of the EVB. This is an excellent example of a **68HC11** system and shows the auxiliary circuits and ICs required to make it function properly.

Glossary

Clock Monitor Enable (CME)   A feature within the **68HC11** that monitors the frequency of its crystal clock. The CME bit in the OPTION register, if set, will cause an interrupt if the clock is too slow.

Computer Operating Properly (COP)   If the COP is invoked, the computer must periodically reset a register to indicate that it is operating properly. Otherwise, a COP interrupt will be generated.

Floating gate A gate with no connections. It can be used to trap charges in response to a high electric field, such as the programming voltage in the EEPROM. Floating gates are used in PROMs and EEPROMs.

IRQE A bit that determines whether an interrupt request is made by a level or an edge on the IRQ pin. Highest Priority Interrupt is the interrupt source that will be awarded the highest priority (be serviced first) as determined by the setting of the PSEL bits.

NOSEC A bit that indicates whether security is invoked in a system (see Security).

Security A feature that prevents a third party from reading the ROM or EEPROM within a **68HC11**.

Problems

Section 12-1.1

12-1 Write a program segment to set IC1 as the highest priority interrupt.

12-2 Set up OC4 as the highest priority interrupt. Then disallow interrupts on OC4 but allow them on OC2, OC3, and OC5. Devise a way to show which register is then awarded the highest priority.

12-3 Set up the EVB so that there is contention between OC4 and IC1. Devise a method to show which register gets the higher priority, and also show how the user can decide which register gets the higher priority.

Section 12-1.2

12-4 On an EVB the CONFIG register contains #$0D. Explain the function of each bit in the register.

Section 12-1.4

12-5 A system is using the COP. Every time there is a COP failure the system is to print the message "COP FAILURE" and then return to the monitor. Write the code to do this.

Section 12-2.1

12-6 A ROM is to occupy locations $A000–$BFFF. Design a decoder so that any addresses from $FFC0 to $FFFF are also decoded as $BFC0–$BFFF. This allows the interrupt vectors to be written into this ROM.

Section 12-2.2

12-7 What must the contents of INIT be if the registers are to occupy locations $D000–$D03F and the internal RAM is to be at $E000–$E0FF?

12-8 Write the instructions to store $AA in $B602, an EEPROM location.

12-9 Write the subroutine to reset the COP timer in EEPROM.

12-10 Write instructions to store the contents of $C000–$C00F in $B610–$B61F. Use row erase.

Section 12-2.4

12-11 List the protected bits and determine what they are on the EVB.

Instruction Set Details

A-1 Introduction

This appendix contains complete detailed information for all M68HC11 instructions. The instructions are arranged in alphabetical order with the instruction mnemonic set in larger type for easy reference.

A-2 Nomenclature

The following nomenclature is used in the subsequent definitions.

(a) Operators

( )	= Contents of Register Shown Inside Parentheses
↑	= Is Transferred to
↟	= Is Pulled from Stack
↡	= Is Pushed onto Stack
•	= Boolean AND
+	= Arithmetic Addition Symbol Except Where Used as Inclusive-OR Symbol in Boolean Formula
⊕	= Exclusive-OR
×	= Multiply
:	= Concatenation
−	= Arithmetic Subtraction Symbol or Negation Symbol (Twos Complement)

(b) Registers in the MPU

ACCA	= Accumulator A
ACCB	= Accumulator B
ACCX	= Accumulator ACCA or ACCB
ACCD	= Double Accumulator — Accumulator A Concatenated with Accumulator B Where A is the Most Significant Byte
CCR	= Condition Code Register
IX	= Index Register X, 16 Bits
IXH	= Index Register X, Higher Order 8 Bits
IXL	= Index Register X, Lower Order 8 Bits
IY	= Index Register Y, 16 Bits
IYH	= Index Register Y, Higher Order 8 Bits
IYL	= Index Register Y, Lower Order 8 Bits
PC	= Program Counter, 16 Bits
PCH	= Program Counter, Higher Order (Most Significant) 8 Bits
PCL	= Program Counter, Lower Order (Least Significant) 8 Bits
SP	= Stack Pointer, 16 Bits
SPH	= Stack Pointer, Higher Order 8 Bits
SPL	= Stack Pointer, Lower Order 8 Bits

* This Appendix courtesy Motorola, Inc. *M68HC11 Reference Manual*, Prentice-Hall, Englewood Cliffs, N.J., 1988.

(c) Memory and Addressing

M   = A Memory Location (One Byte)
M + 1   = The Byte of Memory at $0001 Plus the Address of the Memory Location Indicated by "M"
Rel   = Relative Offset (i.e., the Twos Complement Number Stored in the Last Byte of Machine Code Corresponding to a Branch Instruction)
(opr)   = Operand
(msk)   = Mask Used in Bit Manipulation Instructions
(rel)   = Relative Offset Used in Branch Instructions

(d) Bits 7–0 of the Condition Code Register

S   = Stop Disable, Bit 7
X   = X Interrupt Mask, Bit 6
H   = Half Carry, Bit 5
I   = I Interrupt Mask, Bit 4
N   = Negative Indicator, Bit 3
Z   = Zero Indicator, Bit 2
V   = Twos Complement Overflow Indicator, Bit 1
C   = Carry/Borrow, Bit 0

(e) Status of Individual Bits BEFORE Execution of an Instruction

An   = Bit n of ACCA (n = 7, 6, 5 . . . 0)
Bn   = Bit n of ACCB (n = 7, 6, 5 . . . 0)
Dn   = Bit n of ACCD (n = 15, 14, 13 . . . 0)
             Where Bits 15–8 Refer to ACCA and Bit 7–0 Refer to ACCB
IXn   = Bit n of IX (n = 15, 14, 13 . . . 0)
IXHn   = Bit n of IXH (n = 7, 6, 5 . . . 0)
IXLn   = Bit n of IXL (n = 7, 6, 5 . . . 0)
IYn   = Bit n of IY (n = 15, 14, 13 . . . 0)
IYHn   = Bit n of IYH (n = 7, 6, 5 . . . 0)
IYLn   = Bit n of IYL (n = 7, 6, 5 . . . 0)
Mn   = Bit n of M (n = 7, 6, 5 . . . 0)
SPHn   = Bit n of SPH (n = 7, 6, 5 . . . 0)
SPLn   = Bit n of SPL (n = 7, 6, 5 . . . 0)
Xn   = Bit n of ACCX (n = 7, 6, 5 . . . 0)

(f) Status of Individual Bits of RESULT of Execution of an Instruction

(i) For 8-Bit Results

Rn   = Bit n of the Result (n = 7, 6, 5 . . . 0)
         This applies to instructions which provide a result contained in a single byte of memory or in an 8-bit register.

(ii) For 16-Bit Results

RHn = Bit n of the Most Significant Byte of the Result (n = 7, 6, 5 . . . 0)
RLn = Bit n of the Least Significant Byte of the Result (n = 7, 6, 5 . . . 0)
         This applies to instructions which provide a result contained in two consecutive bytes of memory or in a 16-bit register.
Rn   = Bit n of the Result (n = 15, 14, 13 . . . 0)

(g) Notation Used in CCR Activity Summary Figures
       —   = Bit Not Affected
       0   = Bit Forced to Zero
       1   = Bit Forced to One
       ●   = Bit Set or Cleared According to Results of Operation
       ●   = Bit may change from one to zero, remain zero, or remain one as a result of this operation, but cannot change from zero to one.

(h) Notation Used in Cycle-by-Cycle Execution Tables
       —   = Irrelevant Data
       ii   = One Byte of Immediate Data
       jj   = High-Order Byte of 16-Bit Immediate Data
       kk   = Low-Order Byte of 16-Bit Immediate Data
       hh   = High-Order Byte of 16-Bit Extended Address
       ll   = Low-Order Byte of 16-Bit Extended Address
       dd   = Low-Order 8 Bits of Direct Address $0000–$00FF
                   (High Byte Assumed to be $00)
       mm   = 8-Bit Mask (Set Bits Correspond to Operand Bits Which Will Be Affected)
       ff   = 8-Bit Forward Offset $00 (0) to $FF (255) (Is Added to Index)
       rr   = Signed Relative Offset $80 (− 128) to $7F (+ 127)
                   (Offset Relative to Address Following Machine Code Offset Byte)
       OP   = Address of Opcode Byte
       OP + n   = Address of nth Location after Opcode Byte
       SP   = Address Pointed to by Stack Pointer Value (at the Start of an Instruction)
       SP + n   = Address of nth Higher Address Past That Pointed to by Stack Pointer
       SP − n   = Address of nth Lower Address Before That Pointed to by Stack Pointer
       Sub   = Address of Called Subroutine
       Nxt op   = Opcode of Next Instruction
       Rtn hi   = High-Order Byte of Return Address
       Rtn lo   = Low-Order Byte of Return Address
       Svc hi   = High-Order Byte of Address for Service Routine
       Svc lo   = Low-Order Byte of Address for Service Routine
       Vec hi   = High-Order Byte of Interrupt Vector
       Vec lo   = Low-Order Byte of Interrupt Vector

Add Accumulator B to Accumulator A (ABA)

Operation: ACCA ◀ (ACCA) + (ACCB)

Description: Adds the contents of accumulator B to the contents of accumulator A and places the result in accumulator A. Accumulator B is not changed. This instruction affects the H condition code bit so it is suitable for use in BCD arithmetic operations (see DAA instruction for additional information).

Condition Codes and Boolean Formulae:

S	X	H	I	N	Z	V	C
—	—	✸	—	✸	✸	✸	✸

H $A3 \cdot B3 + B3 \cdot \overline{R3} + \overline{R3} \cdot A3$
   Set if there was a carry from bit 3; cleared otherwise.

N   R7
   Set if MSB of result is set; cleared otherwise.

Z $\overline{R7} \cdot \overline{R6} \cdot \overline{R5} \cdot \overline{R4} \cdot \overline{R3} \cdot \overline{R2} \cdot \overline{R1} \cdot \overline{R0}$
   Set if result is $00; cleared otherwise.

V $A7 \cdot B7 \cdot \overline{R7} + \overline{A7} \cdot \overline{B7} \cdot R7$
   Set if a twos complement overflow resulted from the operation; cleared otherwise.

C $A7 \cdot B7 + B7 \cdot \overline{R7} + \overline{R7} \cdot A7$
   Set if there was a carry from the MSB of the result; cleared otherwise.

Source Forms: ABA

Addressing Modes, Machine Code, and Cycle-by-Cycle Execution:

Cycle	ABA (INH)		
	Addr	Data	R/W̅
1	OP	1B	1
2	OP + 1	—	1

Add Accumulator B to Index Register X (ABX)

Operation: IX ◄ (IX) + (ACCB)

Description: Adds the 8-bit unsigned contents of accumulator B to the contents of index register X (IX) considering the possible carry out of the low-order byte of the index register X; places the result in index register X (IX). Accumulator B is not changed. There is no equivalent instruction to add accumulator A to an index register.

Condition Codes and Boolean Formulae:

S	X	H	I	N	Z	V	C
—	—	—	—	—	—	—	—

None affected

Source Forms: ABX

Addressing Modes, Machine Code, and Cycle-by-Cycle Execution:

Cycle	ABX (INH)		
	Addr	Data	R/W̄
1	OP	3A	1
2	OP + 1	—	1
3	FFFF	—	1

Add Accumulator B to Index Register Y (ABY)

Operation: IY ◀ (IY) + (ACCB)

Description: Adds the 8-bit unsigned contents of accumulator B to the contents of index register Y (IY) considering the possible carry out of the low-order byte of index register Y; places the result in index register Y (IY). Accumulator B is not changed. There is no equivalent instruction to add accumulator A to an index register.

Condition Codes and Boolean Formulae:

S	X	H	I	N	Z	V	C
—	—	—	—	—	—	—	—

None affected

Source Forms: ABY

Addressing Modes, Machine Code, and Cycle-by-Cycle Execution:

Cycle	ABY (INH)		
	Addr	Data	R/W̄
1	OP	18	1
2	OP + 1	3A	1
3	OP + 2	—	1
4	FFFF	—	1

Add with Carry (ADC)

Operation: ACCX ← (ACCX) + (M) + (C)

Description: Adds the contents of the C bit to the sum of the contents of ACCX and M and places the result in ACCX. This instruction affects the H condition code bit so it is suitable for use in BCD arithmetic operations (see DAA instruction for additional information).

Condition Codes and Boolean Formulae:

S	X	H	I	N	Z	V	C
—	—	↕	—	↕	↕	↕	↕

H $X3 \cdot M3 + M3 \cdot \overline{R3} + \overline{R3} \cdot X3$
Set if there was a carry from bit 3; cleared otherwise.

N $R7$
Set if MSB of result is set; cleared otherwise.

Z $\overline{R7} \cdot \overline{R6} \cdot \overline{R5} \cdot \overline{R4} \cdot \overline{R3} \cdot \overline{R2} \cdot \overline{R1} \cdot \overline{R0}$
Set if result is $00; cleared otherwise.

V $X7 \cdot M7 \cdot \overline{R7} + \overline{X7} \cdot \overline{M7} \cdot R7$
Set if a twos complement overflow resulted from the operation; cleared otherwise.

C $X7 \cdot M7 + M7 \cdot \overline{R7} + \overline{R7} \cdot X7$
Set if there was a carry from the MSB of the result; cleared otherwise.

Source Forms: ADCA (opr); ADCB (opr)

Addressing Modes, Machine Code, and Cycle-by-Cycle Execution:

Cycle	ADCA (IMM)			ADCA (DIR)			ADCA (EXT)			ADCA (IND, X)			ADCA (IND, Y)		
	Addr	Data	R/$\overline{W}$	Addr	Data	R/$\overline{W}$	Addr	Data	R/$\overline{W}$	Addr	Data	R/$\overline{W}$	Addr	Data	R/$\overline{W}$
1	OP	89	1	OP	99	1	OP	B9	1	OP	A9	1	OP	18	1
2	OP + 1	ii	1	OP + 1	dd	1	OP + 1	hh	1	OP + 1	ff	1	OP + 1	A9	1
3				00dd	(00dd)	1	OP + 2	ll	1	FFFF	—	1	OP + 2	ff	1
4							hhll	(hhll)	1	X + ff	(X + ff)	1	FFFF	—	1
5													Y + ff	(Y + ff)	1

Cycle	ADCB (IMM)			ADCB (DIR)			ADCB (EXT)			ADCB (IND, X)			ADCB (IND, Y)		
	Addr	Data	R/$\overline{W}$	Addr	Data	R/$\overline{W}$	Addr	Data	R/$\overline{W}$	Addr	Data	R/$\overline{W}$	Addr	Data	R/$\overline{W}$
1	OP	C9	1	OP	D9	1	OP	F9	1	OP	E9	1	OP	18	1
2	OP + 1	ii	1	OP + 1	dd	1	OP + 1	hh	1	OP + 1	ff	1	OP + 1	E9	1
3				00dd	(00dd)	1	OP + 2	ll	1	FFFF	—	1	OP + 2	ff	1
4							hhll	(hhll)	1	X + ff	(X + ff)	1	FFFF	—	1
5													Y + ff	(Y + ff)	1

Add without Carry (ADD)

Operation: ACCX ⬦ (ACCX) + (M)

Description: Adds the contents of M to the contents of ACCX and places the result in ACCX. This instruction affects the H condition code bit so it is suitable for use in the BCD arithmetic operations (see DAA instruction for additional information).

Condition Codes and Boolean Formulae:

S	X	H	I	N	Z	V	C
—	—	✿	—	✿	✿	✿	✿

H $X3 \cdot M3 + M3 \cdot \overline{R3} + \overline{R3} \cdot X3$
Set if there was a carry from bit 3; cleared otherwise.

N   R7
Set if MSB of result is set; cleared otherwise.

Z $\overline{R7} \cdot \overline{R6} \cdot \overline{R5} \cdot \overline{R4} \cdot \overline{R3} \cdot \overline{R2} \cdot \overline{R1} \cdot \overline{R0}$
Set if result is $00; cleared otherwise.

V $X7 \cdot M7 \cdot \overline{R7} + \overline{X7} \cdot \overline{M7} \cdot R7$
Set if a twos complement overflow resulted from the operation; cleared otherwise.

C $X7 \cdot M7 + M7 \cdot \overline{R7} + \overline{R7} \cdot X7$
Set if there was a carry from the MSB of the result; cleared otherwise.

Source Forms: ADDA (opr); ADDB (opr)

Addressing Modes, Machine Code, and Cycle-by-Cycle Execution:

Cycle	ADDA (IMM)			ADDA (DIR)			ADDA (EXT)			ADDA (IND, X)			ADDA (IND, Y)		
	Addr	Data	R/W̄	Addr	Data	R/W̄	Addr	Data	R/W̄	Addr	Data	R/W̄	Addr	Data	R/W̄
1	OP	8B	1	OP	9B	1	OP	BB	1	OP	AB	1	OP	18	1
2	OP+1	ii	1	OP+1	dd	1	OP+1	hh	1	OP+1	ff	1	OP+1	AB	1
3				00dd	(00dd)	1	OP+2	ll	1	FFFF	—	1	OP+2	ff	1
4							hhll	(hhll)	1	X+ff	(X+ff)	1	FFFF	—	1
5													Y+ff	(Y+ff)	1

Cycle	ADDB (IMM)			ADDB (DIR)			ADDB (EXT)			ADDB (IND, X)			ADDB (IND, Y)		
	Addr	Data	R/W̄	Addr	Data	R/W̄	Addr	Data	R/W̄	Addr	Data	R/W̄	Addr	Data	R/W̄
1	OP	CB	1	OP	DB	1	OP	FB	1	OP	EB	1	OP	18	1
2	OP+1	ii	1	OP+1	dd	1	OP+1	hh	1	OP+1	ff	1	OP+1	EB	1
3				00dd	(00dd)	1	OP+2	ll	1	FFFF	—	1	OP+2	ff	1
4							hhll	(hhll)	1	X+ff	(X+ff)	1	FFFF	—	1
5													Y+ff	(Y+ff)	1

Add Double Accumulator (ADDD)

Operation: $ACCD \Leftarrow (ACCD) + (M:M+1)$

Description: Adds the contents of M concatenated with M + 1 to the contents of ACCD and places the results in ACCD. Accumulator A corresponds to the high-order half of the 16-bit double accumulator D.

Condition Codes and Boolean Formulae:

S	X	H	I	N	Z	V	C
—	—	—	—	↕	↕	↕	↕

N R15
    Set if MSB of result is set; cleared otherwise.

Z $\overline{R15} \cdot \overline{R14} \cdot \overline{R13} \cdot \overline{R12} \cdot \overline{R11} \cdot \overline{R10} \cdot \overline{R9} \cdot \overline{R8} \cdot \overline{R7} \cdot \overline{R6} \cdot \overline{R5} \cdot \overline{R4} \cdot \overline{R3} \cdot \overline{R2} \cdot \overline{R1} \cdot \overline{R0}$
    Set if result is $0000; cleared otherwise.

V $D15 \cdot M15 \cdot \overline{R15} + \overline{D15} \cdot \overline{M15} \cdot R15$
    Set if a twos complement overflow resulted from the operation; cleared otherwise.

C $D15 \cdot M15 + M15 \cdot \overline{R15} + \overline{R15} \cdot D15$
    Set if there was a carry from the MSB of the result; cleared otherwise.

Source Form: ADDD (opr)

Addressing Modes, Machine Code, and Cycle-by-Cycle Execution:

Cycle	ADDD (IMM)			ADDD (DIR)			ADDD (EXT)			ADDD (IND, X)			ADDD (IND, Y)		
	Addr	Data	R/W̄	Addr	Data	R/W̄	Addr	Data	R/W̄	Addr	Data	R/W̄	Addr	Data	R/W̄
1	OP	C3	1	OP	D3	1	OP	F3	1	OP	E3	1	OP	18	1
2	OP + 1	jj	1	OP + 1	dd	1	OP + 1	hh	1	OP + 1	ff	1	OP + 1	E3	1
3	OP + 2	kk	1	00dd	(00dd)	1	OP + 2	ll	1	FFFF	—	1	OP + 2	ff	1
4	FFFF	—	1	00dd + 1	(00dd + 1)	1	hhll	(hhll)	1	X + ff	(X + ff)	1	FFFF	—	1
5				FFFF	—	1	hhll + 1	(hhll + 1)	1	X + ff + 1	(X + ff + 1)	1	Y + ff	(Y + ff)	1
6							FFFF	—	1	FFFF	—	1	Y + ff + 1	(Y + ff + 1)	1
7													FFFF	—	1

Logical AND

Operation: ACCX ◀ (ACCX) • (M)

Description: Performs the logical AND between the contents of ACCX and the contents of M and places the result in ACCX. (Each bit of ACCX after the operation will be the logical AND of the corresponding bits of M and of ACCX before the operation.)

Condition Codes and Boolean Formulae:

S	X	H	I	N	Z	V	C
—	—	—	—	↕	↕	0	—

N   R7
　　Set if MSB of result is set; cleared otherwise.

Z $\overline{R7} • \overline{R6} • \overline{R5} • \overline{R4} • \overline{R3} • \overline{R2} • \overline{R1} • \overline{R0}$
　　Set if result is $00; cleared otherwise.

V   0
　　Cleared

Source Forms: ANDA (opr); ANDB (opr)

Addressing Modes, Machine Code, and Cycle-by-Cycle Execution:

Cycle	ANDA (IMM)			ANDA (DIR)			ANDA (EXT)			ANDA (IND, X)			ANDA (IND, Y)		
	Addr	Data	R/W̄	Addr	Data	R/W̄	Addr	Data	R/W̄	Addr	Data	R/W̄	Addr	Data	R/W̄
1	OP	84	1	OP	94	1	OP	B4	1	OP	A4	1	OP	18	1
2	OP+1	ii	1	OP+1	dd	1	OP+1	hh	1	OP+1	ff	1	OP+1	A4	1
3				00dd	(00dd)	1	OP+2	ll	1	FFFF	—	1	OP+2	ff	1
4							hhll	(hhll)	1	X+ff	(X+ff)	1	FFFF	—	1
5													Y+ff	(Y+ff)	1

Cycle	ANDB (IMM)			ANDB (DIR)			ANDB (EXT)			ANDB (IND, X)			ANDB (IND, Y)		
	Addr	Data	R/W̄	Addr	Data	R/W̄	Addr	Data	R/W̄	Addr	Data	R/W̄	Addr	Data	R/W̄
1	OP	C4	1	OP	D4	1	OP	F4	1	OP	E4	1	OP	18	1
2	OP+1	ii	1	OP+1	dd	1	OP+1	hh	1	OP+1	ff	1	OP+1	E4	1
3				00dd	(00dd)	1	OP+2	ll	1	FFFF	—	1	OP+2	ff	1
4							hhll	(hhll)	1	X+ff	(X+ff)	1	FFFF	—	1
5													Y+ff	(Y+ff)	1

Arithmetic Shift Left (ASL) (Same as LSL)

Operation:

$$C \longleftarrow b7 ------ b0 \longleftarrow 0$$

Description: Shifts all bits of the ACCX or M one place to the left. Bit 0 is loaded with a zero. The C bit in the CCR is loaded from the most significant bit of ACCX or M.

Condition Codes and Boolean Formulae:

S	X	H	I	N	Z	V	C
—	—	—	—	↕	↕	↕	↕

N   R7
   Set if MSB of result is set; cleared otherwise.

Z $\overline{R7} \cdot \overline{R6} \cdot \overline{R5} \cdot \overline{R4} \cdot \overline{R3} \cdot \overline{R2} \cdot \overline{R1} \cdot \overline{R0}$
   Set if result is $00; cleared otherwise.

V $N \oplus C = [N \cdot \overline{C}] + [\overline{N} \cdot C]$ (for N and C after the shift)
   Set if (N is set and C is clear) or (N is clear and C is set); cleared otherwise (for values of N and C after the shift).

C   M7
   Set if, before the shift, the MSB of ACCX or M was set; cleared otherwise.

Source Forms: ASLA; ASLB; ASL (opr)

Addressing Modes, Machine Code, and Cycle-by-Cycle Execution:

Cycle	ASLA (INH)			ASLB (INH)			ASL (EXT)			ASL (IND, X)			ASL (IND, Y)		
	Addr	Data	R/W̄	Addr	Data	R/W̄	Addr	Data	R/W̄	Addr	Data	R/W̄	Addr	Data	R/W̄
1	OP	48	1	OP	58	1	OP	78	1	OP	68	1	OP	18	1
2	OP + 1	—	1	OP + 1	—	1	OP + 1	hh	1	OP + 1	ff	1	OP + 1	68	1
3							OP + 2	ll	1	FFFF	—	1	OP + 2	ff	1
4							hhll	(hhll)	1	X + ff	(X + ff)	1	FFFF	—	1
5							FFFF	—	1	FFFF	—	1	Y + ff	(Y + ff)	1
6							hhll	result	0	X + ff	result	0	FFFF	—	1
7													Y + ff	result	0

Arithmetic Shift Left Double Accumulator (ASLD) (Same as LSLD)

Operation:

$$C \leftarrow \boxed{b7 ------ b0} \leftarrow \boxed{b7 ------ b0} \leftarrow 0$$

ACCA   ACCB

Description: Shifts all bits of ACCD one place to the left. Bit 0 is loaded with a zero. The C bit in the CCR is loaded from the most significant bit of ACCD.

Condition Codes and Boolean Formulae:

S	X	H	I	N	Z	V	C
—	—	—	—	↕	↕	↕	↕

N R15
Set if MSB of result is set; cleared otherwise.

Z $\overline{R15} \cdot \overline{R14} \cdot \overline{R13} \cdot \overline{R12} \cdot \overline{R11} \cdot \overline{R10} \cdot \overline{R9} \cdot \overline{R8} \cdot \overline{R7} \cdot \overline{R6} \cdot \overline{R5} \cdot \overline{R4} \cdot \overline{R3} \cdot \overline{R2} \cdot \overline{R1} \cdot \overline{R0}$
Set if result is $0000; cleared otherwise.

V $N \oplus C = [N \cdot \overline{C}] + [\overline{N} \cdot C]$ (for N and C after the shift)
Set if (N is set and C is clear) or (N is clear and C is set); cleared otherwise (for values of N and C after the shift).

C D15
Set if, before the shift, the MSB of ACCD was set; cleared otherwise.

Source Form: ASLD

Addressing Modes, Machine Code, and Cycle-by-Cycle Execution:

Cycle	ASLD (INH)		
	Addr	Data	R/$\overline{W}$
1	OP	05	1
2	OP + 1	—	1
3	FFFF	—	1

Arithmetic Shift Right (ASR)

Operation: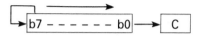

Description: Shifts all of ACCX or M one place to the right. Bit 7 is held constant. Bit 0 is loaded into the C bit of the CCR. This operation effectively divides a twos complement value by two without changing its sign. The carry bit can be used to round the result.

Condition Codes and Boolean Formulae:

S	X	H	I	N	Z	V	C
—	—	—	—	⬍	⬍	⬍	⬍

N R7
Set if MSB of result is set; cleared otherwise.

Z $\overline{R7} \cdot \overline{R6} \cdot \overline{R5} \cdot \overline{R4} \cdot \overline{R3} \cdot \overline{R2} \cdot \overline{R1} \cdot \overline{R0}$
Set if result is $00; cleared otherwise.

V $N \oplus C = [N \cdot \overline{C}] + [\overline{N} \cdot C]$ (for N and C after the shift)
Set if (N is set and C is clear) or (N is clear and C is set); cleared otherwise (for values of N and C after the shift).

C M0
Set if, before the shift, the LSB of ACCX or M was set; cleared otherwise.

Source Forms: ASRA; ASRB; ASR (opr)

Addressing Modes, Machine Code, and Cycle-by-Cycle Execution:

Cycle	ASRA (INH)			ASRB (INH)			ASR (EXT)			ASR (IND, X)			ASR (IND, Y)		
	Addr	Data	R/W̄	Addr	Data	R/W̄	Addr	Data	R/W̄	Addr	Data	R/W̄	Addr	Data	R/W̄
1	OP	47	1	OP	57	1	OP	77	1	OP	67	1	OP	18	1
2	OP+1	—	1	OP+1	—	1	OP+1	hh	1	OP+1	ff	1	OP+1	67	1
3							OP+2	ll	1	FFFF	—	1	OP+2	ff	1
4							hhll	(hhll)	1	X+ff	(X+ff)	1	FFFF	—	1
5							FFFF	—	1	FFFF	—	1	Y+ff	(Y+ff)	1
6							hhll	result	0	X+ff	result	0	FFFF	—	1
7													Y+ff	result	0

Branch if Carry Clear (BCC) (Same as BHS)

Operation: PC ⧫ (PC) + $0002 + Rel if (C) = 0

Description: Tests the state of the C bit in the CCR and causes a branch if C is clear.

See BRA instruction for further details of the execution of the branch.

Condition Codes and Boolean Formulae:

S	X	H	I	N	Z	V	C
—	—	—	—	—	—	—	—

None affected

Source Form: BCC (rel)

Addressing Modes, Machine Code, and Cycle-by-Cycle Execution:

Cycle	BCC (REL)		
	Addr	Data	R/$\overline{W}$
1	OP	24	1
2	OP + 1	rr	1
3	FFFF	—	1

The following table is a summary of all branch instructions.

Test	Boolean	Mnemonic	Opcode	Complementary		Branch	Comment
r>m	$Z+(N \oplus V)=0$	BGT	2E	r≤m	BLE	2F	Signed
r≥m	$N \oplus V=0$	BGE	2C	r<m	BLT	2D	Signed
r=m	$Z=1$	BEQ	27	r≠m	BNE	26	Signed
r≤m	$Z+(N \oplus V)=1$	BLE	2F	r>m	BGT	2E	Signed
r<m	$N \oplus V=1$	BLT	2D	r≥m	BGE	2C	Signed
r>m	$C+Z=0$	BHI	22	r≤m	BLS	23	Unsigned
r≥m	$C=0$	BHS/BCC	24	r<m	BLO/BCS	25	Unsigned
r=m	$Z=1$	BEQ	27	r≠m	BNE	26	Unsigned
r≤m	$C+Z=1$	BLS	23	r>m	BHI	22	Unsigned
r<m	$C=1$	BLO/BCS	25	r≥m	BHS/BCC	24	Unsigned
Carry	$C=1$	BCS	25	No Carry	BCC	24	Simple
Negative	$N=1$	BMI	2B	Plus	BPL	2A	Simple
Overflow	$V=1$	BVS	29	No Overflow	BVC	28	Simple
r=0	$Z=1$	BEQ	27	r≠0	BNE	26	Simple
Always	—	BRA	20	Never	BRN	21	Unconditional

Clear Bit(s) in Memory (BCLR)

Operation: $M \Leftarrow (M) \cdot (\overline{PC + 2})$
$M \Leftarrow (M) \cdot (\overline{PC + 3})$ (for IND, Y address mode only)

Description: Clear multiple bits in location M. The bit(s) to be cleared are specified by ones in the mask byte. All other bits in M are rewritten to their current state.

Condition Codes and Boolean Formulae:

S	X	H	I	N	Z	V	C
—	—	—	—	⬍	⬍	0	—

N   R7
Set if MSB of result is set; cleared otherwise.

Z $\overline{R7} \cdot \overline{R6} \cdot \overline{R5} \cdot \overline{R4} \cdot \overline{R3} \cdot \overline{R2} \cdot \overline{R1} \cdot \overline{R0}$
Set if result is $00; cleared otherwise.

V   0
Cleared

Source Forms: BCLR (opr) (msk)

Addressing Modes, Machine Code, and Cycle-by-Cycle Execution:

Cycle	BCLR (DIR)			BCLR (IND, X)			BCLR (IND, Y)		
	Addr	Data	R/W̄	Addr	Data	R/W̄	Addr	Data	R/W̄
1	OP	15	1	OP	1D	1	OP	18	1
2	OP + 1	dd	1	OP + 1	ff	1	OP + 1	1D	1
3	00dd	(00dd)	1	FFFF	—	1	OP + 2	ff	1
4	OP + 2	mm	1	X + ff	(X + ff)	1	FFFF	—	1
5	FFFF	—	1	OP + 2	mm	1	(IY) + ff	(Y + ff)	1
6	00dd	result	0	FFFF	—	1	OP + 3	mm	1
7				X + ff	result	0	FFFF	—	1
8							Y + ff	result	0

Branch if Carry Set (BCS) (Same as BLO)

Operation: PC ◆ (PC) + $0002 + Rel   if (C) = 1

Description: Tests the state of the C bit in the CCR and causes a branch if C is set.

See BRA instruction for further details of the execution of the branch.

Condition Codes and Boolean Formulae:

S	X	H	I	N	Z	V	C
—	—	—	—	—	—	—	—

None affected

Source Form: BCS (rel)

Addressing Modes, Machine Code, and Cycle-by-Cycle Execution:

Cycle	BCS (REL)		
	Addr	Data	R/W̄
1	OP	25	1
2	OP + 1	rr	1
3	FFFF	—	1

The following table is a summary of all branch instructions.

Test	Boolean	Mnemonic	Opcode	Complementary	Branch	Opcode	Comment
r>m	Z+(N ⊕ V) = 0	BGT	2E	r≤m	BLE	2F	Signed
r≥m	N ⊕ V = 0	BGE	2C	r<m	BLT	2D	Signed
r = m	Z = 1	BEQ	27	r ≠ m	BNE	26	Signed
r≤m	Z+(N ⊕ V) = 1	BLE	2F	r>m	BGT	2E	Signed
r<m	N ⊕ V = 1	BLT	2D	r≥m	BGE	2C	Signed
r>m	C+Z = 0	BHI	22	r≤m	BLS	23	Unsigned
r≥m	C = 0	BHS/BCC	24	r<m	BLO/BCS	25	Unsigned
r = m	Z = 1	BEQ	27	r ≠ m	BNE	26	Unsigned
r≤m	C+Z = 1	BLS	23	r>m	BHI	22	Unsigned
r<m	C = 1	BLO/BCS	25	r≥m	BHS/BCC	24	Unsigned
Carry	C = 1	BCS	25	No Carry	BCC	24	Simple
Negative	N = 1	BMI	2B	Plus	BPL	2A	Simple
Overflow	V = 1	BVS	29	No Overflow	BVC	28	Simple
r = 0	Z = 1	BEQ	27	r ≠ 0	BNE	26	Simple
Always	—	BRA	20	Never	BRN	21	Unconditional

Branch if Equal (BEQ)

Operation: PC ◀ (PC) + $0002 + Rel   if (Z) = 1

Description: Tests the state of the Z bit in the CCR and causes a branch if Z is set.

See BRA instruction for further details of the execution of the branch.

Condition Codes and Boolean Formulae:

S	X	H	I	N	Z	V	C
—	—	—	—	—	—	—	—

None affected

Source Form: BEQ (rel)

Addressing Modes, Machine Code, and Cycle-by-Cycle Execution:

Cycle	BEQ (REL)		
	Addr	Data	R/W̄
1	OP	27	1
2	OP + 1	rr	1
3	FFFF	—	1

The following table is a summary of all branch instructions.

Test	Boolean	Mnemonic	Opcode	Complementary	Branch	Comment
r>m	Z+(N ⊕ V)=0	BGT	2E	r≤m	BLE	2F	Signed
r≥m	N ⊕ V=0	BGE	2C	r<m	BLT	2D	Signed
r=m	Z=1	BEQ	27	r≠m	BNE	26	Signed
r≤m	Z+(N ⊕ V)=1	BLE	2F	r>m	BGT	2E	Signed
r<m	N ⊕ V=1	BLT	2D	r≥m	BGE	2C	Signed
r>m	C+Z=0	BHI	22	r≤m	BLS	23	Unsigned
r≥m	C=0	BHS/BCC	24	r<m	BLO/BCS	25	Unsigned
r=m	Z=1	BEQ	27	r≠m	BNE	26	Unsigned
r≤m	C+Z=1	BLS	23	r>m	BHI	22	Unsigned
r<m	C=1	BLO/BCS	25	r≥m	BHS/BCC	24	Unsigned
Carry	C=1	BCS	25	No Carry	BCC	24	Simple
Negative	N=1	BMI	2B	Plus	BPL	2A	Simple
Overflow	V=1	BVS	29	No Overflow	BVC	28	Simple
r=0	Z=1	BEQ	27	r≠0	BNE	26	Simple
Always	—	BRA	20	Never	BRN	21	Unconditional

Branch if Greater than or Equal to Zero (BGE)

Operation: PC ◀ (PC) + $0002 + Rel   if (N)⊕(V) = 0
i.e., if (ACCX)≥(M)   (twos-complement "signed" numbers)

Description: If the BGE instruction is executed immediately after execution of any of the instructions, CBA, CMP(A, B, or D), CP(X or Y), SBA, SUB(A, B, or D), the branch will occur if and only if the twos-complement number represented by the ACCX was greater than or equal to the two-complement number represented by M.

See BRA instruction for further details of the execution of the branch.

Condition Codes and Boolean Formulae:

S	X	H	I	N	Z	V	C
—	—	—	—	—	—	—	—

None affected

Source Form: BGE (rel)

Addressing Modes, Machine Code, and Cycle-by-Cycle Execution:

Cycle	BGE (REL)		
	Addr	Data	R/W̄
1	OP	2C	1
2	OP + 1	rr	1
3	FFFF	—	1

The following table is a summary of all branch instructions.

Test	Boolean	Mnemonic	Opcode	Complementary	Branch		Comment
r>m	Z + (N ⊕ V) = 0	BGT	2E	r≤m	BLE	2F	Signed
r≥m	N ⊕ V = 0	BGE	2C	r<m	BLT	2D	Signed
r = m	Z = 1	BEQ	27	r ≠ m	BNE	26	Signed
r≤m	Z + (N ⊕ V) = 1	BLE	2F	r>m	BGT	2E	Signed
r<m	N ⊕ V = 1	BLT	2D	r≥m	BGE	2C	Signed
r>m	C + Z = 0	BHI	22	r≤m	BLS	23	Unsigned
r≥m	C = 0	BHS/BCC	24	r<m	BLO/BCS	25	Unsigned
r = m	Z = 1	BEQ	27	r ≠ m	BNE	26	Unsigned
r≤m	C + Z = 1	BLS	23	r>m	BHI	22	Unsigned
r<m	C = 1	BLO/BCS	25	r≥m	BHS/BCC	24	Unsigned
Carry	C = 1	BCS	25	No Carry	BCC	24	Simple
Negative	N = 1	BMI	2B	Plus	BPL	2A	Simple
Overflow	V = 1	BVS	29	No Overflow	BVC	28	Simple
r = 0	Z = 1	BEQ	27	r ≠ 0	BNE	26	Simple
Always	—	BRA	20	Never	BRN	21	Unconditional

Branch if Greater than Zero (BGT)

Operation: PC $\leftarrow$ (PC) + $0002 + Rel   if (Z)•[(N)$\oplus$(V)] = 0
i.e., if (ACCX) > (M)   (twos-complement signed numbers)

Description: If the BGT instruction is executed immediately after execution of any of the instructions, CBA, CMP(A, B, or D), CP(X or Y), SBA, SUB(A, B, or D), the branch will occur if and only if the twos-complement number represented by ACCX was greater than the twos-complement number represented by M.

See BRA instruction for further details of the execution of the branch.

Condition Codes and Boolean Formulae:

S	X	H	I	N	Z	V	C
—	—	—	—	—	—	—	—

None affected

Source Form: BGT (rel)

Addressing Modes, Machine Code, and Cycle-by-Cycle Execution:

Cycle	BGT (REL)		
	Addr	Data	R/$\overline{W}$
1	OP	2E	1
2	OP + 1	rr	1
3	FFFF	—	1

The following table is a summary of all branch instructions.

Test	Boolean	Mnemonic	Opcode	Complementary	Branch	Comment
r > m	Z + (N $\oplus$ V) = 0	BGT	2E	r ≤ m	BLE	2F	Signed
r ≥ m	N $\oplus$ V = 0	BGE	2C	r < m	BLT	2D	Signed
r = m	Z = 1	BEQ	27	r ≠ m	BNE	26	Signed
r ≤ m	Z + (N $\oplus$ V) = 1	BLE	2F	r > m	BGT	2E	Signed
r < m	N $\oplus$ V = 1	BLT	2D	r ≥ m	BGE	2C	Signed
r > m	C + Z = 0	BHI	22	r ≤ m	BLS	23	Unsigned
r ≥ m	C = 0	BHS/BCC	24	r < m	BLO/BCS	25	Unsigned
r = m	Z = 1	BEQ	27	r ≠ m	BNE	26	Unsigned
r ≤ m	C + Z = 1	BLS	23	r > m	BHI	22	Unsigned
r < m	C = 1	BLO/BCS	25	r ≥ m	BHS/BCC	24	Unsigned
Carry	C = 1	BCS	25	No Carry	BCC	24	Simple
Negative	N = 1	BMI	2B	Plus	BPL	2A	Simple
Overflow	V = 1	BVS	29	No Overflow	BVC	28	Simple
r = 0	Z = 1	BEQ	27	r ≠ 0	BNE	26	Simple
Always	—	BRA	20	Never	BRN	21	Unconditional

Branch if Higher (BHI)

Operation: PC ◄ (PC) + $0002 + Rel   if (C) + (Z) = 0
                  i.e., if (ACCX)>(M)   (unsigned binary numbers)

Description: If the BHI instruction is executed immediately after execution of any of the instructions, CBA, CMP(A, B, or D), CP(X or Y), SBA, SUB(A, B, or D), the branch will occur if and only if the unsigned binary number represented by ACCX was greater than the unsigned binary number represented by M. Generally not useful after INC/DEC, LD/ST, TST/CLR/COM because these instructions do not affect the C bit in the CCR.

See BRA instruction for further details of the execution of the branch.

Condition Codes and Boolean Formulae:

S	X	H	I	N	Z	V	C
—	—	—	—	—	—	—	—

None affected

Source Form: BHI (rel)

Addressing Modes, Machine Code, and Cycle-by-Cycle Execution:

Cycle	BHI (REL)		
	Addr	Data	R/W̄
1	OP	22	1
2	OP + 1	rr	1
3	FFFF	—	1

The following table is a summary of all branch instructions.

Test	Boolean	Mnemonic	Opcode	Complementary		Branch	Comment
r>m	Z + (N ⊕ V) = 0	BGT	2E	r≤m	BLE	2F	Signed
r≥m	N ⊕ V = 0	BGE	2C	r<m	BLT	2D	Signed
r = m	Z = 1	BEQ	27	r ≠ m	BNE	26	Signed
r≤m	Z + (N ⊕ V) = 1	BLE	2F	r>m	BGT	2E	Signed
r<m	N ⊕ V = 1	BLT	2D	r≥m	BGE	2C	Signed
r>m	C + Z = 0	BHI	22	r≤m	BLS	23	Unsigned
r≥m	C = 0	BHS/BCC	24	r<m	BLO/BCS	25	Unsigned
r = m	Z = 1	BEQ	27	r ≠ m	BNE	26	Unsigned
r≤m	C + Z = 1	BLS	23	r>m	BHI	22	Unsigned
r<m	C = 1	BLO/BCS	25	r≥m	BHS/BCC	24	Unsigned
Carry	C = 1	BCS	25	No Carry	BCC	24	Simple
Negative	N = 1	BMI	2B	Plus	BPL	2A	Simple
Overflow	V = 1	BVS	29	No Overflow	BVC	28	Simple
r = 0	Z = 1	BEQ	27	r ≠ 0	BNE	26	Simple
Always	—	BRA	20	Never	BRN	21	Unconditional

Branch if Higher or Same (BHS) (Same as BCC)

Operation: PC ◀ (PC) + $0002 + Rel   if (C) = 0
i.e., if (ACCX) ≥ (M)   (unsigned binary numbers)

Description: If the BHI instruction is executed immediately after execution of any of the instructions, CBA, CMP(A, B, or D), CP(X or Y), SBA, SUB(A, B, or D), the branch will occur if and only if the unsigned binary number represented by ACCX was greater than or equal to the unsigned binary number represented by M. Generally not useful after INC/DEC, LD/ST, TST/CLR/COM because these instructions do not affect the C bit in the CCR.

See BRA instruction for further details of the execution of the branch.

Condition Codes and Boolean Formulae:

S	X	H	I	N	Z	V	C
—	—	—	—	—	—	—	—

None affected

Source Form: BHS (rel)

Addressing Modes, Machine Code, and Cycle-by-Cycle Execution:

Cycle	BHS (REL)		
	Addr	Data	R/W̄
1	OP	24	1
2	OP + 1	rr	1
3	FFFF	—	1

The following table is a summary of all branch instructons.

Test	Boolean	Mnemonic	Opcode	Complementary		Branch	Comment
r>m	Z + (N ⊕ V) = 0	BGT	2E	r≤m	BLE	2F	Signed
r≥m	N ⊕ V = 0	BGE	2C	r<m	BLT	2D	Signed
r = m	Z = 1	BEQ	27	r ≠ m	BNE	26	Signed
r≤m	Z + (N ⊕ V) = 1	BLE	2F	r>m	BGT	2E	Signed
r<m	N ⊕ V = 1	BLT	2D	r≥m	BGE	2C	Signed
r>m	C + Z = 0	BHI	22	r≤m	BLS	23	Unsigned
r≥m	C = 0	BHS/BCC	24	r<m	BLO/BCS	25	Unsigned
r = m	Z = 1	BEQ	27	r ≠ m	BNE	26	Unsigned
r≤m	C + Z = 1	BLS	23	r>m	BHI	22	Unsigned
r<m	C = 1	BLO/BCS	25	r≥m	BHS/BCC	24	Unsigned
Carry	C = 1	BCS	25	No Carry	BCC	24	Simple
Negative	N = 1	BMI	2B	Plus	BPL	2A	Simple
Overflow	V = 1	BVS	29	No Overflow	BVC	28	Simple
r = 0	Z = 1	BEQ	27	r ≠ 0	BNE	26	Simple
Always	—	BRA	20	Never	BRN	21	Unconditional

Bit Test

Operation: (ACCX)•(M)

Description: Performs the logical AND operation between the contents of ACCX and the contents of M and modifies the condition codes accordingly. Neither the contents of ACCX or M operands are affected. (Each bit of the result of the AND would be the logical AND of the corresponding bits of ACCX and M.)

Condition Codes and Boolean Formulae:

S	X	H	I	N	Z	V	C
—	—	—	—	⬍	⬍	0	—

N R7
Set if MSB of result is set; cleared otherwise.

Z $\overline{R7} \cdot \overline{R6} \cdot \overline{R5} \cdot \overline{R4} \cdot \overline{R3} \cdot \overline{R2} \cdot \overline{R1} \cdot \overline{R0}$
Set if result is $00; cleared otherwise.

V 0
Cleared

Source Forms: BITA (opr); BITB (opr)

Addressing Modes, Machine Code, and Cycle-by-Cycle Execution:

Cycle	BITA (IMM)			BITA (DIR)			BITA (EXT)			BITA (IND, X)			BITA (IND, Y)		
	Addr	Data	R/W̄	Addr	Data	R/W̄	Addr	Data	R/W̄	Addr	Data	R/W̄	Addr	Data	R/W̄
1	OP	85	1	OP	95	1	OP	B5	1	OP	A5	1	OP	18	1
2	OP+1	ii	1	OP+1	dd	1	OP+1	hh	1	OP+1	ff	1	OP+1	A5	1
3				00dd	(00dd)	1	OP+2	ll	1	FFFF	—	1	OP+2	ff	1
4							hhll	(hhll)	1	X+ff	(X+ff)	1	FFFF	—	1
5													Y+ff	(Y+ff)	1

Cycle	BITB (IMM)			BITB (DIR)			BITB (EXT)			BITB (IND, X)			BITB (IND, Y)		
	Addr	Data	R/W̄	Addr	Data	R/W̄	Addr	Data	R/W̄	Addr	Data	R/W̄	Addr	Data	R/W̄
1	OP	C5	1	OP	D5	1	OP	F5	1	OP	E5	1	OP	18	1
2	OP+1	ii	1	OP+1	dd	1	OP+1	hh	1	OP+1	ff	1	OP+1	E5	1
3				00dd	(00dd)	1	OP+2	ll	1	FFFF	—	1	OP+2	ff	1
4							hhll	(hhll)	1	X+ff	(X+ff)	1	FFFF	—	1
5													Y+ff	(Y+ff)	1

NOOMAIR AHMER ZUBER!

Branch if Less than or Equal to Zero (BLE)

Operation: PC ◆ (PC) + $0002 + Rel   if (Z) + [(N)⊕(V)] = 1
i.e., if (ACCX)≤(M)   (twos-complement signed numbers)

Description: If the BLE instruction is executed immediately after execution of any of the instructions, CBA, CMP(A, B, or D), CP(X, or Y), SBA, SUB(A, B, or D), the branch will occur if and only if the twos-complement number represented by ACCX was less than or equal to the twos-complement number represented by M.

See BRA instruction for further details of the execution of the branch.

Condition Codes and Boolean Formulae:

S	X	H	I	N	Z	V	C
—	—	—	—	—	—	—	—

None affected

Source Form: BLE (rel)

Addressing Modes, Machine Code, and Cycle-by-Cycle Execution:

Cycle	BLE (REL)		
	Addr	Data	R/W̄
1	OP	2F	1
2	OP + 1	rr	1
3	FFFF	—	1

The following table is a summary of all branch instructions.

Test	Boolean	Mnemonic	Opcode	Complementary		Branch	Comment
r>m	Z + (N ⊕ V) = 0	BGT	2E	r≤m	BLE	2F	Signed
r≥m	N ⊕ V = 0	BGE	2C	r<m	BLT	2D	Signed
r = m	Z = 1	BEQ	27	r ≠ m	BNE	26	Signed
r≤m	Z + (N ⊕ V) = 1	BLE	2F	r>m	BGT	2E	Signed
r<m	N ⊕ V = 1	BLT	2D	r≥m	BGE	2C	Signed
r>m	C + Z = 0	BHI	22	r≤m	BLS	23	Unsigned
r≥m	C = 0	BHS/BCC	24	r<m	BLO/BCS	25	Unsigned
r = m	Z = 1	BEQ	27	r ≠ m	BNE	26	Unsigned
r≤m	C + Z = 1	BLS	23	r>m	BHI	22	Unsigned
r<m	C = 1	BLO/BCS	25	r≥m	BHS/BCC	24	Unsigned
Carry	C = 1	BCS	25	No Carry	BCC	24	Simple
Negative	N = 1	BMI	2B	Plus	BPL	2A	Simple
Overflow	V = 1	BVS	29	No Overflow	BVC	28	Simple
r = 0	Z = 1	BEQ	27	r ≠ 0	BNE	26	Simple
Always	—	BRA	20	Never	BRN	21	Unconditional

Branch if Lower (BLO) (Same as BCS)

Operation: PC $\blacklozenge$ (PC) + $0002 + Rel   if (C) = 1
i.e., if (ACCX)<(M)   (unsigned binary numbers)

Description: If the BLO instruction is executed immediately after execution of any of the instructions, CBA, CMP(A, B, or D), CP(X or Y), SBA, SUB(A, B, or D), the branch will occur if and only if the unsigned binary number represented by ACCX was less than or equal to the unsigned binary number represented by M. Generally not useful after INC/DEC, LD/ST, TST/CLR/COM because these instructions do not affect the C bit in the CCR.

See BRA instruction for further details of the execution of the branch.

Condition Codes and Boolean Formulae:

S	X	H	I	N	Z	V	C
—	—	—	—	—	—	—	—

None affected

Source Form: BLO (rel)

Addressing Modes, Machine Code, and Cycle-by-Cycle Execution:

Cycle	BLO (REL)		
	Addr	Data	R/$\overline{W}$
1	OP	25	1
2	OP + 1	rr	1
3	FFFF	—	1

The following table is a summary of all branch instructions.

Test	Boolean	Mnemonic	Opcode	Complementary	Branch		Comment
r>m	Z + (N $\oplus$ V) = 0	BGT	2E	r≤m	BLE	2F	Signed
r≥m	N $\oplus$ V = 0	BGE	2C	r<m	BLT	2D	Signed
r = m	Z = 1	BEQ	27	r ≠ m	BNE	26	Signed
r≤m	Z + (N $\oplus$ V) = 1	BLE	2F	r>m	BGT	2E	Signed
r<m	N $\oplus$ V = 1	BLT	2D	r≥m	BGE	2C	Signed
r>m	C + Z = 0	BHI	22	r≤m	BLS	23	Unsigned
r≥m	C = 0	BHS/BCC	24	r<m	BLO/BCS	25	Unsigned
r = m	Z = 1	BEQ	27	r ≠ m	BNE	26	Unsigned
r≤m	C + Z = 1	BLS	23	r>m	BHI	22	Unsigned
r<m	C = 1	BLO/BCS	25	r≥m	BHS/BCC	24	Unsigned
Carry	C = 1	BCS	25	No Carry	BCC	24	Simple
Negative	N = 1	BMI	2B	Plus	BPL	2A	Simple
Overflow	V = 1	BVS	29	No Overflow	BVC	28	Simple
r = 0	Z = 1	BEQ	27	r ≠ 0	BNE	26	Simple
Always	—	BRA	20	Never	BRN	21	Unconditional

Branch if Lower or Same (BLS)

Operation: PC ◄ (PC) + $0002 + Rel   if (C) + (Z) = 1
i.e., if (ACCX) ≤ (M)   (unsigned binary numbers)

Description: If the BLS instruction is executed immediately after execution of any of the instructions, CBA, CMP(A, B, or D), CP(X or Y), SBA, SUB(A, B, or D), the branch will occur if and only if the unsigned binary number represented by ACCX was less than or equal to the unsigned binary number represented by M. Generally not useful after INC/DEC, LD/ST, TST/CLR/COM because these instructions do not affect the C bit in the CCR.

See BRA instruction for further details of the execution of the branch.

Condition Codes and Boolean Formulae:

S	X	H	I	N	Z	V	C
—	—	—	—	—	—	—	—

None affected

Source Form: BLS (rel)

Addressing Modes, Machine Code, and Cycle-by-Cycle Execution:

Cycle	BLS (REL)		
	Addr	Data	R/W̄
1	OP	23	1
2	OP + 1	rr	1
3	FFFF	—	1

The following table is a summary of all branch instructions.

Test	Boolean	Mnemonic	Opcode	Complementary	Branch	Comment
r > m	Z + (N ⊕ V) = 0	BGT	2E	r ≤ m	BLE	2F	Signed
r ≥ m	N ⊕ V = 0	BGE	2C	r < m	BLT	2D	Signed
r = m	Z = 1	BEQ	27	r ≠ m	BNE	26	Signed
r ≤ m	Z + (N ⊕ V) = 1	BLE	2F	r > m	BGT	2E	Signed
r < m	N ⊕ V = 1	BLT	2D	r ≥ m	BGE	2C	Signed
r > m	C + Z = 0	BHI	22	r ≤ m	BLS	23	Unsigned
r ≥ m	C = 0	BHS/BCC	24	r < m	BLO/BCS	25	Unsigned
r = m	Z = 1	BEQ	27	r ≠ m	BNE	26	Unsigned
r ≤ m	C + Z = 1	BLS	23	r > m	BHI	22	Unsigned
r < m	C = 1	BLO/BCS	25	r ≥ m	BHS/BCC	24	Unsigned
Carry	C = 1	BCS	25	No Carry	BCC	24	Simple
Negative	N = 1	BMI	2B	Plus	BPL	2A	Simple
Overflow	V = 1	BVS	29	No Overflow	BVC	28	Simple
r = 0	Z = 1	BEQ	27	r ≠ 0	BNE	26	Simple
Always	—	BRA	20	Never	BRN	21	Unconditional

Branch if Less than Zero (BLT)

Operation: PC ◆ (PC) + $0002 + Rel   if (N)⊕(V) = 1
i.e., if (ACCX)<(M)   (twos-complement signed numbers)

Description: If the BLT instruction is executed immediately after execution of any of the instructons, CBA, CMP(A, B, or D), CP(X or Y), SBA, SUB(A, B, or D), the branch will occur if and only if the twos-complement number represented by ACCX was less than the twos-complement number represented by M.

See BRA instruction for further details of the execution of the branch.

Condition Codes and Boolean Formulae:

S	X	H	I	N	Z	V	C
—	—	—	—	—	—	—	—

None affected

Source Form: BLT (rel)

Addressing Modes, Machine Code, and Cycle-by-Cycle Execution:

Cycle	BLT (REL)		
	Addr	Data	R/W̄
1	OP	2D	1
2	OP + 1	rr	1
3	FFFF	—	1

The following table is a summary of all branch instructions.

Test	Boolean	Mnemonic	Opcode	Complementary	Branch	Comment
r>m	Z+(N ⊕ V) = 0	BGT	2E	r≤m	BLE	2F	Signed
r≥m	N ⊕ V = 0	BGE	2C	r<m	BLT	2D	Signed
r = m	Z = 1	BEQ	27	r ≠ m	BNE	26	Signed
r≤m	Z+(N ⊕ V) = 1	BLE	2F	r>m	BGT	2E	Signed
r<m	N ⊕ V = 1	BLT	2D	r≥m	BGE	2C	Signed
r>m	C+Z = 0	BHI	22	r≤m	BLS	23	Unsigned
r≥m	C = 0	BHS/BCC	24	r<m	BLO/BCS	25	Unsigned
r = m	Z = 1	BEQ	27	r ≠ m	BNE	26	Unsigned
r≤m	C+Z = 1	BLS	23	r>m	BHI	22	Unsigned
r<m	C = 1	BLO/BCS	25	r≥m	BHS/BCC	24	Unsigned
Carry	C = 1	BCS	25	No Carry	BCC	24	Simple
Negative	N = 1	BMI	2B	Plus	BPL	2A	Simple
Overflow	V = 1	BVS	29	No Overflow	BVC	28	Simple
r = 0	Z = 1	BEQ	27	r ≠ 0	BNE	26	Simple
Always	—	BRA	20	Never	BRN	21	Unconditional

Branch if Minus (BMI)

Operation: PC ◀ (PC) + $0002 + Rel   if (N) = 1

Description: Tests the state of the N bit in the CCR and causes a branch if N is set.

See BRA instruction for further details of the execution of the branch.

Condition Codes and Boolean Formulae:

S	X	H	I	N	Z	V	C
—	—	—	—	—	—	—	—

None affected

Source Form: BMI (rel)

Addressing Modes, Machine Code, and Cycle-by-Cycle Execution:

Cycle	BMI (REL)		
	Addr	Data	R/$\overline{W}$
1	OP	2B	1
2	OP + 1	rr	1
3	FFFF	—	1

The following table is a summary of all branch instructions.

Test	Boolean	Mnemonic	Opcode	Complementary	Branch		Comment
r>m	Z + (N ⊕ V) = 0	BGT	2E	r≤m	BLE	2F	Signed
r≥m	N ⊕ V = 0	BGE	2C	r<m	BLT	2D	Signed
r = m	Z = 1	BEQ	27	r ≠ m	BNE	26	Signed
r≤m	Z + (N ⊕ V) = 1	BLE	2F	r>m	BGT	2E	Signed
r<m	N ⊕ V = 1	BLT	2D	r≥m	BGE	2C	Signed
r>m	C + Z = 0	BHI	22	r≤m	BLS	23	Unsigned
r≥m	C = 0	BHS/BCC	24	r<m	BLO/BCS	25	Unsigned
r = m	Z = 1	BEQ	27	r ≠ m	BNE	26	Unsigned
r≤m	C + Z = 1	BLS	23	r>m	BHI	22	Unsigned
r<m	C = 1	BLO/BCS	25	r≥m	BHS/BCC	24	Unsigned
Carry	C = 1	BCS	25	No Carry	BCC	24	Simple
Negative	N = 1	BMI	2B	Plus	BPL	2A	Simple
Overflow	V = 1	BVS	29	No Overflow	BVC	28	Simple
r = 0	Z = 1	BEQ	27	r ≠ 0	BNE	26	Simple
Always	—	BRA	20	Never	BRN	21	Unconditional

Branch if Not Equal to Zero (BNE)

Operation: PC ◀ (PC) + $0002 + Rel   if (Z) = 0

Description: Tests the state of the Z bit in the CCR and causes a branch if Z is clear.

See BRA instruction for further details of the execution of the branch.

Condition Codes and Boolean Formulae:

S	X	H	I	N	Z	V	C
—	—	—	—	—	—	—	—

None affected

Source Form: BNE (rel)

Addressing Modes, Machine Code, and Cycle-by-Cycle Execution:

Cycle	BNE (REL)		
	Addr	Data	R/W̄
1	OP	26	1
2	OP + 1	rr	1
3	FFFF	—	1

The following table is a summary of all branch instructions.

Test	Boolean	Mnemonic	Opcode	Complementary	Branch	Comment
r>m	Z + (N ⊕ V) = 0	BGT	2E	r≤m	BLE	2F	Signed
r≥m	N ⊕ V = 0	BGE	2C	r<m	BLT	2D	Signed
r = m	Z = 1	BEQ	27	r ≠ m	BNE	26	Signed
r≤m	Z + (N ⊕ V) = 1	BLE	2F	r>m	BGT	2E	Signed
r<m	N ⊕ V = 1	BLT	2D	r≥m	BGE	2C	Signed
r>m	C + Z = 0	BHI	22	r≤m	BLS	23	Unsigned
r≥m	C = 0	BHS/BCC	24	r<m	BLO/BCS	25	Unsigned
r = m	Z = 1	BEQ	27	r ≠ m	BNE	26	Unsigned
r≤m	C + Z = 1	BLS	23	r>m	BHI	22	Unsigned
r<m	C = 1	BLO/BCS	25	r≥m	BHS/BCC	24	Unsigned
Carry	C = 1	BCS	25	No Carry	BCC	24	Simple
Negative	N = 1	BMI	2B	Plus	BPL	2A	Simple
Overflow	V = 1	BVS	29	No Overflow	BVC	28	Simple
r = 0	Z = 1	BEQ	27	r ≠ 0	BNE	26	Simple
Always	—	BRA	20	Never	BRN	21	Unconditional

Branch if Plus (BPL)

Operation: PC ◀ (PC) + $0002 + Rel   if (N) = 0

Description: Tests the state of the N bit in the CCR and causes a branch if N is clear.

See BRA instruction for details of the execution of the branch.

Condition Codes and Boolean Formulae:

S	X	H	I	N	Z	V	C
—	—	—	—	—	—	—	—

None affected

Source Form: BPL (rel)

Addressing Modes, Machine Code, and Cycle-by-Cycle Execution:

Cycle	BPL (REL)		
	Addr	Data	R/W̄
1	OP	2A	1
2	OP + 1	rr	1
3	FFFF	—	1

The following table is a summary of all branch instructions.

Test	Boolean	Mnemonic	Opcode	Complementary		Branch	Comment
r>m	Z + (N ⊕ V) = 0	BGT	2E	r≤m	BLE	2F	Signed
r≥m	N ⊕ V = 0	BGE	2C	r<m	BLT	2D	Signed
r = m	Z = 1	BEQ	27	r ≠ m	BNE	26	Signed
r≤m	Z + (N ⊕ V) = 1	BLE	2F	r>m	BGT	2E	Signed
r<m	N ⊕ V = 1	BLT	2D	r≥m	BGE	2C	Signed
r>m	C + Z = 0	BHI	22	r≤m	BLS	23	Unsigned
r≥m	C = 0	BHS/BCC	24	r<m	BLO/BCS	25	Unsigned
r = m	Z = 1	BEQ	27	r ≠ m	BNE	26	Unsigned
r≤m	C + Z = 1	BLS	23	r>m	BHI	22	Unsigned
r<m	C = 1	BLO/BCS	25	r≥m	BHS/BCC	24	Unsigned
Carry	C = 1	BCS	25	No Carry	BCC	24	Simple
Negative	N = 1	BMI	2B	Plus	BPL	2A	Simple
Overflow	V = 1	BVS	29	No Overflow	BVC	28	Simple
r = 0	Z = 1	BEQ	27	r ≠ 0	BNE	26	Simple
Always	—	BRA	20	Never	BRN	21	Unconditional

Branch Always (BRA)

Operation: PC ◄ (PC) + $0002 + Rel

Description: Unconditional branch to the address given by the foregoing formula, in which Rel is the relative offset stored as a twos complement number in the second byte of machine code corresponding to the branch instruction.

The source program specifies the destination of any branch instruction by its absolute address, either as a numerical value or as a symbol or expression, that can be numerically evaluated by the assembler. The assembler obtains the relative address, Rel, from the absolute address and the current value of the location counter.

Condition Codes and Boolean Formulae:

S	X	H	I	N	Z	V	C
—	—	—	—	—	—	—	—

None affected

Source Form: BRA (rel)

Addressing Modes, Machine Code, and Cycle-by-Cycle Execution:

Cycle	BRA (REL)		
	Addr	Data	R/W̄
1	OP	20	1
2	OP + 1	rr	1
3	FFFF	—	1

The following table is a summary of all branch instructions.

Test	Boolean	Mnemonic	Opcode	Complementary	Branch	Comment
r>m	Z + (N ⊕ V) = 0	BGT	2E	r≤m	BLE	2F	Signed
r≥m	N ⊕ V = 0	BGE	2C	r<m	BLT	2D	Signed
r = m	Z = 1	BEQ	27	r ≠ m	BNE	26	Signed
r≤m	Z + (N ⊕ V) = 1	BLE	2F	r>m	BGT	2E	Signed
r<m	N ⊕ V = 1	BLT	2D	r≥m	BGE	2C	Signed
r>m	C + Z = 0	BHI	22	r≤m	BLS	23	Unsigned
r≥m	C = 0	BHS/BCC	24	r<m	BLO/BCS	25	Unsigned
r = m	Z = 1	BEQ	27	r ≠ m	BNE	26	Unsigned
r≤m	C + Z = 1	BLS	23	r>m	BHI	22	Unsigned
r<m	C = 1	BLO/BCS	25	r≥m	BHS/BCC	24	Unsigned
Carry	C = 1	BCS	25	No Carry	BCC	24	Simple
Negative	N = 1	BMI	2B	Plus	BPL	2A	Simple
Overflow	V = 1	BVS	29	No Overflow	BVC	28	Simple
r = 0	Z = 1	BEQ	27	r ≠ 0	BNE	26	Simple
Always	—	BRA	20	Never	BRN	21	Unconditional

Branch if Bit(s) Clear (BRCLR)

Operation: PC ◀ (PC) + $0004 + Rel   if (M)•(PC + 2) = 0
                   PC ◀ (PC) + $0005 + Rel   if (M)•(PC + 3) = 0 (for IND, Y address mode only)

Description: Performs the logical AND of location M and the mask supplied with the instruction, then branches if the result is zero (only if all bits corresponding to ones in the mask byte are zeros in the tested byte).

Condition Codes and Boolean Formulae:

S	X	H	I	N	Z	V	C
—	—	—	—	—	—	—	—

None affected

Source Form: BRCLR (opr) (msk) (rel)

Addressing Modes, Machine Code, and Cycle-by-Cycle Execution:

Cycle	BRCLR (DIR)			BRCLR (IND, X)			BRCLR (IND, Y)		
	Addr	Data	R/W̄	Addr	Data	R/W̄	Addr	Data	R/W̄
1	OP	13	1	OP	1F	1	OP	18	1
2	OP + 1	dd	1	OP + 1	ff	1	OP + 1	1F	1
3	00dd	(00dd)	1	FFFF	—	1	OP + 2	ff	1
4	OP + 2	mm	1	X + ff	(X + ff)	1	FFFF	—	1
5	OP + 3	rr	1	OP + 2	mm	1	(IY) + ff	(Y + ff)	1
6	FFFF	—	1	OP + 3	rr	1	OP + 3	mm	1
7				FFFF	—	1	OP + 4	rr	1
8							FFFF	—	1

Branch Never (BRN)

Operation: PC ◆ (PC) + $0002

Description: Never branches. In effect, this instruction can be considered as a two-byte NOP (no operation) requiring three cycles for execution. Its inclusion in the instruction set is to provide a complement for the BRA instruction. The instruction is useful during program debug to negate the effect of another branch instruction without disturbing the offset byte. Having a complement for BRA is also useful in compiler implementations.

Condition Codes and Boolean Formulae:

S	X	H	I	N	Z	V	C
—	—	—	—	—	—	—	—

None affected

Source Form: BRN (rel)

Addressing Modes, Machine Code, and Cycle-by-Cycle Execution:

Cycle	BRN (REL)		
	Addr	Data	R/$\overline{W}$
1	OP	21	1
2	OP + 1	rr	1
3	FFFF	—	1

The following table is a summary of all branch instructions.

Test	Boolean	Mnemonic	Opcode	Complementary	Branch		Comment
r>m	Z + (N ⊕ V) = 0	BGT	2E	r≤m	BLE	2F	Signed
r≥m	N ⊕ V = 0	BGE	2C	r<m	BLT	2D	Signed
r = m	Z = 1	BEQ	27	r ≠ m	BNE	26	Signed
r≤m	Z + (N ⊕ V) = 1	BLE	2F	r>m	BGT	2E	Signed
r<m	N ⊕ V = 1	BLT	2D	r≥m	BGE	2C	Signed
r>m	C + Z = 0	BHI	22	r≤m	BLS	23	Unsigned
r≥m	C = 0	BHS/BCC	24	r<m	BLO/BCS	25	Unsigned
r = m	Z = 1	BEQ	27	r ≠ m	BNE	26	Unsigned
r≤m	C + Z = 1	BLS	23	r>m	BHI	22	Unsigned
r<m	C = 1	BLO/BCS	25	r≥m	BHS/BCC	24	Unsigned
Carry	C = 1	BCS	25	No Carry	BCC	24	Simple
Negative	N = 1	BMI	2B	Plus	BPL	2A	Simple
Overflow	V = 1	BVS	29	No Overflow	BVC	28	Simple
r = 0	Z = 1	BEQ	27	r ≠ 0	BNE	26	Simple
Always	—	BRA	20	Never	BRN	21	Unconditional

Branch if Bit(s) Set (BRSET)

Operation: PC ◀ (PC) + $0004 + Rel   if $\overline{(M)} \cdot (PC + 2) = 0$
PC ◀ (PC) + $0005 + Rel   if $\overline{(M)} \cdot (PC + 3) = 0$ (for IND, Y address mode only)

Description: Performs the logical AND of location M inverted and the mask supplied with the instruction, then branches if the result is zero (only if all bits corresponding to ones in the mask byte are ones in the tested byte).

Condition Codes and Boolean Formulae:

S	X	H	I	N	Z	V	C
—	—	—	—	—	—	—	—

None affected

Source Form: BRSET (opr) (msk) (rel)

Addressing Modes, Machine Code, and Cycle-by-Cycle Execution:

Cycle	BRSET (DIR)			BRSET (IND, X)			BRSET (IND, Y)		
	Addr	Data	R/W̄	Addr	Data	R/W̄	Addr	Data	R/W̄
1	OP	12	1	OP	1E	1	OP	18	1
2	OP + 1	dd	1	OP + 1	ff	1	OP + 1	1E	1
3	00dd	(00dd)	1	FFFF	—	1	OP + 2	ff	1
4	OP + 2	mm	1	X + ff	(X + ff)	1	FFFF	—	1
5	OP + 3	rr	1	OP + 2	mm	1	(IY) + ff	(Y + ff)	1
6	FFFF	—	1	OP + 3	rr	1	OP + 3	mm	1
7				FFFF	—	1	OP + 4	rr	1
8							FFFF	—	1

Set Bit(s) in Memory (BSET)

Operation: M ◀ (M) + (PC + 2)
               M ◀ (M) + (PC + 3)   (for IND, Y address mode only)

Description: Set multiple bits in location M. The bit(s) to be set are specified by ones in the mask byte (last machine code byte of the instruction). All other bits in M are unaffected.

Condition Codes and Boolean Formulae:

S	X	H	I	N	Z	V	C
—	—	—	—	↕	↕	0	—

N R7
   Set if MSB of result is set; cleared otherwise.

Z $\overline{R7} \cdot \overline{R6} \cdot \overline{R5} \cdot \overline{R4} \cdot \overline{R3} \cdot \overline{R2} \cdot \overline{R1} \cdot \overline{R0}$
   Set if result is $00; cleared otherwise.

V 0
   Cleared

Source Form: BSET (opr) (msk)

Addressing Modes, Machine Code, and Cycle-by-Cycle Execution:

Cycle	BSET (DIR)			BSET (IND, X)			BSET (IND, Y)		
	Addr	Data	R/W̄	Addr	Data	R/W̄	Addr	Data	R/W̄
1	OP	14	1	OP	1C	1	OP	18	1
2	OP + 1	dd	1	OP + 1	ff	1	OP + 1	1C	1
3	00dd	(00dd)	1	FFFF	—	1	OP + 2	ff	1
4	OP + 2	mm	1	X + ff	(X + ff)	1	FFFF	—	1
5	FFFF	—	1	OP + 2	mm	1	(IY) + ff	(Y + ff)	1
6	00dd	result	0	FFFF	—	1	OP + 3	mm	1
7				X + ff	result	0	FFFF	—	1
8							Y + ff	result	0

Branch to Subroutine (BSR)

Operation: PC ◀ (PC) + $0002   Advance PC to return address
                  ◀(PCL)   Push low-order return onto stack
                  SP ◀ (SP) − 0001
                  ◀(PCH)   Push high-order return onto stack
                  SP ◀ (SP) − $0001
                  PC ◀ (PC) + Rel   Load start address of requested subroutine

Description: The program counter is incremented by two (this will be the return address). The least significant byte of the contents of the program counter (low-order
return address) is pushed onto the stack. The stack pointer is then decremented by
one. The most significant byte of the contents of the program counter (high-order
return address) is pushed onto the stack. The stack pointer is then decremented by
one. A branch then occurs to the location specified by the branch offset.

See BRA instruction for further details of the execution of the branch.

Condition Codes and Boolean Formulae:

S	X	H	I	N	Z	V	C
—	—	—	—	—	—	—	—

None affected

Source Form: BSR (rel)

Addressing Modes, Machine Code, and Cycle-by-Cycle Execution:

Cycle	BSR (INH)		
	Addr	Data	R/W̄
1	OP	8D	1
2	OP + 1	rr	1
3	FFFF	—	1
4	Sub	Nxt op	1
5	SP	Rtn lo	0
6	SP − 1	Rtn hi	0

Branch if Overflow Clear (BVC)

Operation: PC ◀ (PC) + $0002 + Rel   if (V) = 0

Description: Tests the state of the V bit in the CCR and causes a branch if V is clear.

Used after an operation on twos-complement binary values, this instruction will cause a branch if there was NO overflow. That is, branch if the twos-complement result was valid.

See BRA instruction for further details of the execution of the branch.

Condition Codes and Boolean Formulae:

S	X	H	I	N	Z	V	C
—	—	—	—	—	—	—	—

None affected

Source Form: BVC (rel)

Addressing Modes, Machine Code, and Cycle-by-Cycle Execution:

Cycle	BVC (REL)		
	Addr	Data	R/W̄
1	OP	28	1
2	OP + 1	rr	1
3	FFFF	—	1

The following table is a summary of all branch instructions.

Test	Boolean	Mnemonic	Opcode	Complementary	Branch		Comment
r>m	Z+(N ⊕ V) = 0	BGT	2E	r≤m	BLE	2F	Signed
r≥m	N ⊕ V = 0	BGE	2C	r<m	BLT	2D	Signed
r=m	Z=1	BEQ	27	r≠m	BNE	26	Signed
r≤m	Z+(N ⊕ V) = 1	BLE	2F	r>m	BGT	2E	Signed
r<m	N ⊕ V = 1	BLT	2D	r≥m	BGE	2C	Signed
r>m	C+Z = 0	BHI	22	r≤m	BLS	23	Unsigned
r≥m	C = 0	BHS/BCC	24	r<m	BLO/BCS	25	Unsigned
r=m	Z=1	BEQ	27	r≠m	BNE	26	Unsigned
r≤m	C+Z = 1	BLS	23	r>m	BHI	22	Unsigned
r<m	C = 1	BLO/BCS	25	r≥m	BHS/BCC	24	Unsigned
Carry	C = 1	BCS	25	No Carry	BCC	24	Simple
Negative	N = 1	BMI	2B	Plus	BPL	2A	Simple
Overflow	V = 1	BVS	29	No Overflow	BVC	28	Simple
r=0	Z=1	BEQ	27	r≠0	BNE	26	Simple
Always	—	BRA	20	Never	BRN	21	Unconditional

Branch if Overflow Set (BVS)

Operation: PC ◀ (PC) + $0002 + Rel   if (V) = 1

Description: Tests the state of the V bit in the CCR and causes a branch if V is set.

Used after an operation on twos-complement binary values, this instruction will cause a branch if there was an overflow. That is, branch if the twos-complement result was invalid.

See BRA instruction for details of the execution of the branch.

Condition Codes and Boolean Formulae:

S	X	H	I	N	Z	V	C
—	—	—	—	—	—	—	—

None affected

Source Form: BVS (rel)

Addressing Modes, Machine Code, and Cycle-by-Cycle Execution:

Cycle	BVS (REL)		
	Addr	Data	R/$\overline{W}$
1	OP	29	1
2	OP + 1	rr	1
3	FFFF	—	1

The following table is a summary of all branch instructions.

Test	Boolean	Mnemonic	Opcode	Complementary		Branch	Comment
r>m	Z + (N ⊕ V) = 0	BGT	2E	r≤m	BLE	2F	Signed
r≥m	N ⊕ V = 0	BGE	2C	r<m	BLT	2D	Signed
r = m	Z = 1	BEQ	27	r ≠ m	BNE	26	Signed
r≤m	Z + (N ⊕ V) = 1	BLE	2F	r>m	BGT	2E	Signed
r<m	N ⊕ V = 1	BLT	2D	r≥m	BGE	2C	Signed
r>m	C + Z = 0	BHI	22	r≤m	BLS	23	Unsigned
r≥m	C = 0	BHS/BCC	24	r<m	BLO/BCS	25	Unsigned
r = m	Z = 1	BEQ	27	r ≠ m	BNE	26	Unsigned
r≤m	C + Z = 1	BLS	23	r>m	BHI	22	Unsigned
r<m	C = 1	BLO/BCS	25	r≥m	BHS/BCC	24	Unsigned
Carry	C = 1	BCS	25	No Carry	BCC	24	Simple
Negative	N = 1	BMI	2B	Plus	BPL	2A	Simple
Overflow	V = 1	BVS	29	No Overflow	BVC	28	Simple
r = 0	Z = 1	BEQ	27	r ≠ 0	BNE	26	Simple
Always	—	BRA	20	Never	BRN	21	Unconditional

Compare Accumulators (CBA)

Operation: (ACCA) − (ACCB)

Description: Compares the contents of ACCA to the contents of ACCB and sets the condition codes, which may be used for arithmetic and logical conditional branches. Both operands are unaffected.

Condition Codes and Boolean Formulae:

S	X	H	I	N	Z	V	C
—	—	—	—	↕	↕	↕	↕

N   R7
   Set if MSB of result is set; cleared otherwise.

Z $\overline{R7} \cdot \overline{R6} \cdot \overline{R5} \cdot \overline{R4} \cdot \overline{R3} \cdot \overline{R2} \cdot \overline{R1} \cdot \overline{R0}$
   Set if result is $00; cleared otherwise.

V $A7 \cdot \overline{B7} \cdot \overline{R7} + \overline{A7} \cdot B7 \cdot R7$
   Set if a twos complement overflow resulted from the operation; cleared otherwise.

C $\overline{A7} \cdot B7 + B7 \cdot R7 + R7 \cdot \overline{A7}$
   Set if there was a borrow from the MSB of the result; cleared otherwise.

Source Form: CBA

Addressing Modes, Machine Code, and Cycle-by-Cycle Execution:

Cycle	CBA (INH)		
	Addr	Data	R/W̄
1	OP	11	1
2	OP+1	—	1

Clear Carry (CLC)

Operation: C bit ◀ 0

Description: Clears the C bit in the CCR.

CLC may be used to set up the C bit prior to a shift or rotate instruction involving the C bit.

Condition Codes and Boolean Formulae:

S	X	H	I	N	Z	V	C
—	—	—	—	—	—	—	0

C 0
   Cleared

Source Form: CLC

Addressing Modes, Machine Code, and Cycle-by-Cycle Execution:

Cycle	CLC (INH)		
	Addr	Data	R/W̄
1	OP	0C	1
2	OP + 1	—	1

Clear Interrupt Mask (CLI)

Operation: I bit ♦ 0

Description: Clears the interrupt mask bit in the CCR. When the I bit is clear, interrupts are enabled. There is a one E-clock cycle delay in the clearing mechanism for the I bit so that, if interrupts were previously disabled, the next instruction after a CLI will always be executed, even if there was an interrupt pending prior to execution of the CLI instruction.

Condition Codes and Boolean Formulae:

S	X	H	I	N	Z	V	C
—	—	—	0	—	—	—	—

I   0
   Cleared

Source Form: CLI

Addressing Modes, Machine Code, and Cycle-by-Cycle Execution:

Cycle	CLI (INH)		
	Addr	Data	R/W̄
1	OP	0E	1
2	OP+1	—	1

Clear (CLR)

Operation: ACCX ⬧ 0   or: M ⬧ 0

Description;   The contents of ACCX or M are replaced with zeros.

Condition Codes and Boolean Formulae:

S	X	H	I	N	Z	V	C
—	—	—	—	0	1	0	0

N 0
   Cleared

Z 1
   Set

V 0
   Cleared

C 0
   Cleared

Source Forms: CLRA; CLRB; CLR (opr)

Addressing Modes, Machine Code, and Cycle-by-Cycle Execution:

Cycle	CLRA (INH)			CLRB (INH)			CLR (EXT)			CLR (IND, X)			CLR (IND, Y)		
	Addr	Data	R/W̄	Addr	Data	R/W̄	Addr	Data	R/W̄	Addr	Data	R/W̄	Addr	Data	R/W̄
1	OP	4F	1	OP	5F	1	OP	7F	1	OP	6F	1	OP	18	1
2	OP + 1	—	1	OP + 1	—	1	OP + 1	hh	1	OP + 1	ff	1	OP + 1	6F	1
3							OP + 2	ll	1	FFFF	—	1	OP + 2	ff	1
4							hhll	(hhll)	1	X + ff	(X + ff)	1	FFFF	—	1
5							FFFF	—	1	FFFF	—	1	Y + ff	(Y + ff)	1
6							hhll	00	0	X + ff	00	0	FFFF	—	1
7													Y + ff	00	0

Clear Twos-Complement Overflow Bit (CLV)

Operation: V bit ◀ 0

Description: Clears the twos complement overflow bit in the CCR.

Condition Codes and Boolean Formulae:

S	X	H	I	N	Z	V	C
—	—	—	—	—	—	0	—

V   0
   Cleared

Source Form: CLV

Addressing Modes, Machine Code, and Cycle-by-Cycle Execution:

Cycle	CLV (INH)		
	Addr	Data	R/W̄
1	OP	0A	1
2	OP + 1	—	1

Compare (CMP)

Operation: $(ACCX) - (M)$

Description: Compares the contents of ACCX to the contents of M and sets the condition codes, which may be used for arithmetic and logical conditional branching. Both operands are unaffected.

Condition Codes and Boolean Formulae:

S	X	H	I	N	Z	V	C
—	—	—	—	‡	‡	‡	‡

N   R7
  Set if MSB of result is set; cleared otherwise.

Z $\overline{R7} \cdot \overline{R6} \cdot \overline{R5} \cdot \overline{R4} \cdot \overline{R3} \cdot \overline{R2} \cdot \overline{R1} \cdot \overline{R0}$
  Set if result is $00; cleared otherwise.

V $X7 \cdot \overline{M7} \cdot \overline{R7} + \overline{X7} \cdot M7 \cdot R7$
  Set if a twos complement overflow resulted from the operation; cleared otherwise.

C $\overline{X7} \cdot M7 + M7 \cdot R7 + R7 \cdot \overline{X7}$
  Set if there was a borrow from the MSB of the result; cleared otherwise.

Source Forms: CMPA (opr); CMPB (opr)

Addressing Modes, Machine Code, and Cycle-by-Cycle Execution:

Cycle	CMPA (IMM)			CMPA (DIR)			CMPA (EXT)			CMPA (IND, X)			CMPA (IND, Y)		
	Addr	Data	R/W̄	Addr	Data	R/W̄	Addr	Data	R/W̄	Addr	Data	R/W̄	Addr	Data	R/W̄
1	OP	81	1	OP	91	1	OP	B1	1	OP	A1	1	OP	18	1
2	OP+1	ii	1	OP+1	dd	1	OP+1	hh	1	OP+1	ff	1	OP+1	A1	1
3				00dd	(00dd)	1	OP+2	ll	1	FFFF	—	1	OP+2	ff	1
4							hhll	(hhll)	1	X+ff	(X+ff)	1	FFFF	—	1
5													Y+ff	(Y+ff)	1

Cycle	CMPB (IMM)			CMPB (DIR)			CMPB (EXT)			CMPB (IND, X)			CMPB (IND, Y)		
	Addr	Data	R/W̄	Addr	Data	R/W̄	Addr	Data	R/W̄	Addr	Data	R/W̄	Addr	Data	R/W̄
1	OP	C1	1	OP	D1	1	OP	F1	1	OP	E1	1	OP	18	1
2	OP+1	ii	1	OP+1	dd	1	OP+1	hh	1	OP+1	ff	1	OP+1	E1	1
3				00dd	(00dd)	1	OP+2	ll	1	FFFF	—	1	OP+2	ff	1
4							hhll	(hhll)	1	X+ff	(X+ff)	1	FFFF	—	1
5													Y+ff	(Y+ff)	1

Complement (COM)

Operation: ACCX ♦ ($\overline{\text{ACCX}}$) = $FF − (ACCX)   **or:**   M ♦ ($\overline{\text{M}}$) = $FF − (M)

Description: Replaces the contents of ACCX or M with its ones complement. (Each bit of the contents of ACCX or M is replaced with the complement of that bit.) Immediately after a COM operation on unsigned values, only the BEQ and BNE branches can be expected to perform consistently. When operating on twos-complement values, all signed branches are available.

Condition Codes and Boolean Formulae:

S	X	H	I	N	Z	V	C
—	—	—	—	↕	↕	0	1

N R7
   Set if MSB of result is set; cleared otherwise.

Z $\overline{R7} \cdot \overline{R6} \cdot \overline{R5} \cdot \overline{R4} \cdot \overline{R3} \cdot \overline{R2} \cdot \overline{R1} \cdot \overline{R0}$
   Set if result is $00; cleared otherwise.

V 0
   Cleared
C 1
   Set

Source Forms: COMA; COMB; COM (opr)

Addressing Modes, Machine Code, and Cycle-by-Cycle Execution:

Cycle	COMA (INH)			COMB (INH)			COM (EXT)			COM (IND, X)			COM (IND, Y)		
	Addr	Data	R/$\overline{\text{W}}$	Addr	Data	R/$\overline{\text{W}}$	Addr	Data	R/$\overline{\text{W}}$	Addr	Data	R/$\overline{\text{W}}$	Addr	Data	R/$\overline{\text{W}}$
1	OP	43	1	OP	53	1	OP	73	1	OP	63	1	OP	18	1
2	OP+1	—	1	OP+1	—	1	OP+1	hh	1	OP+1	ff	1	OP+1	63	1
3							OP+2	ll	1	FFFF	—	1	OP+2	ff	1
4							hhll	(hhll)	1	X+ff	(X+ff)	1	FFFF	—	1
5							FFFF	—	1	FFFF	—	1	Y+ff	(Y+ff)	1
6							hhll	result	0	X+ff	result	0	FFFF	—	1
7													Y+ff	result	0

Compare Double Accumulator (CPD)

Operation: $(ACCD) - (M:M+1)$

Description: Compares the contents of accumulator D with a 16-bit value at the address specified and sets the condition codes accordingly. The compare is accomplished internally by doing a 16-bit subtract of $(M:M+1)$ from accumulator D without modifying either accumulator D or $(M:M+1)$.

Condition Codes and Boolean Formulae:

S	X	H	I	N	Z	V	C
—	—	—	—	✱	✱	✱	✱

N R15
   Set if MSB of result is set; cleared otherwise.

Z $\overline{R15} \cdot \overline{R14} \cdot \overline{R13} \cdot \overline{R12} \cdot \overline{R11} \cdot \overline{R10} \cdot \overline{R9} \cdot \overline{R8} \cdot \overline{R7} \cdot \overline{R6} \cdot \overline{R5} \cdot \overline{R4} \cdot \overline{R3} \cdot \overline{R2} \cdot \overline{R1} \cdot \overline{R0}$
   Set if result is \$0000; cleared otherwise.

V $D15 \cdot \overline{M15} \cdot \overline{R15} + \overline{D15} \cdot M15 \cdot R15$
   Set if a twos complement overflow resulted from the operation; cleared otherwise.

C $\overline{D15} \cdot M15 + M15 \cdot R15 + R15 \cdot \overline{D15}$
   Set if the absolute value of the contents of memory is larger than the absolute value of the accumulator; cleared otherwise.

Source Form: CPD (opr)

Addressing Modes, Machine Code, and Cycle-by-Cycle Execution:

Cycle	CPD (IMM)			CPD (DIR)			CPD (EXT)			CPD (IND, X)			CPD (IND, Y)		
	Addr	Data	R/W̄	Addr	Data	R/W̄	Addr	Data	R/W̄	Addr	Data	R/W̄	Addr	Data	R/W̄
1	OP	1A	1	OP	1A	1	OP	1A	1	OP	1A	1	OP	CD	1
2	OP+1	83	1	OP+1	93	1	OP+1	B3	1	OP+1	A3	1	OP+1	A3	1
3	OP+2	jj	1	OP+2	dd	1	OP+2	hh	1	OP+2	ff	1	OP+2	ff	1
4	OP+3	kk	1	00dd	(00dd)	1	OP+3	ll	1	FFFF	—	1	FFFF	—	1
5	FFFF	—	1	00dd+1	(00dd+1)	1	hhll	(hhll)	1	X+ff	(X+ff)	1	Y+ff	(Y+ff)	1
6				FFFF	—	1	hhll+1	(hhll+1)	1	X+ff+1	(X+ff+1)	1	Y+ff+1	(Y+ff+1)	1
7							FFFF	—	1	FFFF	—	1	FFFF	—	1

Compare Index Register X (CPX)

Operation: $(IX) - (M:M+1)$

Description: Compares the contents of the index register X with a 16-bit value at the address specified and sets the condition codes accordingly. The compare is accomplished internally by doing a 16-bit subtract of $(M:M+1)$ from index register X without modifying either index register X or $(M:M+1)$.

Condition Codes and Boolean Formulae:

S	X	H	I	N	Z	V	C
—	—	—	—	↕	↕	↕	↕

N R15
   Set if MSB of result is set; cleared otherwise.

Z $\overline{R15} \cdot \overline{R14} \cdot \overline{R13} \cdot \overline{R12} \cdot \overline{R11} \cdot \overline{R10} \cdot \overline{R9} \cdot \overline{R8} \cdot \overline{R7} \cdot \overline{R6} \cdot \overline{R5} \cdot \overline{R4} \cdot \overline{R3} \cdot \overline{R2} \cdot \overline{R1} \cdot \overline{R0}$
   Set if result is \$0000; cleared otherwise.

V $IX15 \cdot \overline{M15} \cdot \overline{R15} + \overline{IX15} \cdot M15 \cdot R15$
   Set if a twos complement overflow resulted from the operation; cleared otherwise.

C $\overline{IX15} \cdot M15 + M15 \cdot R15 + R15 \cdot \overline{IX15}$
   Set if the absolute value of the contents of memory is larger than the absolute value of the index register; cleared otherwise.

Source Form: CPX (opr)

Addressing Modes, Machine Code, and Cycle-by-Cycle Execution:

Cycle	CPX (IMM)			CPX (DIR)			CPX (EXT)			CPX (IND, X)			CPX (IND, Y)		
	Addr	Data	R/W̄	Addr	Data	R/W̄	Addr	Data	R/W̄	Addr	Data	R/W̄	Addr	Data	R/W̄
1	OP	8C	1	OP	9C	1	OP	BC	1	OP	AC	1	OP	CD	1
2	OP+1	jj	1	OP+1	dd	1	OP+1	hh	1	OP+1	ff	1	OP+1	AC	1
3	OP+2	kk	1	00dd	(00dd)	1	OP+2	ll	1	FFFF	—	1	OP+2	ff	1
4	FFFF	—	1	00dd+1	(00dd+1)	1	hhll	(hhll)	1	X+ff	(X+ff)	1	FFFF	—	1
5				FFFF	—	1	hhll+1	(hhll+1)	1	X+ff+1	(X+ff+1)	1	Y+ff	(Y+ff)	1
6							FFFF	—	1	FFFF	—	1	Y+ff+1	(Y+ff+1)	1
7													FFFF	—	1

Compare Index Register Y (CPY)

Operation: $(IY) - (M:M + 1)$

Description: Compares the contents of the index register Y with a 16-bit value at the address specified and sets the condition codes accordingly. The compare is accomplished internally by doing a 16-bit subtract of $(M:M + 1)$ from index register Y without modifying either index register Y or $(M:M + 1)$.

Condition Codes and Boolean Formulae:

S	X	H	I	N	Z	V	C
—	—	—	—	⬦	⬦	⬦	⬦

N R15
   Set if MSB of result is set; cleared otherwise.

Z $\overline{R15} \cdot \overline{R14} \cdot \overline{R13} \cdot \overline{R12} \cdot \overline{R11} \cdot \overline{R10} \cdot \overline{R9} \cdot \overline{R8} \cdot \overline{R7} \cdot \overline{R6} \cdot \overline{R5} \cdot \overline{R4} \cdot \overline{R3} \cdot \overline{R2} \cdot \overline{R1} \cdot \overline{R0}$
   Set if result is $0000; cleared otherwise.

V $IY15 \cdot \overline{M15} \cdot \overline{R15} + \overline{IY15} \cdot M15 \cdot R15$
   Set if a twos complement overflow resulted from the operation; cleared otherwise.

C $\overline{IY15} \cdot M15 + M15 \cdot R15 + R15 \cdot \overline{IY15}$
   Set if the absolute value of the contents of memory is larger than the absolute value of the index register; cleared otherwise.

Source Form: CPY (opr)

Addressing Modes, Machine Code, and Cycle-by-Cycle Execution:

Cycle	CPY (IMM)			CPY (DIR)			CPY (EXT)			CPY (IND, X)			CPY (IND, Y)		
	Addr	Data	R/W̄	Addr	Data	R/W̄	Addr	Data	R/W̄	Addr	Data	R/W̄	Addr	Data	R/W̄
1	OP	18	1	OP	18	1	OP	18	1	OP	1A	1	OP	18	1
2	OP+1	8C	1	OP+1	9C	1	OP+1	BC	1	OP+1	AC	1	OP+1	AC	1
3	OP+2	jj	1	OP+2	dd	1	OP+2	hh	1	OP+2	ff	1	OP+2	ff	1
4	OP+3	kk	1	00dd	(00dd)	1	OP+3	ll	1	FFFF	—	1	FFFF	—	1
5	FFFF	—	1	00dd+1	(00dd+1)	1	hhll	(hhll)	1	X+ff	(X+ff)	1	Y+ff	(Y+ff)	1
6				FFFF	—	1	hhll+1	(hhll+1)	1	X+ff+1	(X+ff+1)	1	Y+ff+1	(Y+ff+1)	1
7							FFFF	—	1	FFFF	—	1	FFFF	—	1

Decimal Adjust ACCA (DAA)

Operation: The following table summarizes the operation of the DAA instruction for all legal combinations of input operands. A correction factor (column 5 in the following table) is added to ACCA to restore the result of an addition of two BCD operands to a valid BCD value and set or clear the carry bit.

State of C Bit Before DAA (Column 1)	Upper Half-Byte of ACCA (Bits 7–4) (Column 2)	Initial Half-Carry H Bit from CCR (Column 3)	Lower Half-Byte of ACCA (Bits 3–0) (Column 4)	Number Added of ACCA by DAA (Column 5)	State of C Bit After DAA (Column 6)
0	0-9	0	0-9	00	0
0	0-8	0	A-F	06	0
0	0-9	1	0-3	06	0
0	A-F	0	0-9	60	1
0	9-F	0	A-F	66	1
0	A-F	1	0-3	66	1
1	0-2	0	0-9	60	1
1	0-2	0	A-F	66	1
1	0-3	1	0-3	66	1

NOTE

Columns (1) through (4) of the above table represent all possible cases which can result from any of the operations ABA, ADD, or ADC, with initial carry either set or clear, applied to two binary-coded-decimal operands. The table shows hexadecimal values.

Description: If the contents of ACCA and the state of the carry/borrow bit C and the state of the half-carry bit H are all the result of applying any of the operations ABA, ADD, or ADC to binary-coded-decimal operands, with or without an initial carry, the DAA operation will adjust the contents of ACCA and the carry bit C in the CCR to represent the correct binary-coded-decimal sum and the correct state of the C bit.

Condition Codes and Boolean Formulae:

S	X	H	I	N	Z	V	C
—	—	—	—	↕	↕	?	↕

N R7
   Set if MSB of result is set; cleared otherwise.

Z $\overline{R7} \cdot \overline{R6} \cdot \overline{R5} \cdot \overline{R4} \cdot \overline{R3} \cdot \overline{R2} \cdot \overline{R1} \cdot \overline{R0}$
   Set if result is $00; cleared otherwise.

V ?
   Not defined

C See table above.

DAA (*Continued*)

Source Form: DAA

Addressing Modes, Machine Code, and Cycle-by-Cycle Execution:

Cycle	DAA (INH)		
	Addr	Data	R/W̄
1	OP	19	1
2	OP+1	—	1

For the purpose of illustration, consider the case where the BCD value $99 was just added to the BCD value $22. The add instruction is a binary operation, which yields the result $BB with no carry (C) or half carry (H). This corresponds to the fifth row of the table on the previous page. The DAA instruction will therefore add the correction factor $66 to the result of the addition, giving a result of $21 with the carry bit set. This result corresponds to the BCD value $121, which is the expected BCD result.

Decrement (DEC)

Operation: ACCX ◆ (ACCX) − $01   or: M ◆ (M) − $01

Description: Subtract one from the contents of ACCX or M.

The N, Z, and V bits in the CCR are set or cleared according to the results of the operation. The C bit in the CCR is not affected by the operation, thus allowing the DEC instruction to be used as a loop counter in multiple-precision computations.

When operating on unsigned values, only BEQ and BNE branches can be expected to perform consistently. When operating on twos-complement values, all signed branches are available.

Condition Codes and Boolean Formulae:

S	X	H	I	N	Z	V	C
—	—	—	—	◆	◆	◆	—

N   R7
   Set if MSB of result is set; cleared otherwise.

Z $\overline{R7} \cdot \overline{R6} \cdot \overline{R5} \cdot \overline{R4} \cdot \overline{R3} \cdot \overline{R2} \cdot \overline{R1} \cdot \overline{R0}$
   Set if result is $00; cleared otherwise

V $X7 \cdot \overline{X6} \cdot \overline{X5} \cdot \overline{X4} \cdot X3 \cdot \overline{X2} \cdot \overline{X1} \cdot \overline{X0} = \overline{R7} \cdot R6 \cdot R5 \cdot R4 \cdot R3 \cdot R2 \cdot R1 \cdot R0$
   Set if there was a twos complement overflow as a result of the operation; cleared otherwise. Twos complement overflow occurs if and only if (ACCX) or (M) was $80 before the operation.

Source Forms: DECA; DECB; DEC (opr)

Addressing Modes, Machine Code, and Cycle-by-Cycle Execution:

Cycle	DECA (INH)			DECB (INH)			DEC (EXT)			DEC (IND, X)			DEC (IND, Y)		
	Addr	Data	R/W̄	Addr	Data	R/W̄	Addr	Data	R/W̄	Addr	Data	R/W̄	Addr	Data	R/W̄
1	OP	4A	1	OP	5A	1	OP	7A	1	OP	6A	1	OP	18	1
2	OP+1	—	1	OP+1	—	1	OP+1	hh	1	OP+1	ff	1	OP+1	6A	1
3							OP+2	ll	1	FFFF	—	1	OP+2	ff	1
4							hhll	(hhll)	1	X+ff	(X+ff)	1	FFFF	—	1
5							FFFF	—	1	FFFF	—	1	Y+ff	(Y+ff)	1
6							hhll	result	0	X+ff	result	0	FFFF	—	1
7													Y+ff	result	0

Decrement Stack Pointer (DES)

Operation: SP ◀ (SP) − $0001

Description: Subtract one from the stack pointer.

Condition Codes and Boolean Formulae:

S	X	H	I	N	Z	V	C
—	—	—	—	—	—	—	—

None affected

Source Form: DES

Addressing Modes, Machine Code, and Cycle-by-Cycle Execution:

Cycle	DES (REL)		
	Addr	Data	R/W̄
1	OP	34	1
2	OP + 1	—	1
3	SP	—	1

Decrement Index Register X (DEX)

Operation: IX ◀ (IX) − $0001

Description: Subtract one from the index register X.

Only the Z bit is set or cleared according to the result of this operation.

Condition Codes and Boolean Formulae:

S	X	H	I	N	Z	V	C
—	—	—	—	—	✱	—	—

Z $\overline{R15} \cdot \overline{R14} \cdot \overline{R13} \cdot \overline{R12} \cdot \overline{R11} \cdot \overline{R10} \cdot \overline{R9} \cdot \overline{R8} \cdot \overline{R7} \cdot \overline{R6} \cdot \overline{R5} \cdot \overline{R4} \cdot \overline{R3} \cdot \overline{R2} \cdot \overline{R1} \cdot \overline{R0}$
Set if result is $0000; cleared otherwise.

Source Form: DEX

Addressing Modes, Machine Code, and Cycle-by-Cycle Execution:

Cycle	DEX (REL)		
	Addr	Data	R/W̄
1	OP	09	1
2	OP+1	—	1
3	FFFF	—	1

Decrement Index Register Y (DEY)

Operation: IY ◀ (IY) − $0001

Description: Subtract one from the index register Y.

Only the Z bit is set or cleared according to the result of this operation.

Condition Codes and Boolean Formulae:

S	X	H	I	N	Z	V	C
—	—	—	—	—	✿	—	—

Z $\overline{R15} \cdot \overline{R14} \cdot \overline{R13} \cdot \overline{R12} \cdot \overline{R11} \cdot \overline{R10} \cdot \overline{R9} \cdot \overline{R8} \cdot \overline{R7} \cdot \overline{R6} \cdot \overline{R5} \cdot \overline{R4} \cdot \overline{R3} \cdot \overline{R2} \cdot \overline{R1} \cdot \overline{R0}$
Set if result is $0000; cleared otherwise.

Source Form: DEY

Addressing Modes, Machine Code, and Cycle-by-Cycle Execution:

Cycle	DEY (INH)		
	Addr	Data	R/W̄
1	OP	18	1
2	OP + 1	09	1
3	OP + 2	—	1
4	FFFF	—	1

Exclusive-OR (EOR)

Operation: ACCX ⬧ (ACCX) ⊕ (M)

Description: Performs the logical exclusive-OR between the contents of ACCX and the contents of M and places the result in ACCX. (Each bit of ACCX after the operation will be the logical exclusive-OR of the corresponding bits of M and ACCX before the operation.)

Condition Codes and Boolean Formulae:

S	X	H	I	N	Z	V	C
—	—	—	—	⬦	⬦	0	—

N R7
   Set if MSB of result is set; cleared otherwise.

Z $\overline{R7} \cdot \overline{R6} \cdot \overline{R5} \cdot \overline{R4} \cdot \overline{R3} \cdot \overline{R2} \cdot \overline{R1} \cdot \overline{R0}$
   Set if result is $00; cleared otherwise

V 0
   Cleared

Source Forms: EORA (opr); EORB (opr)

Addressing Modes, Machine Code, and Cycle-by-Cycle Execution:

Cycle	EORA (IMM)			EORA (DIR)			EORA (EXT)			EORA (IND, X)			EORA (IND, Y)		
	Addr	Data	R/W̄	Addr	Data	R/W̄	Addr	Data	R/W̄	Addr	Data	R/W̄	Addr	Data	R/W̄
1	OP	88	1	OP	98	1	OP	B8	1	OP	A8	1	OP	18	1
2	OP+1	ii	1	OP+1	dd	1	OP+1	hh	1	OP+1	ff	1	OP+1	A8	1
3				00dd	(00dd)	1	OP+2	ll	1	FFFF	—	1	OP+2	ff	1
4							hhll	(hhll)	1	X+ff	(X+ff)	1	FFFF	—	1
5													Y+ff	(Y+ff)	1

Cycle	EORB (IMM)			EORB (DIR)			EORB (EXT)			EORB (IND, X)			EORB (IND, Y)		
	Addr	Data	R/W̄	Addr	Data	R/W̄	Addr	Data	R/W̄	Addr	Data	R/W̄	Addr	Data	R/W̄
1	OP	C8	1	OP	D8	1	OP	F8	1	OP	E8	1	OP	18	1
2	OP+1	ii	1	OP+1	dd	1	OP+1	hh	1	OP+1	ff	1	OP+1	E8	1
3				00dd	(00dd)	1	OP+2	ll	1	FFFF	—	1	OP+2	ff	1
4							hhll	(hhll)	1	X+ff	(X+ff)	1	FFFF	—	1
5													Y+ff	(Y+ff)	1

Fractional Divide (FDIV)

Operation: (ACCD)/(IX); IX ◀ Quotient, ACCD ◀ Remainder

Description: Performs an unsigned fractional divide of the 16-bit numerator in the D accumulator by the 16-bit denominator in the index register X and sets the condition codes accordingly. The quotient is placed in the index register X, and the remainder is placed in the D accumulator. The radix point is assumed to be in the same place for both the numerator and the denominator. The radix point is to the left of bit 15 for the quotient. The numerator is assumed to be less than the denominator. In the case of overflow (denominator is less than or equal to the numerator) or divide by zero, the quotient is set to $FFFF, and the remainder is indeterminate.

FDIV is equivalent to multiplying the numerator by 2^{16} and then performing a 32×16-bit integer divide. The result is interpreted as a binary-weighted fraction, which resulted from the division of a 16-bit integer by a larger 16-bit integer. A result of $0001 corresponds to 0.000015, and $FFFF corresponds to 0.99998. The remainder of an IDIV instruction can be resolved into a binary-weighted fraction by an FDIV instruction. The remainder of an FDIV instruction can be resolved into the next 16-bits of binary-weighted fraction by another FDIV instruction.

Condition Codes and Boolean Formulae:

S	X	H	I	N	Z	V	C
—	—	—	—	—	↕	↕	↕

Z $\overline{R15} \cdot \overline{R14} \cdot \overline{R13} \cdot \overline{R12} \cdot \overline{R11} \cdot \overline{R10} \cdot \overline{R9} \cdot \overline{R8} \cdot \overline{R7} \cdot \overline{R6} \cdot \overline{R5} \cdot \overline{R4} \cdot \overline{R3} \cdot \overline{R2} \cdot \overline{R1} \cdot \overline{R0}$
Set if quotient is $0000; cleared otherwise.

V 1 if IX≤D
Set if denominator was less than or equal to the numerator; cleared otherwise.

C $\overline{IX15} \cdot \overline{IX14} \cdot \overline{IX13} \cdot \overline{IX12} \cdot \overline{IX11} \cdot \overline{IX10} \cdot \overline{IX9} \cdot \overline{IX8} \cdot$
$\overline{IX7} \cdot \overline{IX6} \cdot \overline{IX5} \cdot \overline{IX4} \cdot \overline{IX3} \cdot \overline{IX2} \cdot \overline{IX1} \cdot \overline{IX0}$
Set if denominator was $0000; cleared otherwise.

Source Form: FDIV

Addressing Modes, Machine Code, and Cycle-by-Cycle Execution:

Cycle	FDIV (INH)		
	Addr	Data	R/W̄
1	OP	03	1
2	OP + 1	—	1
3–41	FFFF	—	1

Integer Divide (IDIV)

Operation: (ACCD)/(IX); IX ◀ Quotient, ACCD ◀ Remainder

Description: Performs an unsigned integer divide of the 16-bit numerator in D accumulator by the 16-bit denominator in index register X and sets the condition codes accordingly. The quotient is placed in index register X, and the remainder is placed in accumulator D. The radix point is assumed to be in the same place for both the numerator and the denominator. The radix point is to the right of bit zero for the quotient. In the case of divide by zero, the quotient is set to $FFFF, and the remainder is indeterminate.

Condition Codes and Boolean Formulae:

S	X	H	I	N	Z	V	C
—	—	—	—	—	✱	0	✱

Z $\overline{R15} \cdot \overline{R14} \cdot \overline{R13} \cdot \overline{R12} \cdot \overline{R11} \cdot \overline{R10} \cdot \overline{R9} \cdot \overline{R8} \cdot \overline{R7} \cdot \overline{R6} \cdot \overline{R5} \cdot \overline{R4} \cdot \overline{R3} \cdot \overline{R2} \cdot \overline{R1} \cdot \overline{R0}$
Set if result is $0000; cleared otherwise.

V   0
Cleared.

C $\overline{IX15} \cdot \overline{IX14} \cdot \overline{IX13} \cdot \overline{IX12} \cdot \overline{IX11} \cdot \overline{IX10} \cdot \overline{IX9} \cdot \overline{IX8} \cdot$
$\overline{IX7} \cdot \overline{IX6} \cdot \overline{IX5} \cdot \overline{IX4} \cdot \overline{IX3} \cdot \overline{IX2} \cdot \overline{IX1} \cdot \overline{IX0}$
Set if denominator was $0000; cleared otherwise.

Source Form: IDIV

Addressing Modes, Machine Code, and Cycle-by-Cycle Execution:

Cycle	IDIV (INH)		
	Addr	Data	R/W̄
1	OP	02	1
2	OP + 1	—	1
3–41	FFFF	—	1

Increment (INC)

Operation: ACCX ◀ (ACCX) + $01   **or:**   M ◀ (M) + $01

Description: Add one to the contents of ACCX or M.

The N, Z, and V bits in the CCR are set or cleared according to the results of the operation. The C bit in the CCR is not affected by the operation, thus allowing the INC instruction to be used as a loop counter in multiple-precision computations.

When operating on unsigned values, only BEQ and BNE branches can be expected to perform consistently. When operating on twos-complement values, all signed branches are available.

Condition Codes and Boolean Formulae:

S	X	H	I	N	Z	V	C
—	—	—	—	↕	↕	↕	—

N R7
   Set if MSB of result is set; cleared otherwise.

Z $\overline{R7} \cdot \overline{R6} \cdot \overline{R5} \cdot \overline{R4} \cdot \overline{R3} \cdot \overline{R2} \cdot \overline{R1} \cdot \overline{R0}$
   Set if result is $00; cleared otherwise.

V $\overline{X7} \cdot X6 \cdot X5 \cdot X4 \cdot X3 \cdot X2 \cdot X1 \cdot X0$
   Set if there is a twos complement overflow as a result of the operation; cleared otherwise. Twos complement overflow occurs if and only if (ACCX) or (M) was $7F before the operation.

Source Forms: INCA; INCB; INC (opr)

Addressing Modes, Machine Code, and Cycle-by-Cycle Execution:

Cycle	INCA (INH)			INCB (INH)			INC (EXT)			INC (IND, X)			INC (IND, Y)		
	Addr	Data	R/W̄	Addr	Data	R/W̄	Addr	Data	R/W̄	Addr	Data	R/W̄	Addr	Data	R/W̄
1	OP	4C	1	OP	5C	1	OP	7C	1	OP	6C	1	OP	18	1
2	OP+1	—	1	OP+1	—	1	OP+1	hh	1	OP+1	ff	1	OP+1	6C	1
3							OP+2	ll	1	FFFF	—	1	OP+2	ff	1
4							hhll	(hhll)	1	X+ff	(X+ff)	1	FFFF	—	1
5							FFFF	—	1	FFFF	—	1	Y+ff	(Y+ff)	1
6							hhll	result	0	X+ff	result	0	FFFF	—	1
7													Y+ff	result	0

Increment Stack Pointer (INS)

Operation: SP ◀ (SP) + $0001

Description: Add one to the stack pointer.

Condition Codes and Boolean Formulae:

S	X	H	I	N	Z	V	C
—	—	—	—	—	—	—	—

None affected

Source Form: INS

Addressing Modes, Machine Code, and Cycle-by-Cycle Execution:

Cycle	INS (INH)		
	Addr	Data	R/W̄
1	OP	31	1
2	OP+1	—	1
3	SP	—	1

Increment Index Register X (INX)

Operation: IX ◆ (IX) + $0001

Description: Add one to index register X.

Only the Z bit is set or cleared according to the result of this operation.

Condition Codes and Boolean Formulae:

S	X	H	I	N	Z	V	C
—	—	—	—	—	↕	—	—

Z $\overline{R15} \cdot \overline{R14} \cdot \overline{R13} \cdot \overline{R12} \cdot \overline{R11} \cdot \overline{R10} \cdot \overline{R9} \cdot \overline{R8} \cdot \overline{R7} \cdot \overline{R6} \cdot \overline{R5} \cdot \overline{R4} \cdot \overline{R3} \cdot \overline{R2} \cdot \overline{R1} \cdot \overline{R0}$
Set if result is $0000; cleared otherwise.

Source Form: INX

Addressing Modes, Machine Code, and Cycle-by-Cycle Execution:

Cycle	INX (INH)		
	Addr	Data	R/$\overline{\text{W}}$
1	OP	08	1
2	OP + 1	—	1
3	FFFF	—	1

Increment Index Register Y (INY)

Operation: IY ⬦ (IY) + $0001

Description: Add one to index register Y.

Only the Z bit is set or cleared according to the result of this operation.

Condition Codes and Boolean Formulae:

S	X	H	I	N	Z	V	C
—	—	—	—	—	⬍	—	—

Z $\overline{R15} \cdot \overline{R14} \cdot \overline{R13} \cdot \overline{R12} \cdot \overline{R11} \cdot \overline{R10} \cdot \overline{R9} \cdot \overline{R8} \cdot \overline{R7} \cdot \overline{R6} \cdot \overline{R5} \cdot \overline{R4} \cdot \overline{R3} \cdot \overline{R2} \cdot \overline{R1} \cdot \overline{R0}$
Set if result is $0000; cleared otherwise.

Source Form: INY

Addressing Modes, Machine Code, and Cycle-by-Cycle Execution:

Cycle	INY (INH)		
	Addr	Data	R/W̄
1	OP	18	1
2	OP + 1	08	1
3	OP + 2	—	1
4	FFFF	—	1

Jump (JMP)

Operation: PC ◀ Effective Address

Description: A jump occurs to the instruction stored at the effective address. The effective address is obtained according to the rules for EXTended or INDexed addressing.

Condition Codes and Boolean Formulae:

S	X	H	I	N	Z	V	C
—	—	—	—	—	—	—	—

None affected

Source Form: JMP (opr)

Addressing Modes, Machine Code, and Cycle-by-Cycle Execution:

Cycle	JMP (EXT)			JMP (IND, X)			JMP (IND, Y)		
	Addr	Data	R/W̅	Addr	Data	R/W̅	Addr	Data	R/W̅
1	OP	7E	1	OP	6E	1	OP	18	1
2	OP+1	hh	1	OP+1	ff	1	OP+1	6E	1
3	OP+2	ll	1	FFFF	—	1	OP+2	ff	1
4							FFFF	—	1

Jump to Subroutine (JSR)

Operation: PC ◀ (PC) + $0003   (for EXTended or INDexed, Y addressing)   **or:**
PC ◀ (PC) + $0002   (for DIRect or INDexed, X addressing)
◀(PCL)   Push low-order return address onto stack
SP ◀ (SP) − $0001
◀(PCH)   Push high-order return address onto stack
SP ◀ (SP) − $0001
PC ◀ Effective Addr   Load start address of requested subroutine

Description: The program counter is incremented by three or by two, depending on the addressing mode, and is then pushed onto the stack, eight bits at a time, least significant byte first. The stack pointer points to the next empty location in the stack. A jump occurs to the instruction stored at the effective address. The effective address is obtained according to the rules for EXTended, DIRect, or INDexed addressing.

Condition Codes and Boolean Formulae:

S	X	H	I	N	Z	V	C
—	—	—	—	—	—	—	—

None affected

Source Form: JSR (opr)

Addressing Modes, Machine Code, and Cycle-by-Cycle Execution:

Cycle	JSR (DIR)			JSR (EXT)			JSR (IND, X)			JSR (IND, Y)		
	Addr	Data	R/W̄	Addr	Data	R/W̄	Addr	Data	R/W̄	Addr	Data	R/W̄
1	OP	9D	1	OP	BD	1	OP	AD	1	OP	18	1
2	OP + 1	dd	1	OP + 1	hh	1	OP + 1	ff	1	OP + 1	AD	1
3	00dd	(00dd)	1	OP + 2	ll	1	FFFF	—	1	OP + 2	ff	1
4	SP	Rtn lo	0	hhll	(hhll)	1	X + ff	(X + ff)	1	FFFF	—	1
5	SP − 1	Rtn hi	0	SP	Rtn lo	0	SP	Rtn lo	0	Y + ff	(Y + ff)	1
6				SP − 1	Rtn hi	0	SP − 1	Rtn hi	0	SP	Rtn lo	0
7										SP − 1	Rtn hi	0

Load Accumulator (LDA)

Operation: ACCX ⬅ (M)

Description: Loads the contents of memory into the 8-bit accumulator. The condition codes are set according to the data.

Condition Codes and Boolean Formulae:

S	X	H	I	N	Z	V	C
—	—	—	—	⬍	⬍	0	—

N R7
   Set if MSB of result is set; cleared otherwise.

Z $\overline{R7} \cdot \overline{R6} \cdot \overline{R5} \cdot \overline{R4} \cdot \overline{R3} \cdot \overline{R2} \cdot \overline{R1} \cdot \overline{R0}$
   Set if result is $00; cleared otherwise

V 0
   Cleared

Source Form: LDAA (opr); LDAB (opr)

Addressing Modes, Machine Code, and Cycle-by-Cycle Execution:

Cycle	LDAA (IMM) Addr	Data	R/W̄	LDAA (DIR) Addr	Data	R/W̄	LDAA (EXT) Addr	Data	R/W̄	LDAA (IND, X) Addr	Data	R/W̄	LDAA (IND, Y) Addr	Data	R/W̄
1	OP	86	1	OP	96	1	OP	B6	1	OP	A6	1	OP	18	1
2	OP+1	ii	1	OP+1	dd	1	OP+1	hh	1	OP+1	ff	1	OP+1	A6	1
3				00dd	(00dd)	1	OP+2	ll	1	FFFF	—	1	OP+2	ff	1
4							hhll	(hhll)	1	X+ff	(X+ff)	1	FFFF	—	1
5													Y+ff	(Y+ff)	1

Cycle	LDAB (IMM) Addr	Data	R/W̄	LDAB (DIR) Addr	Data	R/W̄	LDAB (EXT) Addr	Data	R/W̄	LDAB (IND, X) Addr	Data	R/W̄	LDAB (IND, Y) Addr	Data	R/W̄
1	OP	C6	1	OP	D6	1	OP	F6	1	OP	E6	1	OP	18	1
2	OP+1	ii	1	OP+1	dd	1	OP+1	hh	1	OP+1	ff	1	OP+1	E6	1
3				00dd	(00dd)	1	OP+2	ll	1	FFFF	—	1	OP+2	ff	1
4							hhll	(hhll)	1	X+ff	(X+ff)	1	FFFF	—	1
5													Y+ff	(Y+ff)	1

Load Double Accumulator (LDD)

Operation: ACCD ◄ (M:M + 1); ACCA ◄ (M), ACCB ◄ (M + 1)

Description: Loads the contents of memory locations M and M + 1 into the double accumulator D. The condition codes are set according to the data. The information from location M is loaded into accumulator A, and the information from location M + 1 is loaded into accumulator B.

Condition Codes and Boolean Formulae:

S	X	H	I	N	Z	V	C
—	—	—	—	↕	↕	0	—

N R15
   Set if MSB of result is set; cleared otherwise.

Z $\overline{R15} \cdot \overline{R14} \cdot \overline{R13} \cdot \overline{R12} \cdot \overline{R11} \cdot \overline{R10} \cdot \overline{R9} \cdot \overline{R8} \cdot \overline{R7} \cdot \overline{R6} \cdot \overline{R5} \cdot \overline{R4} \cdot \overline{R3} \cdot \overline{R2} \cdot \overline{R1} \cdot \overline{R0}$
   Set if result is $0000; cleared otherwise.

V 0
   Cleared

Source Form: LDD (opr)

Addressing Modes, Machine Code, and Cycle-by-Cycle Execution:

Cycle	LDD (IMM)			LDD (DIR)			LDD (EXT)			LDD (IND, X)			LDD (IND, Y)		
	Addr	Data	R/W̅	Addr	Data	R/W̅	Addr	Data	R/W̅	Addr	Data	R/W̅	Addr	Data	R/W̅
1	OP	CC	1	OP	DC	1	OP	FC	1	OP	EC	1	OP	18	1
2	OP + 1	jj	1	OP + 1	dd	1	OP + 1	hh	1	OP + 1	ff	1	OP + 1	EC	1
3	OP + 2	kk	1	00dd	(00dd)	1	OP + 2	ll	1	FFFF	—	1	OP + 2	ff	1
4				00dd + 1	(00dd + 1)	1	hhll	(hhll)	1	X + ff	(X + ff)	1	FFFF	—	1
5							hhll + 1	(hhll + 1)	1	X + ff + 1	(X + ff + 1)	1	Y + ff	(Y + ff)	1
6													Y + ff + 1	(Y + ff + 1)	1

Load Stack Pointer (LDS)

Operation: SPH ◀ (M), SPL ◀ (M + 1)

Description: Loads the most significant byte of the stack pointer from the byte of memory at the address specified by the program, and loads the least significant byte of the stack pointer from the next byte of memory at one plus the address specified by the program.

Condition Codes and Boolean Formulae:

S	X	H	I	N	Z	V	C
—	—	—	—	↕	↕	0	—

N R15
Set if MSB of result is set; cleared otherwise.

Z $\overline{R15} \cdot \overline{R14} \cdot \overline{R13} \cdot \overline{R12} \cdot \overline{R11} \cdot \overline{R10} \cdot \overline{R9} \cdot \overline{R8} \cdot \overline{R7} \cdot \overline{R6} \cdot \overline{R5} \cdot \overline{R4} \cdot \overline{R3} \cdot \overline{R2} \cdot \overline{R1} \cdot \overline{R0}$
Set if result is $0000; cleared otherwise.

V 0
Cleared

Source Form: LDS (opr)

Addressing Modes, Machine Code, and Cycle-by-Cycle Execution:

Cycle	LDS (IMM)			LDS (DIR)			LDS (EXT)			LDS (IND. X)			LDS (IND, Y)		
	Addr	Data	R/W̄	Addr	Data	R/W̄	Addr	Data	R/W̄	Addr	Data	R/W̄	Addr	Data	R/W̄
1	OP	8E	1	OP	9E	1	OP	BE	1	OP	AE	1	OP	18	1
2	OP + 1	jj	1	OP + 1	dd	1	OP + 1	hh	1	OP + 1	ff	1	OP + 1	AE	1
3	OP + 2	kk	1	00dd	(00dd)	1	OP + 2	ll	1	FFFF	—	1	OP + 2	ff	1
4				00dd + 1	(00dd + 1)	1	hhll	(hhll)	1	X + ff	(X + ff)	1	FFFF	—	1
5							hhll + 1	(hhll + 1)	1	X + ff + 1	(X + ff + 1)	1	Y + ff	(Y + ff)	1
6													Y + ff + 1	(Y + ff + 1)	1

Load Index Register X (LDX)

Operation;   IXH ⬦ (M), IXL ⬦ (M + 1)

Description: Loads the most significant byte of index register X from the byte of memory at the address specified by the program, and loads the least significant byte of index register X from the next byte of memory at one plus the address specified by the program.

Condition Codes and Boolean Formulae:

S	X	H	I	N	Z	V	C
—	—	—	—	⬥	⬥	0	—

N R15
   Set if MSB of result is set; cleared otherwise.

Z $\overline{R15} \cdot \overline{R14} \cdot \overline{R13} \cdot \overline{R12} \cdot \overline{R11} \cdot \overline{R10} \cdot \overline{R9} \cdot \overline{R8} \cdot \overline{R7} \cdot \overline{R6} \cdot \overline{R5} \cdot \overline{R4} \cdot \overline{R3} \cdot \overline{R2} \cdot \overline{R1} \cdot \overline{R0}$
   Set if result is $0000; cleared otherwise.

V 0
   Cleared

Source Form: LDX (opr)

Addressing Modes, Machine Code, and Cycle-by-Cycle Execution:

Cycle	LDX (IMM) Addr	Data	R/W̄	LDX (DIR) Addr	Data	R/W̄	LDX (EXT) Addr	Data	R/W̄	LDX (IND, X) Addr	Data	R/W̄	LDX (IND, Y) Addr	Data	R/W̄
1	OP	CE	1	OP	DE	1	OP	FE	1	OP	EE	1	OP	CD	1
2	OP + 1	jj	1	OP + 1	dd	1	OP + 1	hh	1	OP + 1	ff	1	OP + 1	EE	1
3	OP + 2	kk	1	00dd	(00dd)	1	OP + 2	ll	1	FFFF	—	1	OP + 2	ff	1
4				00dd + 1	(00dd + 1)	1	hhll	(hhll)	1	X + ff	(X + ff)	1	FFFF	—	1
5							hhll + 1	(hhll + 1)	1	X + ff + 1	(X + ff + 1)	1	Y + ff	(Y + ff)	1
6													Y + ff + 1	(Y + ff + 1)	1

Load Index Register Y (LDY)

Operation: $IYH \blacktriangleleft (M), IYL \blacktriangleleft (M+1)$

Description: Loads the most significant byte of index register Y from the byte of memory at the address specified by the program, and loads the least significant byte of index register Y from the next byte of memory at one plus the address specified by the program.

Condition Codes and Boolean Formulae:

S	X	H	I	N	Z	V	C
—	—	—	—	$\updownarrow$	$\updownarrow$	0	—

N   R15
   Set if MSB of result is set; cleared otherwise.

Z $\overline{R15} \cdot \overline{R14} \cdot \overline{R13} \cdot \overline{R12} \cdot \overline{R11} \cdot \overline{R10} \cdot \overline{R9} \cdot \overline{R8} \cdot \overline{R7} \cdot \overline{R6} \cdot \overline{R5} \cdot \overline{R4} \cdot \overline{R3} \cdot \overline{R2} \cdot \overline{R1} \cdot \overline{R0}$
   Set if result is $0000; cleared otherwise.

V   0
   Cleared

Source Form: LDY (opr)

Addressing Modes, Machine Code, and Cycle-by-Cycle Execution:

Cycle	LDY (IMM)			LDY (DIR)			LDY (EXT)			LDY (IND, X)			LDY (IND, Y)		
	Addr	Data	R/W̄	Addr	Data	R/W̄	Addr	Data	R/W̄	Addr	Data	R/W̄	Addr	Data	R/W̄
1	OP	18	1	OP	18	1	OP	18	1	OP	1A	1	OP	18	1
2	OP+1	CE	1	OP+1	DE	1	OP+1	FE	1	OP+1	EE	1	OP+1	EE	1
3	OP+2	jj	1	OP+2	dd	1	OP+2	hh	1	OP+2	ff	1	OP+2	ff	1
4	OP+3	kk	1	00dd	(00dd)	1	OP+3	ll	1	FFFF	—	1	FFFF	—	1
5				00dd+1	(00dd+1)	1	hhll	(hhll)	1	X+ff	(X+ff)	1	Y+ff	(Y+ff)	1
6							hhll+1	(hhll+1)	1	X+ff+1	(X+ff+1)	1	Y+ff+1	(Y+ff+1)	1

Logical Shift Left (LSL) (Same as ASL)

Operation: $C \longleftarrow$ | b7 ------ b0 | $\longleftarrow 0$

Description: Shifts all bits of the ACCX or M one place to the left. Bit 0 is loaded with zero. The C bit is loaded from the most significant bit of ACCX or M.

Condition Codes and Boolean Formulae:

S	X	H	I	N	Z	V	C
—	—	—	—	$\updownarrow$	$\updownarrow$	$\updownarrow$	$\updownarrow$

N   R7
  Set if MSB of result is set; cleared otherwise.

Z $\overline{R7} \cdot \overline{R6} \cdot \overline{R5} \cdot \overline{R4} \cdot \overline{R3} \cdot \overline{R2} \cdot \overline{R1} \cdot \overline{R0}$
  Set if result is $00; cleared otherwise.

V $N \oplus C = [N \cdot \overline{C}] + [\overline{N} \cdot C]$ (for N and C after the shift)
  Set if (N is set and C is clear) or (N is clear and C is set); cleared otherwise (for values of N and C after the shift).

C   M7
  Set if, before the shift, the MSB of ACCX or M was set; cleared otherwise.

Source Forms: LSLA; LSLB; LSL (opr)

Addressing Modes, Machine Code, and Cycle-by-Cycle Execution:

Cycle	LSLA (INH)			LSLB (INH)			LSL (EXT)			LSL (IND, X)			LSL (IND, Y)		
	Addr	Data	R/$\overline{W}$	Addr	Data	R/$\overline{W}$	Addr	Data	R/$\overline{W}$	Addr	Data	R/$\overline{W}$	Addr	Data	R/$\overline{W}$
1	OP	48	1	OP	58	1	OP	78	1	OP	68	1	OP	18	1
2	OP+1	—	1	OP+1	—	1	OP+1	hh	1	OP+1	ff	1	OP+1	68	1
3							OP+2	ll	1	FFFF	—	1	OP+2	ff	1
4							hhll	(hhll)	1	X+ff	(X+ff)	1	FFFF	—	1
5							FFFF	—	1	FFFF	—	1	Y+ff	(Y+ff)	1
6							hhll	result	0	X+ff	result	0	FFFF	—	1
7													Y+ff	result	0

Logical Shift Left Double (LSLD) (Same as ASLD)

Operation:

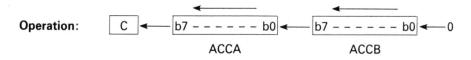

Description: Shifts all of ACCD one place to the left. Bit 0 is loaded with zero. The C bit is loaded from the most significant bit of ACCD.

Condition Codes and Boolean Formulae:

S	X	H	I	N	Z	V	C
—	—	—	—	$\updownarrow$	$\updownarrow$	$\updownarrow$	$\updownarrow$

N R15
   Set if MSB of result is set; cleared otherwise.

Z $\overline{R15} \cdot \overline{R14} \cdot \overline{R13} \cdot \overline{R12} \cdot \overline{R11} \cdot \overline{R10} \cdot \overline{R9} \cdot \overline{R8} \cdot \overline{R7} \cdot \overline{R6} \cdot \overline{R5} \cdot \overline{R4} \cdot \overline{R3} \cdot \overline{R2} \cdot \overline{R1} \cdot \overline{R0}$
   Set if result is $0000; cleared otherwise.

V $N \oplus C = [N \cdot \overline{C}] + [\overline{N} \cdot C]$ (for N and C after the shift)
   Set if (N is set and C is clear) or (N is clear and C is set); cleared otherwise (for values of N and C after the shift).

C D15
   Set if, before the shift, the MSB of ACCD was set; cleared otherwise.

Source Form: LSLD

Addressing Modes, Machine Code, and Cycle-by-Cycle Execution:

Cycle	LSLD (INH)		
	Addr	Data	R/$\overline{W}$
1	OP	05	1
2	OP + 1	—	1
3	FFFF	—	1

Logical Shift Right (LSR)

Operation:

$$0 \longrightarrow \boxed{b7 - - - - - - b0} \longrightarrow \boxed{C}$$

Description: Shifts all bits of ACCX or M one place to the right. Bit 7 is loaded with zero. The C bit is loaded from the least significant bit of ACCX or M.

Condition Codes and Boolean Formulae:

S	X	H	I	N	Z	V	C
—	—	—	—	0	↕	↕	↕

N 0
   Cleared.

Z $\overline{R7} \cdot \overline{R6} \cdot \overline{R5} \cdot \overline{R4} \cdot \overline{R3} \cdot \overline{R2} \cdot \overline{R1} \cdot \overline{R0}$
   Set if result is $00; cleared otherwise.

V $N \oplus C = [N \cdot \overline{C}] + [\overline{N} \cdot C]$ (for N and C after the shift)
   Since $N = 0$, this simplifies to C (after the shift).

C M0
   Set if, before the shift, the LSB of ACCX or M was set; cleared otherwise.

Source Forms: LSRA; LSRB; LSR (opr)

Addressing Modes, Machine Code, and Cycle-by-Cycle Execution:

Cycle	LSRA (INH)			LSRB (INH)			LSR (EXT)			LSR (IND, X)			LSR (IND, Y)		
	Addr	Data	R/W̄	Addr	Data	R/W̄	Addr	Data	R/W̄	Addr	Data	R/W̄	Addr	Data	R/W̄
1	OP	44	1	OP	54	1	OP	74	1	OP	64	1	OP	18	1
2	OP+1	—	1	OP+1	—	1	OP+1	hh	1	OP+1	ff	1	OP+1	64	1
3							OP+2	ll	1	FFFF	—	1	OP+2	ff	1
4							hhll	(hhll)	1	X+ff	(X+ff)	1	FFFF	—	1
5							FFFF	—	1	FFFF	—	1	Y+ff	(Y+ff)	1
6							hhll	result	0	X+ff	result	0	FFFF	—	1
7													Y+ff	result	0

Logical Shift Right Double Accumulator (LSRD)

Operation:

$$0 \rightarrow \boxed{b7 \text{ -- -- -- -- } b0} \rightarrow \boxed{b7 \text{ -- -- -- -- } b0} \rightarrow \boxed{C}$$
$$\qquad\qquad\quad \text{ACCA} \qquad\qquad\quad \text{ACCB}$$

Description: Shifts all bits of ACCD one place to the right. Bit 15 (MSB of ACCA) is loaded with zero. The C bit is loaded from the least significant bit of ACCD (LSB of ACCB).

Condition Codes and Boolean Formulae:

S	X	H	I	N	Z	V	C
—	—	—	—	0	↕	↕	↕

N 0
  Cleared

Z $\overline{R15} \cdot \overline{R14} \cdot \overline{R13} \cdot \overline{R12} \cdot \overline{R11} \cdot \overline{R10} \cdot \overline{R9} \cdot \overline{R8} \cdot \overline{R7} \cdot \overline{R6} \cdot \overline{R5} \cdot \overline{R4} \cdot \overline{R3} \cdot \overline{R2} \cdot \overline{R1} \cdot \overline{R0}$
  Set if result is $0000; cleared otherwise.

V D0
  Set if, after the shift operaton, C is set; cleared otherwise.

C D0
  Set if, before the shift, the least significant bit of ACCD was set; cleared otherwise.

Source Form: LSRD

Addressing Modes, Machine Code, and Cycle-by-Cycle Execution:

Cycle	LSRD (INH)		
	Addr	Data	R/W̄
1	OP	04	1
2	OP + 1	—	1
3	FFFF	—	1

Multiply Unsigned (MUL)

Operation: ACCD ◀ (ACCA) × (ACCB)

Description: Multiplies the 8-bit unsigned binary value in accumulator A by the 8-bit unsigned binary value in accumulator B to obtain a 16-bit unsigned result in the double accumulator D. Unsigned multiply allows multiple-precision operations. The carry flag allows rounding the most significant byte of the result through the sequence: MUL, ADCA #0.

Condition Codes and Boolean Formulae:

S	X	H	I	N	Z	V	C
—	—	—	—	—	—	—	⬥

C   R7
    Set if bit 7 of the result (ACCB bit 7) is set; cleared otherwise.

Source Form: MUL

Addressing Modes, Machine Code, and Cycle-by-Cycle Execution:

Cycle	MUL (INH)		
	Addr	Data	R/W̄
1	OP	3D	1
2	OP+1	—	1
3–10	FFFF	—	1

Negate (NEG)

Operation: $(ACCX) \leftarrow -(ACCX) = \$00 - (ACCX)$ **or:** $(M) \leftarrow -(M) = \$00 - (M)$

Description: Replaces the contents of ACCX or M with its twos complement; the value $80 is left unchanged.

Condition Codes and Boolean Formulae:

S	X	H	I	N	Z	V	C
—	—	—	—	↕	↕	↕	↕

N R7
   Set if MSB of result is set; cleared otherwise.

Z $\overline{R7} \cdot \overline{R6} \cdot \overline{R5} \cdot \overline{R4} \cdot \overline{R3} \cdot \overline{R2} \cdot \overline{R1} \cdot \overline{R0}$
   Set if result is $00; cleared otherwise.

V $R7 \cdot \overline{R6} \cdot \overline{R5} \cdot \overline{R4} \cdot \overline{R3} \cdot \overline{R2} \cdot \overline{R1} \cdot \overline{R0}$
   Set if there is a twos complement overflow from the implied subtraction from zero; cleared otherwise. A twos complement overflow will occur if and only if the contents of ACCX or M is $80.

C $R7 + R6 + R5 + R4 + R3 + R2 + R1 + R0$
   Set if there is a borrow in the implied subtraction from zero; cleared otherwise. The C bit will be set in all cases except when the contents of ACCX or M is $00.

Source Forms: NEGA; NEGB; NEG (opr)

Addressing Modes, Machine Code, and Cycle-by-Cycle Execution:

Cycle	NEGA (INH)			NEGB (INH)			NEG (EXT)			NEG (IND, X)			NEG (IND, Y)		
	Addr	Data	R/W̄	Addr	Data	R/W̄	Addr	Data	R/W̄	Addr	Data	R/W̄	Addr	Data	R/W̄
1	OP	40	1	OP	50	1	OP	70	1	OP	60	1	OP	18	1
2	OP+1	—	1	OP+1	—	1	OP+1	hh	1	OP+1	ff	1	OP+1	60	1
3							OP+2	ll	1	FFFF	—	1	OP+2	ff	1
4							hhll	(hhll)	1	X+ff	(X+ff)	1	FFFF	—	1
5							FFFF	—	1	FFFF	—	1	Y+ff	(Y+ff)	1
6							hhll	result	0	X+ff	result	0	FFFF	—	1
7													Y+ff	result	0

No Operation (NOP)

Description: This is a single-byte instruction that causes only the program counter to be incremented. No other registers are affected. This instruction is typically used to produce a time delay although some software disciplines discourage CPU frequency-based time delays. During debug, NOP instructions are sometimes used to temporarily replace other machine code instructions, thus disabling the replaced instruction(s).

Condition Codes and Boolean Formulae:

S	X	H	I	N	Z	V	C
—	—	—	—	—	—	—	—

None affected

Source Form: NOP

Addressing Modes, Machine Code, and Cycle-by-Cycle Execution:

Cycle	NOP (INH)		
	Addr	Data	R/$\overline{\text{W}}$
1	OP	01	1
2	OP + 1	—	1

Inclusive-OR (ORA)

Operation: ACCX ◀ (ACCX) + (M)

Description: Performs the logical inclusive-OR between the contents of ACCX and the contents of M and places the result in ACCX. (Each bit of ACCX after the operation will be the logical inclusive-OR of the corresponding bits of M and of ACCX before the operation.)

Condition Codes and Boolean Formulae:

S	X	H	I	N	Z	V	C
—	—	—	—	⬍	⬍	0	—

N   R7
   Set if MSB of result is set; cleared otherwise.

Z $\overline{R7} \cdot \overline{R6} \cdot \overline{R5} \cdot \overline{R4} \cdot \overline{R3} \cdot \overline{R2} \cdot \overline{R1} \cdot \overline{R0}$
   Set if result is $00; cleared otherwise

V   0
   Cleared

Source Forms: ORAA (opr); ORAB (opr)

Addressing Modes, Machine Code, and Cycle-by-Cycle Execution:

Cycle	ORAA (IMM)			ORAA (DIR)			ORAA (EXT)			ORAA (IND, X)			ORAA (IND, Y)		
	Addr	Data	R/W̄	Addr	Data	R/W̄	Addr	Data	R/W̄	Addr	Data	R/W̄	Addr	Data	R/W̄
1	OP	8A	1	OP	9A	1	OP	BA	1	OP	AA	1	OP	18	1
2	OP+1	ii	1	OP+1	dd	1	OP+1	hh	1	OP+1	ff	1	OP+1	AA	1
3				00dd	(00dd)	1	OP+2	ll	1	FFFF	—	1	OP+2	ff	1
4							hhll	(hhll)	1	X+ff	(X+ff)	1	FFFF	—	1
5													Y+ff	(Y+ff)	1

Cycle	ORAB (IMM)			ORAB (DIR)			ORAB (EXT)			ORAB (IND, X)			ORAB (IND, Y)		
	Addr	Data	R/W̄	Addr	Data	R/W̄	Addr	Data	R/W̄	Addr	Data	R/W̄	Addr	Data	R/W̄
1	OP	CA	1	OP	DA	1	OP	FA	1	OP	EA	1	OP	18	1
2	OP+1	ii	1	OP+1	dd	1	OP+1	hh	1	OP+1	ff	1	OP+1	EA	1
3				00dd	(00dd)	1	OP+2	ll	1	FFFF	—	1	OP+2	ff	1
4							hhll	(hhll)	1	X+ff	(X+ff)	1	FFFF	—	1
5													Y+ff	(Y+ff)	1

Push Data onto Stack (PSH)

Operation: ➡ACCX, SP ⬇ (SP) − $0001

Description: The contents of ACCX are stored on the stack at the address contained in the stack pointer. The stack pointer is then decremented.

Push instructions are commonly used to save the contents of one or more CPU registers at the start of a subroutine. Just before returning from the subroutine, corresponding pull instructions are used to restore the saved CPU registers so the subroutine will appear not to have affected these registers.

Condition Codes and Boolean Formulae:

S	X	H	I	N	Z	V	C
—	—	—	—	—	—	—	—

None affected

Source Forms: PSHA; PSHB

Addressing Modes, Machine Code, and Cycle-by-Cycle Execution:

Cycle	PSHA (INH)			PSHB (INH)		
	Addr	Data	R/W̄	Addr	Data	R/W̄
1	OP	36	1	OP	37	1
2	OP+1	—	1	OP+1	—	1
3	SP	(A)	0	SP	(B)	0

Push Index Register X onto Stack (PSHX)

Operation: ➡(IXL), SP ⬥ (SP) − $0001
⬥(IXH), SP ⬥ (SP) − $0001

Description: The contents of the index register X are pushed onto the stack (low-order byte first) at the address contained in the stack pointer. The stack pointer is then decremented by two.

Push instructions are commonly used to save the contents of one or more CPU registers at the start of a subroutine. Just before returning from the subroutine, corresponding pull instructions are used to restore the saved CPU registers so the subroutine will appear not to have affected these registers.

Condition Codes and Boolean Formulae:

S	X	H	I	N	Z	V	C
—	—	—	—	—	—	—	—

None affected

Source Form: PSHX

Addressing Modes, Machine Code, and Cycle-by-Cycle Execution:

Cycle	PSHX (INH)		
	Addr	Data	R/W̄
1	OP	3C	1
2	OP + 1	—	1
3	SP	(IXL)	0
4	SP − 1	(IXH)	0

Push Index Register Y onto Stack (PSHY)

Operation: ➡(IYL), SP ⬇ (SP) − $0001
➡(IYH), SP ⬇ (SP) − $0001

Description: The contents of the index register Y are pushed onto the stack (low-order byte first) at the address contained in the stack pointer. The stack pointer is then decremented by two.

Push instructions are commonly used to save the contents of one or more CPU registers at the start of a subroutine. Just before returning from the subroutine, corresponding pull instructions are used to restore the saved CPU registers so the subroutine will appear not to have affected these registers.

Condition Codes and Boolean Formulae:

S	X	H	I	N	Z	V	C
—	—	—	—	—	—	—	—

None affected

Source Form: PSHY

Addressing Modes, Machine Code, and Cycle-by-Cycle Execution:

Cycle	PSHY (INH)		
	Addr	Data	R/W̄
1	OP	18	1
2	OP+1	3C	1
3	OP+2	—	1
4	SP	(IYL)	0
5	SP−1	(IYH)	0

Pull Data from Stack (PUL)

Operation: SP ♦ (SP) + $0001, ☛(ACCX)

Description: The stack pointer is incremented. The ACCX is then loaded from the stack at the address contained in the stack pointer.

Push instructions are commonly used to save the contents of one or more CPU registers at the start of a subroutine. Just before returning from the subroutine, corresponding pull instructions are used to restore the saved CPU registers so the subroutine will appear not to have affected these registers.

Condition Codes and Boolean Formulae:

S	X	H	I	N	Z	V	C
—	—	—	—	—	—	—	—

None affected

Source Forms: PULA; PULB

Addressing Modes, Machine Code, and Cycle-by-Cycle Execution:

Cycle	PULA (INH)			PULB (INH)		
	Addr	Data	R/W̄	Addr	Data	R/W̄
1	OP	32	1	OP	33	1
2	OP + 1	—	1	OP + 1	—	1
3	SP	—	1	SP	—	1
4	SP + 1	get A	1	SP + 1	get B	1

Pull Index Register X from Stack (PULX)

Operation:　　SP ◀ (SP) + $0001; ◀(IXH)
　　　　　　　　SP ◀ (SP) + $0001; ◀(IXL)

Description:　　The index register X is pulled from the stack (high-order byte first), beginning at the address contained in the stack pointer plus one. The stack pointer is incremented by two in total.

Push instructions are commonly used to save the contents of one or more CPU registers at the start of a subroutine. Just before returning from the subroutine, corresponding pull instructions are used to restore the saved CPU registers so the subroutine will appear not to have affected these registers.

Condition Codes and Boolean Formulae:

S	X	H	I	N	Z	V	C
—	—	—	—	—	—	—	—

None affected

Source Form:　PULX

Addressing Modes, Machine Code, and Cycle-by-Cycle Execution:

Cycle	PULX (INH)		
	Addr	Data	R/W̄
1	OP	38	1
2	OP + 1	—	1
3	SP	—	1
4	SP + 1	get IXH	1
5	SP + 2	get IXL	1

Pull Index Register Y from Stack (PULY)

Operation: SP ◆ (SP) + $0001; ←(IYH)
                    SP ◆ (SP) + $0001; ←(IYL)

Description: The index register Y is pulled from the stack (high-order byte first) beginning at the address contained in the stack pointer plus one. The stack pointer is
incremented by two in total.

Push instructions are commonly used to save the contents of one or more CPU registers at the start of a subroutine. Just before returning from the subroutine, corresponding pull instructions are used to restore the saved CPU registers so the subroutine
will appear not to have affected these registers.

Condition Codes and Boolean Formulae:

S	X	H	I	N	Z	V	C
—	—	—	—	—	—	—	—

None affected

Source Form: PULY

Addressing Modes, Machine Code, and Cycle-by-Cycle Execution:

Cycle	PULY (INH)		
	Addr	Data	R/$\overline{W}$
1	OP	18	1
2	OP + 1	38	1
3	OP + 2	—	1
4	SP	—	1
5	SP + 1	get IYH	1
6	SP + 2	get IYL	1

Rotate Left (ROL)

Operation:

$$\text{C} \leftarrow \boxed{\text{b7} - - - - - - \text{b0}} \leftarrow \boxed{\text{C}}$$

Description: Shifts all bits of ACCX or M one place to the left. Bit 0 is loaded from the C bit. The C bit is loaded from the most significant bit of ACCX or M. The rotate operations include the carry bit to allow extension of the shift and rotate operations to multiple bytes. For example, to shift a 24-bit value left one bit, the sequence ASL LOW, ROL MID, ROL HIGH could be used where LOW, MID, and HIGH refer to the low-order, middle, and high-order bytes of the 24-bit value, respectively.

Condition Codes and Boolean Formulae:

S	X	H	I	N	Z	V	C
—	—	—	—	$\updownarrow$	$\updownarrow$	$\updownarrow$	$\updownarrow$

N R7
Set if MSB of result is set; cleared otherwise.

Z $\overline{R7} \cdot \overline{R6} \cdot \overline{R5} \cdot \overline{R4} \cdot \overline{R3} \cdot \overline{R2} \cdot \overline{R1} \cdot \overline{R0}$
Set if result is $00; cleared otherwise.

V $N \oplus C = [N \cdot \overline{C}] + [\overline{N} \cdot C]$ (for N and C after the rotate)
Set if (N is set and C is clear) or (N is clear and C is set); cleared otherwise (for values of N and C after the rotate).

C M7
Set if, before the rotate, the MSB of ACCX or M was set; cleared otherwise.

Source Forms: ROLA; ROLB; ROL (opr)

Addressing Modes, Machine Code, and Cycle-by-Cycle Execution:

Cycle	ROLA (INH)			ROLB (INH)			ROL (EXT)			ROL (IND, X)			ROL (IND, Y)		
	Addr	Data	R/W̄	Addr	Data	R/W̄	Addr	Data	R/W̄	Addr	Data	R/W̄	Addr	Data	R/W̄
1	OP	49	1	OP	59	1	OP	79	1	OP	69	1	OP	18	1
2	OP + 1	—	1	OP + 1	—	1	OP + 1	hh	1	OP + 1	ff	1	OP + 1	69	1
3							OP + 2	ll	1	FFFF	—	1	OP + 2	ff	1
4							hhll	(hhll)	1	X + ff	(X + ff)	1	FFFF	—	1
5							FFFF	—	1	FFFF	—	1	Y + ff	(Y + ff)	1
6							hhll	result	0	X + ff	result	0	FFFF	—	1
7													Y + ff	result	0

Rotate Right (ROR)

Operation:

$$\boxed{C} \blacktriangleright \boxed{b7\text{ -- -- -- -- }b0} \blacktriangleright \boxed{C}$$

Description: Shift all bits of ACCX or M one place to the right. Bit 7 is loaded from the C bit. The C bit is loaded from the least significant bit of ACCX or M. The rotate operations include the carry bit to allow extension of the shift and rotate operations to multiple bytes. For example, to shift a 24-bit value right one bit, the sequence LSR LOW, ROR MID, ROR HIGH could be used where LOW, MID, and HIGH refer to the low-order, middle, and high-order bytes of the 24-bit value, respectively. The first LSR could be replaced by ASR to maintain the original value of the sign bit (MSB of high-order byte) of the 24-bit value.

Condition Codes and Boolean Formulae:

S	X	H	I	N	Z	V	C
—	—	—	—	$\updownarrow$	$\updownarrow$	$\updownarrow$	$\updownarrow$

N R7
   Set if MSB of result is set; cleared otherwise.

Z $\overline{R7} \cdot \overline{R6} \cdot \overline{R5} \cdot \overline{R4} \cdot \overline{R3} \cdot \overline{R2} \cdot \overline{R1} \cdot \overline{R0}$
   Set if result is $00; cleared otherwise.

V $N \oplus C = [N \cdot \overline{C}] + [\overline{N} \cdot C]$ (for N and C after the rotate)
   Set if (N is set and C is clear) or (N is clear and C is set); cleared otherwise (for values of N and C after the rotate).

C M0
   Set if, before the rotate, the LSB of ACCX or M was set; cleared otherwise.

Source Forms: RORA; RORB; ROR (opr)

Addressing Modes, Machine Code, and Cycle-by-Cycle Execution:

Cycle	RORA (INH)			RORB (INH)			ROR (EXT)			ROR (IND, X)			ROR (IND, Y)		
	Addr	Data	R/W̄	Addr	Data	R/W̄	Addr	Data	R/W̄	Addr	Data	R/W̄	Addr	Data	R/W̄
1	OP	46	1	OP	56	1	OP	76	1	OP	66	1	OP	18	1
2	OP+1	—	1	OP+1	—	1	OP+1	hh	1	OP+1	ff	1	OP+1	66	1
3							OP+2	ll	1	FFFF	—	1	OP+2	ff	1
4							hhll	(hhll)	1	X+ff	(X+ff)	1	FFFF	—	1
5							FFFF	—	1	FFFF	—	1	Y+ff	(Y+ff)	1
6							hhll	result	0	X+ff	result	0	FFFF	—	1
7													Y+ff	result	0

Return from Interrupt (RTI)

Operation: SP ◆ (SP) + $0001, ◄(CCR)
SP ◆ (SP) + $0001, ◄(ACCB)
SP ◆ (SP) + $0001, ◄(ACCA)
SP ◆ (SP) + $0001, ◄(IXH)
SP ◆ (SP) + $0001, ◄(IXL)
SP ◆ (SP) + $0001, ◄(IYH)
SP ◆ (SP) + $0001, ◄(IYL)
SP ◆ (SP) + $0001, ◄(PCH)
SP ◆ (SP) + $0001, ◄(PCL)

Description: The condition code, accumulators B and A, index registers X and Y, and the program counter will be restored to a state pulled from the stack. The X bit in the CCR may be cleared as a result of an RTI instruction but may not be set if it was cleared prior to execution of the RTI instruction.

Condition Codes and Boolean Formulae:

S	X	H	I	N	Z	V	C
↕	←	↕	↕	↕	↕	↕	↕

Condition code bits take on the value of the corresponding bit of accumulator A except that the X bit may not change from a zero to a one. Software can leave X set, leave X clear, or change X from one to zero. The XIRQ interrupt mask can only become set as a $\overline{\text{RESET}}$ of a reset or recognition of an XIRQ interrupt.

Source Form: RTI

Addressing Modes, Machine Code, and Cycle-by-Cycle Execution:

Cycle	RTI (INH)		
	Addr	Data	R/W̄
1	OP	3B	1
2	OP + 1	—	1
3	SP	—	1
4	SP + 1	get CC	1
5	SP + 2	get B	1
6	SP + 3	get A	1
7	SP + 4	get IXH	1
8	SP + 5	get IXL	1
9	SP + 6	get IXH	1
10	SP + 7	get IXL	1
11	SP + 8	Rtn hi	1
12	SP + 9	Rtn lo	1

Return from Subroutine (RTS)

Operation: SP ↓ (SP) + $0001, ↖(PCH)
SP ↓ (SP) + $0001, ↖(PCL)

Description: The stack pointer is incremented by one. The contents of the byte of memory, at the address now contained in the stack pointer, are loaded into the high-order eight bits of the program counter. The stack pointer is again incremented by one. The contents of the byte of memory, at the address now contained in the stack pointer, are loaded into the low-order eight bits of the program counter.

Condition Codes and Boolean Formulae:

S	X	H	I	N	Z	V	C
—	—	—	—	—	—	—	—

None affected

Source Form: RTS

Addressing Modes, Machine Code, and Cycle-by-Cycle Execution:

Cycle	RTS (INH)		
	Addr	Data	R/W̄
1	OP	39	1
2	OP + 1	—	1
3	SP	—	1
4	SP + 1	Rtn hi	1
5	SP + 2	Rtn lo	1

Subtract Accumulators (SBA)

Operation: ACCA ⬦ (ACCA) − (ACCB)

Description: Subtracts the contents of ACCB from the contents of ACCA and places the result in ACCA. The contents of ACCB are not affected. For subtract instructions, the C bit in the CCR represents a borrow.

Condition Codes and Boolean Formulae:

S	X	H	I	N	Z	V	C
—	—	—	—	✿	✿	✿	✿

N   R7
  Set if MSB of result is set; cleared otherwise.

Z $\overline{R7} \cdot \overline{R6} \cdot \overline{R5} \cdot \overline{R4} \cdot \overline{R3} \cdot \overline{R2} \cdot \overline{R1} \cdot \overline{R0}$
  Set if result is $00; cleared otherwise.

V $A7 \cdot \overline{B7} \cdot \overline{R7} + \overline{A7} \cdot B7 \cdot R7$
  Set if a twos complement overflow resulted from the operation; cleared otherwise.

C $\overline{A7} \cdot B7 + B7 \cdot R7 + R7 \cdot \overline{A7}$
  Set if the absolute value of ACCB is larger than the absolute value of ACCA; cleared otherwise.

Source Form: SBA

Addressing Modes, Machine Code, and Cycle-by-Cycle Execution:

Cycle	SBA (INH)		
	Addr	Data	R/$\overline{\text{W}}$
1	OP	10	1
2	OP + 1	—	1

Subtract with Carry (SBC)

Operation: ACCX ◀ (ACCX) – (M) – (C)

Description: Subtracts the contents of M and the contents of C from the contents of ACCX and places the result in ACCX. For subtract instructions the C bit in the CCR represents a borrow.

Condition Codes and Boolean Formulae:

S	X	H	I	N	Z	V	C
—	—	—	—	↕	↕	↕	↕

N R7
Set if MSB of result is set; cleared otherwise.

Z $\overline{R7} \cdot \overline{R6} \cdot \overline{R5} \cdot \overline{R4} \cdot \overline{R3} \cdot \overline{R2} \cdot \overline{R1} \cdot \overline{R0}$
Set if result is $00; cleared otherwise.

V $X7 \cdot \overline{M7} \cdot \overline{R7} + \overline{X7} \cdot M7 \cdot R7$
Set if a twos complement overflow resulted from the operation; cleared otherwise.

C $\overline{X7} \cdot M7 + M7 \cdot R7 + R7 \cdot \overline{X7}$
Set if the absolute value of the contents of memory plus previous carry is larger than the absolute value of the accumulator; cleared otherwise.

Source Forms: SBCA (opr); SBCB (opr)

Addressing Modes, Machine Code, and Cycle-by-Cycle Execution:

Cycle	SBCA (IMM)			SBCA (DIR)			SBCA (EXT)			SBCA (IND, X)			SBCA (IND, Y)		
	Addr	Data	R/W̄	Addr	Data	R/W̄	Addr	Data	R/W̄	Addr	Data	R/W̄	Addr	Data	R/W̄
1	OP	82	1	OP	92	1	OP	B2	1	OP	A2	1	OP	18	1
2	OP + 1	ii	1	OP + 1	dd	1	OP + 1	hh	1	OP + 1	ff	1	OP + 1	A2	1
3				00dd	(00dd)	1	OP + 2	ll	1	FFFF	—	1	OP + 2	ff	1
4							hhll	(hhll)	1	X + ff	(X + ff)	1	FFFF	—	1
5													Y + ff	(Y + ff)	1

Cycle	SBCB (IMM)			SBCB (DIR)			SBCB (EXT)			SBCB (IND, X)			SBCB (IND, Y)		
	Addr	Data	R/W̄	Addr	Data	R/W̄	Addr	Data	R/W̄	Addr	Data	R/W̄	Addr	Data	R/W̄
1	OP	C2	1	OP	D2	1	OP	F2	1	OP	E2	1	OP	18	1
2	OP + 1	ii	1	OP + 1	dd	1	OP + 1	hh	1	OP + 1	ff	1	OP + 1	E2	1
3				00dd	(00dd)	1	OP + 2	ll	1	FFFF	—	1	OP + 2	ff	1
4							hhll	(hhll)	1	X + ff	(X + ff)	1	FFFF	—	1
5													Y + ff	(Y + ff)	1

Set Carry (SEC)

Operation: C bit ◀ 1

Description: Sets the C bit in the CCR.

Condition Codes and Boolean Formulae:

S	X	H	I	N	Z	V	C
—	—	—	—	—	—	—	1

C 1
  Set

Source Form: SEC

Addressing Modes, Machine Code, and Cycle-by-Cycle Execution:

Cycle	SEC (INH)		
	Addr	Data	R/W̄
1	OP	0D	1
2	OP + 1	—	1

Set Interrupt Mask (SEI)

Operation: I bit ♦ 1

Description: Sets the interrupt mask bit in the CCR. When the I bit is set, all maskable interrupts are inhibited, and the MPU will recognize only non-maskable interrupt sources or an SWI.

Condition Codes and Boolean Formulae:

S	X	H	I	N	Z	V	C
—	—	—	1	—	—	—	—

I   1
   Set

Source Form: SEI

Addressing Modes, Machine Code, and Cycle-by-Cycle Execution:

Cycle	SEI (INH)		
	Addr	Data	R/W̄
1	OP	0F	1
2	OP + 1	—	1

Set Twos Complement Overflow Bit (SEV)

Operation: V bit ◄ 1

Description: Sets the twos complement overflow bit in the CCR.

Condition Codes and Boolean Formulae:

S	X	H	I	N	Z	V	C
—	—	—	—	—	—	1	—

V   1
   Set

Source Form: SEV

Addressing Modes, Machine Code, and Cycle-by-Cycle Execution:

Cycle	SEV (INH)		
	Addr	Data	R/W̄
1	OP	0B	1
2	OP + 1	—	1

Store Accumulator (STA)

Operation: M ◀ (ACCX)

Description: Stores the contents of ACCX in memory. The contents of ACCX remain unchanged.

Condition Codes and Boolean Formulae:

S	X	H	I	N	Z	V	C
—	—	—	—	⬦	⬦	0	—

N   X7
   Set if MSB of result is set; cleared otherwise.

Z $\overline{X7} \cdot \overline{X6} \cdot \overline{X5} \cdot \overline{X4} \cdot \overline{X3} \cdot \overline{X2} \cdot \overline{X1} \cdot \overline{X0}$
   Set if result is $00; cleared otherwise.

V   0
   Cleared

Source Forms: STAA (opr); STAB (opr)

Addressing Modes, Machine Code, and Cycle-by-Cycle Execution:

Cycle	STAA (DIR)			STAA (EXT)			STAA (IND, X)			STAA (IND, Y)		
	Addr	Data	R/W̄	Addr	Data	R/W̄	Addr	Data	R/W̄	Addr	Data	R/W̄
1	OP	97	1	OP	B7	1	OP	A7	1	OP	18	1
2	OP+1	dd	1	OP+1	hh	1	OP+1	ff	1	OP+1	A7	1
3	00dd	(A)	0	OP+2	ll	1	FFFF	—	1	OP+2	ff	1
4				hhll	(A)	0	X+ff	(A)	0	FFFF	—	1
5										Y+ff	(A)	0

Cycle	STAB (DIR)			STAB (EXT)			STAB (IND, X)			STAB (IND, Y)		
	Addr	Data	R/W̄	Addr	Data	R/W̄	Addr	Data	R/W̄	Addr	Data	R/W̄
1	OP	D7	1	OP	F7	1	OP	E7	1	OP	18	1
2	OP+1	dd	1	OP+1	hh	1	OP+1	ff	1	OP+1	E7	1
3	00dd	(B)	0	OP+2	ll	1	FFFF	—	1	OP+2	ff	1
4				hhll	(B)	0	X+ff	(B)	0	FFFF	—	1
5										Y+ff	(B)	0

Store Double Accumulator (STD)

Operation: M:M + 1 ◆ (ACCD); M ◆ (ACCA), M + 1 ◆ (ACCB)

Description: Stores the contents of double accumulator ACCD in memory. The contents of ACCD remain unchanged.

Condition Codes and Boolean Formulae:

S	X	H	I	N	Z	V	C
—	—	—	—	↕	↕	0	—

N   D15
Set if MSB of result is set; cleared otherwise.

Z $\overline{D15} \cdot \overline{D14} \cdot \overline{D13} \cdot \overline{D12} \cdot \overline{D11} \cdot \overline{D10} \cdot \overline{D9} \cdot \overline{D8} \cdot \overline{D7} \cdot \overline{D6} \cdot \overline{D5} \cdot \overline{D4} \cdot \overline{D3} \cdot \overline{D2} \cdot \overline{D1} \cdot \overline{D0}$
Set if result is $0000; cleared otherwise.

V   0
Cleared

Source Form: STD (opr)

Addressing Modes, Machine Code, and Cycle-by-Cycle Execution:

Cycle	STD (DIR) Addr	Data	R/W̄	STD (EXT) Addr	Data	R/W̄	STD (IND, X) Addr	Data	R/W̄	STD (IND, Y) Addr	Data	R/W̄
1	OP	DD	1	OP	FD	1	OP	ED	1	OP	18	1
2	OP + 1	dd	1	OP + 1	hh	1	OP + 1	ff	1	OP + 1	ED	1
3	00dd	(A)	0	OP + 2	ll	1	FFFF	—	1	OP + 2	ff	1
4	00dd + 1	(B)	0	hhll	(A)	0	X + ff	(A)	0	FFFF	—	1
5				hhll + 1	(B)	0	X + ff + 1	(B)	0	Y + ff	(A)	0
6										Y + ff + 1	(B)	0

Stop Processing (STOP)

Description: If the S bit in the CCR is set, then the STOP instruction is disabled and operates like the NOP instruction. If the S bit in the CCR is clear, the STOP instruction causes all system clocks to halt, and the system is placed in a minimum-power standby mode. All CPU registers remain unchanged. I/O pins also remain unaffected.

Recovery from STOP may be accomplished by $\overline{\text{RESET}}$, $\overline{\text{XIRQ}}$, or an unmasked $\overline{\text{IRQ}}$. When recovering from STOP with $\overline{\text{XIRQ}}$, if the X bit in the CCR is clear, execution will resume with the stacking operations for the $\overline{\text{XIRQ}}$ interrupt. If the X bit in the CCR is set, masking $\overline{\text{XIRQ}}$ interrupts, execution will resume with the opcode fetch for the instruction which follows the STOP instruction (continue).

An error in some mask sets of the M68HC11 caused incorrect recover from STOP under very specific unusual conditions. If the opcode of the instruction before the STOP instruction came from column 4 or 5 of the opcode map, the STOP instruction was incorrectly interpreted as a two-byte instruction. A simple way to avoid this potential problem is to put a NOP instruction (which is a column 0 opcode) immediately before any STOP instruction.

Condition Codes and Boolean Formulae:

S	X	H	I	N	Z	V	C
—	—	—	—	—	—	—	—

None affected

Source Form: STOP

Addressing Modes, Machine Code, and Cycle-by-Cycle Execution:

Cycle	STOP (INH)		
	Addr	Data	R/$\overline{\text{W}}$
1	OP	CF	1
2	OP+1	—	1

Store Stack Pointer (STS)

Operation: M ↓ (SPH), M + 1 ↓ (SPL)

Description: Stores the most significant byte of the stack pointer in memory at the address specified by the program and stores the least significant byte of the stack pointer at the next location in memory, at one plus the address specified by the program.

Condition Codes and Boolean Formulae:

S	X	H	I	N	Z	V	C
—	—	—	—	⬍	⬍	0	—

N SP15
   Set if MSB of result is set; cleared otherwise.

Z $\overline{SP15} \cdot \overline{SP14} \cdot \overline{SP13} \cdot \overline{SP12} \cdot \overline{SP11} \cdot \overline{SP10} \cdot \overline{SP9} \cdot \overline{SP8} \cdot$
   $\overline{SP7} \cdot \overline{SP6} \cdot \overline{SP5} \cdot \overline{SP4} \cdot \overline{SP3} \cdot \overline{SP2} \cdot \overline{SP1} \cdot \overline{SP0}$
   Set if result is $0000; cleared otherwise.

V 0
   Cleared

Source Form: STS (opr)

Addressing Modes, Machine Code, and Cycle-by-Cycle Execution:

Cycle	STS (DIR)			STS (EXT)			STS (IND, X)			STS (IND, Y)		
	Addr	Data	R/W̄	Addr	Data	R/W̄	Addr	Data	R/W̄	Addr	Data	R/W̄
1	OP	9F	1	OP	BF	1	OP	AF	1	OP	18	1
2	OP + 1	dd	1	OP + 1	hh	1	OP + 1	ff	1	OP + 1	AF	1
3	00dd	(SPH)	0	OP + 2	ll	1	FFFF	—	1	OP + 2	ff	1
4	oodd + 1	(SPL)	0	hhll	(SPH)	0	X + ff	(SPH)	0	FFFF	—	1
5				hhll + 1	(SPL)	0	X + ff + 1	(SPL)	0	Y + ff	(SPH)	0
6										Y + ff + 1	(SPL)	0

Store Index Register X (STX)

Operation: M ◀ (IXH), M + 1 ◀ (IXL)

Description: Stores the most significant byte of index register X in memory at the address specified by the program and stores the least significant byte of index register X at the next location in memory, at one plus the address specified by the program.

Condition Codes and Boolean Formulae:

S	X	H	I	N	Z	V	C
—	—	—	—	↕	↕	0	—

N IX15
   Set if MSB of result is set; cleared otherwise.

Z $\overline{IX15} \cdot \overline{IX14} \cdot \overline{IX13} \cdot \overline{IX12} \cdot \overline{IX11} \cdot \overline{IX10} \cdot \overline{IX9} \cdot \overline{IX8} \cdot$
   $\overline{IX7} \cdot \overline{IX6} \cdot \overline{IX5} \cdot \overline{IX4} \cdot \overline{IX3} \cdot \overline{IX2} \cdot \overline{IX1} \cdot \overline{IX0}$
   Set if result is $0000; cleared otherwise.

V 0
   Cleared

Source Form: STX (opr)

Addressing Modes, Machine Code, and Cycle-by-Cycle Execution:

Cycle	STX (DIR)			STX (EXT)			STX (IND, X)			STX (IND, Y)		
	Addr	Data	R/$\overline{W}$	Addr	Data	R/$\overline{W}$	Addr	Data	R/$\overline{W}$	Addr	Data	R/$\overline{W}$
1	OP	DF	1	OP	FF	1	OP	EF	1	OP	CD	1
2	OP + 1	dd	1	OP + 1	hh	1	OP + 1	ff	1	OP + 1	EF	1
3	00dd	(IXH)	0	OP + 2	ll	1	FFFF	—	1	OP + 2	ff	1
4	oodd + 1	(IXL)	0	hhll	(IXH)	0	X + ff	(IXH)	0	FFFF	—	1
5				hhll + 1	(IXL)	0	X + ff + 1	(IXL)	0	Y + ff	(IXH)	0
6										Y + ff + 1	(IXL)	0

Store Index Register Y (STY)

Operation: M ◀ (IYH), M + 1 ◀ (IYL)

Description: Stores the most significant byte of index register Y in memory at the address specified by the program and stores the least significant byte of index register Y at the next location in memory, at one plus the address specified by the program.

Condition Codes and Boolean Formulae:

S	X	H	I	N	Z	V	C
—	—	—	—	↕	↕	0	—

N   IY15
   Set if MSB of result is set; cleared otherwise.

Z $\overline{IY15} \cdot \overline{IY14} \cdot \overline{IY13} \cdot \overline{IY12} \cdot \overline{IY11} \cdot \overline{IY10} \cdot \overline{IY9} \cdot \overline{IY8} \cdot$
   $\overline{IY7} \cdot \overline{IY6} \cdot \overline{IY5} \cdot \overline{IY4} \cdot \overline{IY3} \cdot \overline{IY2} \cdot \overline{IY1} \cdot \overline{IY0}$
   Set if result is $0000; cleared otherwise.

V   0
   Cleared

Source Form: STY (opr)

Addressing Modes, Machine Code, and Cycle-by-Cycle Execution:

Cycle	STY (DIR) Addr	Data	R/W̄	STY (EXT) Addr	Data	R/W̄	STY (IND, X) Addr	Data	R/W̄	STY (IND, Y) Addr	Data	R/W̄
1	OP	18	1	OP	18	1	OP	1A	1	OP	18	1
2	OP + 1	DF	1	OP + 1	FF	1	OP + 1	EF	1	OP + 1	EF	1
3	OP + 2	dd	1	OP + 2	hh	1	OP + 2	ff	1	OP + 2	ff	1
4	00dd	(IYH)	0	OP + 3	ll	1	FFFF	—	1	FFFF	—	1
5	00dd + 1	(IYL)	0	hhll	(IYH)	0	X + ff	(IYH)	0	Y + ff	(IYH)	0
6				hhll + 1	(IYL)	0	X + ff + 1	(IYL)	0	Y + ff + 1	(IYL)	0

Subtract (SUB)

Operation: ACCX ◀ (ACCX) – (M)

Description: Subtracts the contents of M from the contents of ACCX and places the result in ACCX. For subtract instructions, the C bit in the CCR represents a borrow.

Condition Codes and Boolean Formulae:

S	X	H	I	N	Z	V	C
—	—	—	—	↕	↕	↕	↕

N R7
    Set if MSB of result is set; cleared otherwise.

Z $\overline{R7} \cdot \overline{R6} \cdot \overline{R5} \cdot \overline{R4} \cdot \overline{R3} \cdot \overline{R2} \cdot \overline{R1} \cdot \overline{R0}$
    Set if result is $00; cleared otherwise.

V $X7 \cdot \overline{M7} \cdot \overline{R7} + \overline{X7} \cdot M7 \cdot R7$
    Set if a twos complement overflow resulted from the operation; cleared otherwise.

C $\overline{X7} \cdot M7 + M7 \cdot R7 + R7 \cdot \overline{X7}$
    Set if the absolute value of the contents of memory is larger than the absolute value of the contents of the accumulator; cleared otherwise.

Source Forms: SUBA (opr); SUBB (opr)

Addressing Modes, Machine Code, and Cycle-by-Cycle Execution:

Cycle	SUBA (IMM)			SUBA (DIR)			SUBA (EXT)			SUBA (IND, X)			SUBA (IND, Y)		
	Addr	Data	R/W̄	Addr	Data	R/W̄	Addr	Data	R/W̄	Addr	Data	R/W̄	Addr	Data	R/W̄
1	OP	80	1	OP	90	1	OP	B0	1	OP	A0	1	OP	18	1
2	OP+1	ii	1	OP+1	dd	1	OP+1	hh	1	OP+1	ff	1	OP+1	A0	1
3				00dd	(00dd)	1	OP+2	ll	1	FFFF	—	1	OP+2	ff	1
4							hhll	(hhll)	1	X+ff	(X+ff)	1	FFFF	—	1
5													Y+ff	(Y+ff)	1

Cycle	SUBB (IMM)			SUBB (DIR)			SUBB (EXT)			SUBB (IND, X)			SUBB (IND, Y)		
	Addr	Data	R/W̄	Addr	Data	R/W̄	Addr	Data	R/W̄	Addr	Data	R/W̄	Addr	Data	R/W̄
1	OP	C0	1	OP	D0	1	OP	F0	1	OP	E0	1	OP	18	1
2	OP+1	ii	1	OP+1	dd	1	OP+1	hh	1	OP+1	ff	1	OP+1	E0	1
3				00dd	(00dd)	1	OP+2	ll	1	FFFF	—	1	OP+2	ff	1
4							hhll	(hhll)	1	X+ff	(X+ff)	1	FFFF	—	1
5													Y+ff	(Y+ff)	1

Subtract Double Accumulator (SUBD)

Operation: ACCD ◀ (ACCD) − (M:M + 1)

Description: Subtracts the contents of M:M + 1 from the contents of double accumulator D and places the result in ACCD. For subtract instructions, the C bit in the CCR represents a borrow.

Condition Codes and Boolean Formulae:

S	X	H	I	N	Z	V	C
—	—	—	—	‡	‡	‡	‡

N   R15
Set if MSB of result is set; cleared otherwise.

Z $\overline{R15} \cdot \overline{R14} \cdot \overline{R13} \cdot \overline{R12} \cdot \overline{R11} \cdot \overline{R10} \cdot \overline{R9} \cdot \overline{R8} \cdot \overline{R7} \cdot \overline{R6} \cdot \overline{R5} \cdot \overline{R4} \cdot \overline{R3} \cdot \overline{R2} \cdot \overline{R1} \cdot \overline{R0}$
Set if result is $0000; cleared otherwise.

V $D15 \cdot \overline{M15} \cdot \overline{R15} + \overline{D15} \cdot M15 \cdot R15$
Set if a twos complement overflow resulted from the operation; cleared otherwise.

C $\overline{D15} \cdot M15 + M15 \cdot R15 + R15 \cdot \overline{D15}$
Set if the absolute value of the contents of memory is larger than the absolute value of the accumulator; cleared otherwise.

Source Form: SUBD (opr)

Addressing Modes, Machine Code, and Cycle-by-Cycle Execution:

Cycle	SUBD (IMM)			SUBD (DIR)			SUBD (EXT)			SUBD (IND, X)			SUBD (IND, Y)		
	Addr	Data	R/W̄	Addr	Data	R/W̄	Addr	Data	R/W̄	Addr	Data	R/W̄	Addr	Data	R/W̄
1	OP	83	1	OP	93	1	OP	B3	1	OP	A3	1	OP	18	1
2	OP + 1	jj	1	OP + 1	dd	1	OP + 1	hh	1	OP + 1	ff	1	OP + 1	A3	1
3	OP + 2	kk	1	00dd	(00dd)	1	OP + 2	ll	1	FFFF	—	1	OP + 2	ff	1
4	FFFF	—	1	00dd + 1	(00dd + 1)	1	hhll	(hhll)	1	X + ff	(X + ff)	1	FFFF	—	1
5				FFFF	—	1	hhll + 1	(hhll + 1)	1	X + ff + 1	(X + ff + 1)	1	Y + ff	(Y + ff)	1
6							FFFF	—	1	FFFF	—	1	Y + ff + 1	(Y + ff + 1)	1
7													FFFF	—	1

Software Interrupt (SWI)

Operation: PC ◀ (PC) + $0001
 ➡(PCL), SP ◀ (SP) − $0001
 ➡(PCH), SP ◀ (SP) − $0001
 ➡(IYL), SP ◀ (SP) − $0001
 ➡(IYH), SP ◀ (SP) − $0001
 ➡(IXL), SP ◀ (SP) − $0001
 ➡(IXH), SP ◀ (SP) − $0001
 ➡(ACCA), SP ◀ (SP) − $0001
 ➡(ACCB), SP ◀ (SP) − $0001
 ➡(CCR), SP ◀ (SP) − $0001
 I ◀ 1, PC ◀ (SWI vector)

Description: The program counter is incremented by one. The program counter, index registers Y and X, and accumulators A and B are pushed onto the stack. The CCR is then pushed onto the stack. The stack pointer is decremented by one after each byte of data is stored on the stack. The I bit in the CCR is then set. The program counter is loaded with the address stored at the SWI vector, and instruction execution resumes at this location. This instruction is not maskable by the I bit.

Condition Codes and Boolean Formulae:

S	X	H	I	N	Z	V	C
—	—	—	1	—	—	—	—

I   1
    Set

Source Form: SWI

Addressing Modes, Machine Code, and Cycle-by-Cycle Execution:

Cycle	SWI (INH)		
	Addr	Data	R/W̄
1	OP	3F	1
2	OP + 1	—	1
3	SP	Rtn lo	0
4	SP − 1	Rtn hi	0
5	SP − 2	(IYL)	0
6	SP − 3	(IYH)	0
7	SP − 4	(IXL)	0
8	SP − 5	(IXH)	0
9	SP − 6	(A)	0
10	SP − 7	(B)	0
11	SP − 8	(CCR)	0
12	SP − 8	(CCR)	1
13	Vec hi	Svc hi	1
14	Vec lo	Svc lo	1

Transfer from Accumulator A to Accumulator B (TAB)

Operation: ACCB ◀ (ACCA)

Description: Moves the contents of ACCA to ACCB. The former contents of ACCB are lost; the contents of ACCA are not affected.

Condition Codes and Boolean Formulae:

S	X	H	I	N	Z	V	C
—	—	—	—	✸	✸	0	—

N R7
   Set if MSB of result is set; cleared otherwise.

Z $\overline{R7} \cdot \overline{R6} \cdot \overline{R5} \cdot \overline{R4} \cdot \overline{R3} \cdot \overline{R2} \cdot \overline{R1} \cdot \overline{R0}$
   Set if result is $00; cleared otherwise

V 0
   Cleared

Source Form: TAB

Addressing Modes, Machine Code, and Cycle-by-Cycle Execution:

Cycle	TAB (INH)		
	Addr	Data	R/W̄
1	OP	16	1
2	OP+1	—	1

Transfer from Accumulator A to Condition Code Register (TAP)

Operation: CCR ◀ (ACCA)

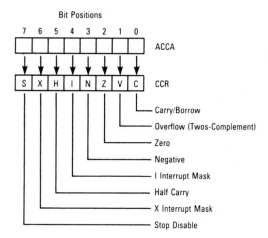

Description: Transfers the contents of bit positions 7–0 of accumulator A to the corresponding bit positions of the CCR. The contents of accumulator A remain unchanged. The X bit in the CCR may be cleared as a result of a TAP instruction but may not be set if it was clear prior to execution of the TAP instruction.

Condition Codes and Boolean Formulae:

S	X	H	I	N	Z	V	C
✿	◄	✿	✿	✿	✿	✿	✿

Condition code bits take on the value of the corresponding bit of accumulator A except that the X bit may not change from a zero to a one. Software can leave X set, leave X clear or change X from one to zero. The $\overline{\text{XIRQ}}$ interrupt mask can only become set as a result of a $\overline{\text{RESET}}$ or recognition of an $\overline{\text{XIRQ}}$ interrupt.

Source Form: TAP

Addressing Modes, Machine Code, and Cycle-by-Cycle Execution:

Cycle	TAP (INH)		
	Addr	Data	R/$\overline{\text{W}}$
1	OP	06	1
2	OP + 1	—	1

Transfer from Accumulator B to Accumulator A (TBA)

Operation: ACCA ◀ (ACCB)

Description: Moves the contents of ACCB to ACCA. The former contents of ACCA are lost; the contents of ACCB are not affected.

Condition Codes and Boolean Formulae:

S	X	H	I	N	Z	V	C
—	—	—	—	↕	↕	0	—

N R7
   Set if MSB of result is set; cleared otherwise.

Z $\overline{R7} \cdot \overline{R6} \cdot \overline{R5} \cdot \overline{R4} \cdot \overline{R3} \cdot \overline{R2} \cdot \overline{R1} \cdot \overline{R0}$
   Set if result is $00; cleared otherwise.

V 0
   Cleared

Source Form: TBA

Addressing Modes, Machine Code, and Cycle-by-Cycle Execution:

Cycle	TBA (INH)		
	Addr	Data	R/$\overline{W}$
1	OP	17	1
2	OP+1	—	1

Test Operation (Test Mode Only) (TEST)

Description: This is a single-byte instruction that causes the program counter to be continuously incremented. It can only be executed while in the test mode. The MPU must be reset to exit this instruction. Code execution is suspended during this instruction. This is an illegal opcode when not in test mode.

Condition Codes and Boolean Formulae:

S	X	H	I	N	Z	V	C
—	—	—	—	—	—	—	—

None affected

Source Form: TEST

Addressing Modes, Machine Code, and Cycle-by-Cycle Execution:

Cycle	TEST (INH)		
	Addr	Data	R/$\overline{\text{W}}$
1	OP	00	1
2	OP + 1	—	1
3	OP + 2	—	1
4	OP + 3	—	1
5 − n	PREV − 1 (PREV − 1)		1

Transfer from Condition Code Register to Accumulator A (TPA)

Operation: (ACCA) ◂ (CCR)

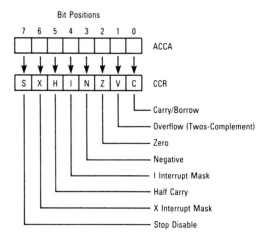

Description: Transfers the contents of the CCR to corresponding bit positions of accumulator A. The CCR remains unchanged.

Condition Codes and Boolean Formulae:

S	X	H	I	N	Z	V	C
—	—	—	—	—	—	—	—

None affected

Source Form: TPA

Addressing Modes, Machine Code, and Cycle-by-Cycle Execution:

Cycle	TPA (INH)		
	Addr	Data	R/W̄
1	OP	07	1
2	OP+1	—	1

Test (TST)

Operation: (ACCX) − $00   **or:**   (M) − $00

Description: Subtracts $00 from the contents of ACCX or M and sets the condition codes accordingly.

The subtraction is accomplished internally without modifying either ACCX or M.

The TST instruction provides only minimum information when testing unsigned values. Since no unsigned value is less than zero, BLO and BLS have no utility. While BHI could be used after TST, it provides exactly the same control as BNE, which is preferred. After testing signed values, all signed branches are available.

Condition Codes and Boolean Formulae:

S	X	H	I	N	Z	V	C
—	—	—	—	⬍	⬍	0	0

N M7
   Set if MSB of result is set; cleared otherwise.

Z $\overline{M7} \cdot \overline{M6} \cdot \overline{M5} \cdot \overline{M4} \cdot \overline{M3} \cdot \overline{M2} \cdot \overline{M1} \cdot \overline{M0}$
   Set if result is $00; cleared otherwise

V 0
   Cleared

C 0
   Cleared

Source Forms: TSTA; TSTB; TST (opr)

Addressing Modes, Machine Code, and Cycle-by-Cycle Execution:

Cycle	TSTA (INH)			TSTB (INH)			TST (EXT)			TST (IND, X)			TST (IND, Y)		
	Addr	Data	R/W̅	Addr	Data	R/W̅	Addr	Data	R/W̅	Addr	Data	R/W̅	Addr	Data	R/W̅
1	OP	4D	1	OP	5D	1	OP	7D	1	OP	6D	1	OP	18	1
2	OP + 1	—	1	OP + 1	—	1	OP + 1	hh	1	OP + 1	ff	1	OP + 1	6D	1
3							OP + 2	ll	1	FFFF	—	1	OP + 2	ff	1
4							hhll	(hhll)	1	X + ff	(X + ff)	1	FFFF	—	1
5							FFFF	—	1	FFFF	—	1	Y + ff	(Y + ff)	1
6							FFFF	—	1	FFFF	—	1	FFFF	—	1
7													FFFF	—	1

Transfer from Stack Pointer to Index Register X (TSX)

Operation: IX ◄ (SP) + $0001

Description: Loads the index register X with one plus the contents of the stack pointer. The contents of the stack pointer remain unchanged. After a TSX instruction the index register X points at the last value that was stored on the stack.

Condition Codes and Boolean Formulae:

S	X	H	I	N	Z	V	C
—	—	—	—	—	—	—	—

None affected

Source Form: TSX

Addressing Modes, Machine Code, and Cycle-by-Cycle Execution:

Cycle	TSX (INH)		
	Addr	Data	R/W̄
1	OP	30	1
2	OP + 1	—	1
3	SP	—	1

Transfer from Stack Pointer to Index Register Y (TSY)

Operation: IY ◄ (SP) + $0001

Description: Loads the index register Y with one plus the contents of the stack pointer. The contents of the stack pointer remain unchanged. After a TSY instruction the index register Y points at the last value that was stored on the stack.

Condition Codes and Boolean Formulae:

S	X	H	I	N	Z	V	C
—	—	—	—	—	—	—	—

None affected

Source Form: TSY

Addressing Modes, Machine Code, and Cycle-by-Cycle Execution:

Cycle	TSY (INH)		
	Addr	Data	R/W̄
1	OP	18	1
2	OP+1	30	1
3	OP+2	—	1
4	SP	—	1

Transfer from Index Register X to Stack Pointer (TXS)

Operation: SP ◀ (IX) − $0001

Description: Loads the stack pointer with the contents of the index register X minus one. The contents of the index register X remain unchanged.

Condition Codes and Boolean Formulae:

S	X	H	I	N	Z	V	C
—	—	—	—	—	—	—	—

None affected

Source Form: TXS

Addressing Modes, Machine Code, and Cycle-by-Cycle Execution:

Cycle	TXS (INH)		
	Addr	Data	R/W̄
1	OP	35	1
2	OP + 1	—	1
3	FFFF	—	1

Transfer from Index Register Y to Stack Pointer (TYS)

Operation;   SP ∮ (IY) − $0001

Description: Loads the stack pointer with the contents of the index register Y minus one. The contents of the index register Y remain unchanged.

Condition Codes and Boolean Formulae:

S	X	H	I	N	Z	V	C
—	—	—	—	—	—	—	—

None affected

Source Form: TYS

Addressing Modes, Machine Code, and Cycle-by-Cycle Execution:

Cycle	TYS (INH)		
	Addr	Data	R/W̄
1	OP	18	1
2	OP + 1	35	1
3	OP + 2	—	1
4	FFFF	—	1

Wait for Interrupt (WAI)

Operation;   PC ◀ (PC) + $0001
◀ (PCL), SP ◀ (SP) − $0001
◀ (PCH), SP ◀ (SP) − $0001
◀ (IYL), SP ◀ (SP) − $0001
◀ (IYH), SP ◀ (SP) − $0001
◀ (IXL), SP ◀ (SP) − $0001
◀ (IXH), SP ◀ (SP) − $0001
◀ (ACCA), SP ◀ (SP) − $0001
◀ (ACCB), SP ◀ (SP) − $0001
◀ (CCR), SP ◀ (SP) − $0001

Description: The program counter is incremented by one. The program counter, index registers Y and X, and accumulators A and B are pushed onto the stack. The CCR is then pushed onto the stack. The stack pointer is decremented by one after each byte of data is stored on the stack.

The MPU then enters a wait state for an integer number of MPU E-clock cycles. While in the wait state, the address/data bus repeatedly runs read bus cycles to the address where the CCR contents were stacked. The MPU leaves the wait state when it senses any interrupt that has not been masked.

Upon leaving the wait state, the MPU sets the I bit in the CCR, fetches the vector (address) corresponding to the interrupt sensed, and instruction execution is resumed at this location.

Condition Codes and Boolean Formulae:

S	X	H	I	N	Z	V	C
—	—	—	—	—	—	—	—

Although the WAI instruction itself does not alter the condition code bits, the interrupt which causes the MCU to resume processing causes the I bit (and the X bit if the interrupt was XIRQ) to be set as the interrupt vector is being fetched.

Source Form: WAI

Addressing Modes, Machine Code, and Cycle-by-Cycle Execution:

Cycle	WAI (INH)		
	Addr	Data	R/W̄
1	OP	3E	1
2	OP + 1	—	1
3	SP	Rtn lo	0
4	SP − 1	Rtn hi	0
5	SP − 2	(IYL)	0
6	SP − 3	(IYH)	0
7	SP − 4	(IXL)	0
8	SP − 5	(IXH)	0
9	SP − 6	(A)	0
10	SP − 7	(B)	0
11	SP − 8	(CCR)	0
12 to 12 + n	SP − 8	(CCR)	1
13 + n	Vec hi	Svc hi	1
14 + n	Vec lo	Svc lo	1

Exchange Double Accumulator and Index Register X (XGDX)

Operation: (IX) ⟺ (ACCD)

Description: Exchanges the contents of double accumulator ACCD and the contents of index register X. A common use for XGDX is to move an index value into the double accumulator to allow 16-bit arithmetic calculations on the index value before exchanging the updated index value back into the X index register.

Condition Codes and Boolean Formulae:

S	X	H	I	N	Z	V	C
—	—	—	—	—	—	—	—

None affected

Source Form: XGDX

Addressing Modes, Machine Code, and Cycle-by-Cycle Execution:

Cycle	XGDX (INH)		
	Addr	Data	R/W̄
1	OP	8F	1
2	OP + 1	—	1
3	FFFF	—	1

Exchange Double Accumulator and Index Register Y (XGDY)

Operation: (IY) ⬌ (ACCD)

Description: Exchanges the contents of double accumulator ACCD and the contents of index register Y. A common use for XGDY is to move an index value into the double accumulator to allow 16-bit arithmetic calculations on the index value before exchanging the updated index value back into the Y index register.

Condition Codes and Boolean Formulae:

S	X	H	I	N	Z	V	C
—	—	—	—	—	—	—	—

None affected

Source Form: XGDY

Addressing Modes, Machine Code, and Cycle-by-Cycle Execution:

Cycle	XGDY (INH)		
	Addr	Data	R/W̄
1	OP	18	1
2	OP + 1	8F	1
3	OP + 2	—	1
4	FFFF	—	1

APPENDIX B

References

Greenfield, J. D. and Wray, W. C., *Using Microprocessors and Microcomputers: The Motorola Family,* 2nd ed. Prentice-Hall, Englewood Cliffs, NJ., 1988. (Originally published by John Wiley, 1988.)

Greenfield, J. D., *Practical Digital Design Using ICs* (latest ed.). Prentice-Hall, Englewood Cliffs, NJ.

Lipovski, G. J., *Single and Multiple-Chip Microcomputer Interfacing.* Prentice-Hall, Englewood Cliffs, NJ, 1988.

Peatman, J. B. *Design with Microcontrollers.* McGraw-Hill, New York, 1988.

Motorola *M68HC11 Reference Manual*, Motorola, Phoenix, AZ, 1990.

Motorola *M68HC11EVB Evaluation Board User's Manual.* Motorola, Phoenix, AZ, 1986.

The Motorola Corporation, Phoenix, AZ, provides a series of application notes on its products. The following pertain to the **68HC11**:

AN997   The CONFIG Register
AN1010   Programming the EEPROM

Motorola will probably develop additional application notes as time proceeds.

Motorola also provides an *electronic bulletin board* designed to help the users of its products. The telephone number is 512-891-3733.

The following companies provide assemblers and simulators for the **68HC11**:

Avocet Systems, Inc.
120 Union Street
Rockport, ME 04856
1-800-448-8500

P&E Microcomputer Systems, Inc.
P.O. Box 2044
Woburn, MA 01888-2044
617-944-7585

Information on Motorola's academic support and on the EVB and EVBU can be obtained by contacting:

Motorola, Inc.
Judy Racino
6501 William Cannon Drive West
Austin, TX 78735
Mail Drop OE39

Information on PROCOMM, a serial communications package, can be obtained from:

Datastorm Technologies
P.O. Box 1471
Columbia, MO 65205
314-443-3282

APPENDIX C

ASCII Conversion Chart

The conversion chart listed below is helpful in converting from a two-digit (2-byte) hexadecimal number to an ASCII character or from an ASCII character to a two-digit hexadecimal number. The example provided below shows the method of using this conversion chart.

Example

		Bits						
		MSB ←					→ LSB	
ASCII	Hex #	6	5	4	3	2	1	0
T	54	1	0	1	0	1	0	0
?	3F	0	1	1	1	1	1	1
+	2B	0	1	0	1	0	1	1

Bits 0 to 3 Second Hex Digit (LSB)	Bits 4 to 6 First Hex Digit (MSB)							
	0	**1**	**2**	**3**	**4**	**5**	**6**	**7**
0	NUL	DLE	SP	0	@	P	(	p
1	SOH	DC1	!	1	A	Q	a	q
2	STX	DC2	''	2	B	R	b	r
3	ETX	DC3	#	3	C	S	c	s
4	EOT	DC4	$	4	D	T	d	t
5	ENQ	NAK	%	5	E	U	e	u
6	ACK	SYN	&	6	F	V	f	v
7	BEL	ETB	'	7	G	W	g	w
8	BS	CAN	(	8	H	X	h	x
9	HT	EM	)	9	I	Y	i	y
A	LF	SUB	*	:	J	Z	j	z
B	VT	ESC	+	;	K	[	k	{
C	FF	FS	,	<	L	/	l	/
D	CR	GS	–	=	M	]	m	}
E	SO	RS	.	>	N	∧	n	≈
F	SI	US	/	?	O	—	o	DEL

APPENDIX D

Answers to Selected Problems

Chapter 1

1-1 a. 23   **b.** 413

1-3 b. 7   **d.** 14

1-4 b. 1100001   **d.** 10011100010011

1-5 a. A + B = 100001, A − B = 10001
   c. A + B = 101001010111, A − B = 1111100011

1-6 The B number in parts (a) and (c) is divisible by 2. The B number in part (a) is the only number divisible by 4; actually it is divisible by 8.

1-7 b. $(43)_{10}$ = $(2B)_{16}$. The negative of 2B is D5.
   d. $(-77)_{10}$ = $(B3)_{16}$. The negative of B3 is 4D or $(+77)_{10}$.

1-8 b. 75 − 23 = 52   (decimal); 4B − 17 = 34   (hex)
   d. −77 − (−87) = +10   (decimal); B3 − A9 = 0A   (hex)
   e. 77 + (−12) = 65   (decimal); 4D + F4 = 41   (hex)

1-9 The numbers are −32,768 to +32,767.

1-10 b. CFE   **d.** 33F9E

1-11 b. 11100000000111111010

1-12 b. 918020

1-13 a. 62   **d.** 1A123

1-14 b. 472AB

1-15 b. 3B17   **d.** 543A

1-16 b, d, f, and g are positive. a, c, e, and g are divisible by 4.

1-17 a. 5C   **c.** 10   **e.** 2224   **g.** C600

Chapter 2

2-1 a. 18   **b.** 8   **c.** 1

2-4 LOAD   A

   SUB   B

   ADD   C

   ADD   D

   ADD   D

2-6 See Figure P2-6.

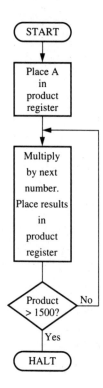

Figure P2-6

Chapter 3

3-1 a. 65 is loaded into B.
   c. 6576 is loaded into X.
   e. 87AB is loaded into X.
   f. C100 is stored in C104 and C105.

3-4 a. Results are CC. N and V are both set to 1.
   c. Results are FF (-1). N and C are 1.
   d. Results are 0. Z and C are 1.

3-5 The numbers add up to 139, which is out of bounds.

3-6 a. C and V are set.   **c.** V is set.   **d.** N and C are set.

3-7 a. C016   **b.** (1) BFCE   **b.** (2) C0CE

3-8 a. EB   **b.** CC

3-11

```
START   CLR    $C600    Clear the counter that will count
                        the numbers of ABs.
        LDX    $C100    Point X to the start of the buffer.
LOOP    LDAA   0,X      Get a byte from memory.
        CMPA   #$AB     Is it AB?
        BNE    CONT
        INC    $C600    Yes. Increment counter.
CONT    INX             Point to next memory location.
        CPX    #$C500   Done?
        BNE    LOOP     If no, go back and get next
                        location.
        BRA    *        Yes. Halt.
```

Figure P3-11

Chapter 4

No problems

Chapter 5

5-1 After the program is run, A will contain 01 and B will contain 00.

5-2
```
        LDX    #C100
LOOP    CLR    0,X
        INX
        CPX    #C300
        BNE    LOOP
```

5-3 b. After the instruction, B will contain 79F2.

5-5 and **5-6** The problem is

```
  01    A0    14
+ 6C    AC    8D
```

The program is

Program		A	N	Z	V	C
LDAA	2C	14	0	0	0	0
ADDA	2F	A1	1	0	0	0
STAA	3C	A1	1	0	0	0
LDAA	2B	A0	1	0	0	0
ADCA	2E	4C	0	0	1	1
STAA	3B	4C	0	0	0	1
LDAA	2A	01	0	0	0	1
ADCA	2D	6E	0	0	0	0
STAA	3A	6E	0	0	0	0

5-7 Part 2.

$$(34)_{10} = (22)_{16} = \quad 0022 \qquad \text{extended}$$
$$(45)_{10} = (2D)_{16} = \quad 002D \qquad \text{extended}$$
$$(-113)_{10} = (8F)_{16} = \quad \underline{FF8F} \qquad \text{extended}$$
$$FFDE = (-22)_{16} = (-34)_{10}$$

5-9

```
        LDX     #C047
        ADDA    #00
```
This instruction is performed only to clear the carry flag at the start. It can be replaced by the CLC (Clear Carry) instruction discussed in Chapter 6.

```
LOOP    LDAA    0,X
        ADCA    10,X
```
Note that the first time through, when the least significant bytes are added, the carry is clear.

```
        STAA    20,X
        DEX
        CPX     #C03F
        BNE     LOOP
```

5-10 c. Location C122 contains 41.
    f. The Z bit is 1.

5-11 The correct answer is c.

5-13 A = 0C, B = 80

5-15 X = 002B, A = 00, B = 1A

5-17 b. B = E6   **d.** A = 5D, B = E6

5-19 b. A = 77   **d.** A = 56, Z = 1

5-20 BITB #60. If Z = 1, both bits 5 and 6 are 0.

5-22 To determine if the parity of the byte is even or odd, all 8 bits must be examined. EOR a byte in another location each time a 1 is encountered. If the results are 1, the parity is odd. The coding is left to the student.

5-23 Part c.

```
        LDX     #C100
LOOP    BSET    0,X 80
        INX
        CPX     #C300
        BNE     LOOP
```

5-24 C100 contains 4F; C101 contains E0.

5-27	Instruction		A	H	N	Z	V	C
	LDAA	60	41	0	0	0	0	0
	ORA	08	49	0	0	0	0	0
	ADDA	#57	A0	1	1	0	1	0
	DAA		06	1	0	0	0	1
	CMPA	#80	06	1	1	0	1	1

Chapter 6

6-1 a. All codes are 0.   **b.** Z = 1   **d.** N = 1, C = 1, V = 1

6-2 If the Op code is BLE, it will branch if A contains A0 or 90.

6-3

START	LDX	#C080	
	LDY	#C500	
LOOP	LDAA	0,X	
	CMPA	#A0	
	BHI	LOOP2	
CONT	INX		
	CPX	C3CD	
	BNE	LOOP	
	SWI		
LOOP2	STX	0,Y	This loop lists the addresses
	INY		where numbers greater than
	INY		A0 are found.
	BRA	CONT	

6-6

C110	LDX	#D400
C113	BRSET	0,X 4A D7

6-8 *Hint:* A BRCLR 0F will branch if the number is divisible by 16.

6-9 c.

	LDAA	#20
	TAP	

6-10 Location C103

6-11 b. Stack C5FF 52

            C5FE C0

The subroutine starts at C02D.

c. Stack C5FF 53

            C5FE C0

The subroutine starts at C023.

e. The subroutine starts at C223.

6-12 The locations that changed are

C12E	79
C12F	2A
C130	A1
C131	C0
C132	2B

6-14 The byte list goes from C000 to C009.
The ASCII list goes from C110 to C123.

6-15 The stack is

C4FA	C3
C4FB	46
C4FC	X
C4FD	X
C4FE	A
C4FF	C0
C500	58

6-18 18 cycles, 9 μs

6-19 The program is

LOOP1	LDY	#$(3600)_{10}$
	JSR	To the 1-s delay routine
	DEY	
	BNE	LOOP1
	LDY	#$(5400)_{10}$
LOOP2	JSR	To the 1-s delay routine
	DEY	
	BNE	LOOP2
	BRA	LOOP1

6-21 It takes 20,031 cycles or 10.0155 ms.

6-24 The program is

LDS	
JSR	INCHAR
ANDA	#0F
STAA	SAM
JSR	INCHAR
ANDA	#0F

```
ADDA     SAM
DAA
JSR      OUTRHLF
JSR      OUTLHLF
```

Chapter 7

7-1 Labels a, d, and f are proper.

7-2
```
BNE      BILL     26    03
LDAA     MAX      B6    C020
         INCA     4A
```

7-5 a.
```
     LDS     #$C500
```

   b.
```
          ORG     $C500
          RMB     $42
JANE      RMB     1
```

7-8 The code on the first line begins at C202. It is 86 02 CE C3 00. . . .

Chapter 8

8-1 It is pin 6 of port B in the single-chip mode.
It is address bit 14 in the expanded mode.

8-3 During the JSR

   S = C01C, Y = C222, X = C155, A = AB, B = 57, CC = C1

   During the STAA

   S = C01B, Y = C222, X = C156, A = 14, B = 57, CC = C1.

   The stack is

```
C01E     2E
C01D     C0
C01C     AB
C01B     57
C01A     C0
C019     22
C018     C2
C017     56
C016     C1
C015     14
C014     57
C013     C1
```

8-4 The CCR receives $CA, B receives $12 . . .

8-5

FFF0	00	
FFF1	EB	
00EB	7E	Jump
00EC	D4	
00ED	00	

8-8 If the SEI were omitted, a 60 could appear in the memory list if an interrupt occurred between the SECONDS = 60? and the CLEAR SECONDS instructions in the program.

8-10 JACK-3 contains the low byte of Y in the main program at the time of the interrupt.

JACK-14 contains the low byte of X in the lower priority service routine when it is interrupted.

8-12 To clear the bits

TPA

ANDA #2F

TAP

8-14 The original stack is

C400	24
C3FF	C2

The new stack is

D000	44
CFFF	C2

Chapter 9

9-1

PORTB	EQU	$1004	
START	LDX	#$C100	
LOOP	LDAA	0,X	
	STAA	PORTB	
	JSR		Jump to a 10-ms delay subroutine.
	INX		
	CPX	#C124	
	BNE	LOOP	
	CONTINUE		

9-5

LDAA	#$C1
STAA	DDRC

9-7 Part of the program is

```
START   LDAA    #C0
        STAA    DDRC
        LDAA    PORTC     Load the data.
        ANDA    #$3F      Zero out the two MSBs.
        CMPA    #$25
        BLS     LOOP2
CONT


LOOP2   LDAA    #$00
        STAA    PORTC
        BRA     CONT
```

9-8 The connections between the μP and the memory are shown in Figure P9-8.

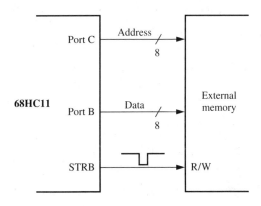

Figure P9-8

The program is

```
START   LDAA    #$FF
        STAA    DDRC      Set port C for output.
        LDX     #$C200
        CLR     PORTC     Set the initial address to 0.
LOOP    LDAA    0,X
        STAA    PORTB     Send out the first byte. This
                          also provides a write pulse
                          from STRB.
        INC     PORTC     Increment the address to the
                          memory.
        INX
        CPX     #$C300
        BNE     LOOP
```

```
9-10  PIOC    EQU     $1002
      PORTC   EQU     $1003
      PORTCL  EQU     $1005
      DDRC    EQU     $1007
      START   LDAA    #02         Set DDRC so that bit 1 is
              STAA    DDRC        output.
      LOOP    LDAA    PIOC
              BPL     LOOP
              LDAA    #$02
              STAA    PORTC       Turn the light on.
              JSR                 Jump to 10-ms delay.
              CLR     PORTC       Turn off the light.
              JSR                 Jump to 10-ms delay.
              LDAA    PORTCL      Clear STAF.
              BRA     LOOP

9-14  START   LDX     #BUFFER
              LDAA    #1C         Set up the PIOC for
              STAA    PIOC        handshaking output in pulse
                                  mode.

              BRA     LOOP2       Send out the first character.
      LOOP    LDAA    PIOC
              BPL     LOOP
      LOOP2   LDAA    0,X
              STAA    PORTB       This sends the character to
                                  the printer and provides
                                  the strobe.

              INX
              CPX     BUFFER+1
              BNE     LOOP
```

Chapter 10

10-1 The program is simply

```
LDX     TCNT
LDY     TCNT
```

The time difference will be six counts or 3 μs.

10-3 a. The delay will be 4.096 ms.
c. There will be no delay; the BGT will never branch.

10-4 The program cannot change the value in A.

10-5 The time delay is 69.64 ms.

10-6 a. N1 = 0, DIFF = $04E2
  c. N1 = 15, DIFF = $4204

10-7 The maximum delay is approximately 2^{21} ms.

10-8 b. Y = 4, D = 100

10-11 The program should count 18,310 RTIs.

10-12 LDAA   #$13
    STAA $1021

10-13 LDAA   #$7E
    STAA $00E5
    LDD   #$C400
    STD $00E6
    LDAA   #$40
    STAA $1021

10-14 a. There are two TOFs. The value in TCNT at the time of the second edge is $CF20.

10-17 The outline of the program is
  a. Put the waveform into one of the IC pins.
  b. Read the time of the first transition.
  c. Read the time of the next transition.
  d. Compare the difference to the upper and lower limits for 5 ms. Jump to ERROR if the difference is out of bounds.
  e. Go to a similar routine for the 3-ms pulse and repeat.

10-18 Connect the input line to PA0, PA1, and PA2.
    The program is

RESUL1	FDB			Set aside space
RESUL2	FDB			for the results.
	LDX	#$1000		
	LDAA	#$16		Set TCTL2 to
	STAA	21,X		detect rising edges on IC1 and IC2 and a falling edge on IC3.
	LDAA	#$07		
	STAA	23,X		Clear the flags.
LOOP1	BRCLR	23,X 04	LOOP1	Wait for the first rising edge.

(program continues)

LOOP2	BRCLR	23,X 01	LOOP2	Wait for the falling edge.
	LDD	14,X		Find the time of
	SUBD	10,X		the first pulse and
	STD	RESUL1		store it in
	LDAA	#$07		RESUL1.
	STAA	23,X		Clear the flags.
LOOP3	BRCLR	23,X 02	LOOP3	
LOOP4	BRCLR	23,X 01	LOOP4	
	LDD	14,X		Repeat for the
	SUBD	12,X		second pulse.
	STD	RESUL2		

10-19 An outline of the program is

a. Set up an input capture register to detect a negative edge and clear its flag.

b. Poll it until a negative edge appears.

c. Read TCNT, add 23 ms, and store it in an output compare register.

d. Allow the output compare register to interrupt.

10-22 Outline:

a. Set OM3 and OL3 to 11 in TCTL1 so that PA5 goes high.

b. Add 4 ms to the time in TOC3 and store it in TOC3.

c. Change TCTL1 so that PA5 will go low on the next compare.

d. Clear the flag in TFLG1 and wait for it to set.

e. Change TCTL1 so that the PA5 will go high on the next compare.

e. Add 7 ms to the time in TOC3.

f. Clear the flag in TFLG1 and wait for it to set.

g. Go back to step b.

10-23 Refer to Figure 10-16.

a. An address for the interrupt routine would have to be selected and the interrupt vector would have to be set up in the initialization.

b. The program could go into an infinite loop after the STAA $20,X instruction.

c. The interrupt routine could start at LOOP1.

10-26 Change the INC instruction in Figure 10-18 to LDD #$0064 followed by an STD $101D.

10-28 The program segment is

```
LDAA    #$F8
STAA    $100C
```

```
LDAA      #$F0
STAA      #$100D
```

10-31 The program must
 a. Set PA5 high.
 b. Set up TCTL1 to bring it low.
 c. Wait 20 ms.
 d. Set TCTL1 to make PA5 set.
 e. Wait 5 ms.
 f. Go to step b.
   In addition, it must monitor PA1. If it finds a positive transition, it should do a CFORC to cause PA5 to go low.
   The effectiveness of this program can be monitored in the laboratory by
 a. Applying no pulses to PA1. The waveform should be up for 20 ms and down for 5 ms.
 b. Applying a square wave with a period between 5 and 20 ms. The output should go negative on every active edge of the square wave.

10-33 **a.** Place $C700 in the SP.
 b. Place a JMP $C600 in $00CD–00CF.
 c. Write #$E7 into PACNT at $1027. This will wait 25 decimal pulses before rolling over.
 d. Write #$50 into PACTL at $1026. This will enable the pulse accumulator and cause it to respond to positive edges. It also sets PA7 for input.
 e. Write #$20 into TMSK2 at 1024. This enables pulse accumulator overflow interrupts.

10-35 **a.** The counter will roll over approximately every 88 ms.

Chapter 11

11-1 **a.** $80

11-2 **b.** 3.89 V

11-3 3.25 V

11-5 For parts (a) and (b) the program might be

```
LDS       #STACK        Set the stack pointer.
LDX       #$1030        Point X to ADCTL.
LDAA      #$22          Set up for continuous conversions
STAA      0,X           on channel 2.
CLR       LOC1          Clear a memory location.
CLRA
CLRB                    Clear the double registers.
```

(*program continues*)

LOOP1	BRCLR	0,X 80	LOOP1	Wait until conversions are complete.
LOOP2	PSHB			Preserve B
	LDAB	LOC1 + 1		Store it as the LS byte of a 16-bit number.
	PULB			
	ADDD	LOC1		Add the four converted
	INX			results.
	CMPX	#$1034		
	BNE	LOOP2		Done?
	ASRA			
	RORB			Yes, Divide the results
	ASRA			in the double
	RORB			register by 4.

11-6 The program must perform an A/D conversion on PE4 and turn the light on whenever the result is greater than $66.

11-7 The program should
   a. Write to ADCTL.
   b. Store TCNT.
   c. Wait for bit 7 of ADCTL to set.
   d. Get the new value in TCNT and take the difference.

11-9 The conversions should be made at $-86.4°$, $-28.8°$, $+28.8°$, and $+86.4°$.

11-11 The message is HI5.

11-12 c. See Figure P11-12.

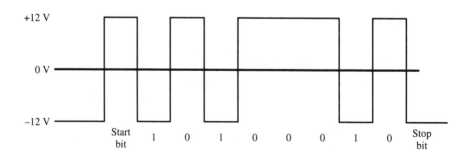

Figure P11-12

11-13 See Figure P11-13.

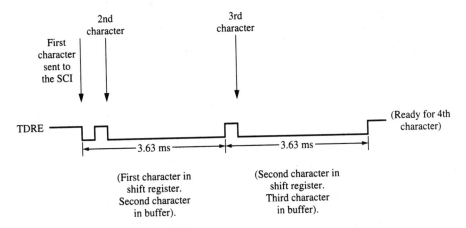

Figure P11-13

11-15 The program could be

CLR	$102D	Clear the TE and RE bits in SCCR2.
LDAA	#$02	Set DDRD so that PD0 is input
STAA	$1009	and PD1 is output.
LDAA	$1008	Bring in the data on PD0.
LSLA		Shift it.
STAA	$1008	Send it out on PD1.

11-16 $32

11-17 a. 1302 baud

11-19 The program might be

	LDX	#$C100	
	LDS	#STACK	
LOOP1	LDAA	0,X	Get character.
	CMPA	#$0D	Is it a RETURN?
	BEQ	DONE	
	JSR	E461	Send character when TDRE = 1.
	INX		Prepare for the next character.
	CPX	$C150	
	BNE	LOOP1	Done?
DONE	. . . .		

11-21 The core of the program could be

```
             LDX     #$C000   Point X to the list.
             LDY     #$1000   Point Y to the registers.
      LOOP1  LDAA    1,X      Get MSB.
             BITA    #$01     Test it.
             BEQ     SET0     Branch if it is a 0.
             BSET    2C,Y 40  Set T8 in SCCR1 if it is a 1.
             BRA     CONT
      SET0   BCLR    2C,Y 40  Clear T8 in SCCR1 if it is a 0.
      CONT   LDAA    0,X      Put the rest of the character
             STAA    2F,Y     in SCDR.
             INX              Set up for the next 9-bit
             INX              character.
             CPX     #C00A
             BNE     LOOP1
```

11-24 363 ms

11-26 Set the SCI for 9 bits in receive. A framing error on the second STOP bit will be signaled by the framing error bit. A framing error on the first STOP bit can be detected if bit 7 of the SCDR is 0.

Chapter 12

12-1 IC1 requires an 8 in the priority table. The code

```
LDAA     #$08
STAA     $103C
```

should store the proper priority in the HPRIO register.

12-2 Set the interrupt vectors for OC2, OC3, and OC5 to go to different locations in RAM. Then put a breakpoint at each of these locations. They can then all be enabled at the same time by storing a $68 into TMSK1 at $1022. The program should break in the routine for the highest priority interrupt.

12-5 **1.** The user must decide where the COP failure interrupt service routine is to start, and set the COP vectors (at FFFA and FFFB, or OOFA-OOFC on the EVB) to point to the start of the COP routine.

2. The message 'COP FAILURE' must be placed in memory, preferably by using an FCB.

3. The program must write the message out to the screen by placing the characters in the message, one at a time, into A, and then performing a JSR to the OUTA (Output accumulator A) subroutine in the monitor.

4. The interrupt routine can now return to the monitor by performing a JMP to the start of the SWI routine.

12-7 $EB

12-10 A suggested program is

```
* This is the erase loop. Because it can erase all locations
* by doing a row erase, it is only done once.
START   LDX    #$B610   Point X to the EEPROM locations.
        LDY    #$C000   Point Y to the source list.
        LDAA   #$0E     Set up for row erase.
        STAA   $103B    Put it in PPROG.
        STAA   $0,X     Write anything to the proper
                        EEPROM address.
        LDAA   #$0F     Turn on high voltage.
        STAA   $103B
        JSR             Delay for 10 ms.
        CLR    $103B    End of erase.
* This is the writing part of the program.
LOOP    LDAB   #$02                    Set EELAT bit.
        STAB   $103B
        LDAA   0,Y                     Load source byte.
        STAA   0,X                     Write it into the EEPROM.
        LDAB   #$03                    Turn on high voltage.
        STAB   $103B (EEPGM = 1)
        JSR                            Delay 10 ms.
        CLR    $103B                   End of write.
        INX
        INY
        CPX    #$B620                  More data to be written?
        BNE    LOOP                    Yes, branch.
```

Index